THE LOEB CLASSICAL LIBRARY

FOUNDED BY JAMES LOEB, LL.D.

AELIAN

ON THE CHARACTERISTICS OF ANIMALS

III

BOOKS XII—XVII

AELIAN

ON THE CHARACTERISTICS OF ANIMALS

WITH AN ENGLISH TRANSLATION BY

A. F. SCHOLFIELD

FELLOW OF KING'S COLLEGE, CAMBRIDGE

IN THREE VOLUMES

III

BOOKS XII—XVII

CAMBRIDGE, MASSACHUSETTS

HARVARD UNIVERSITY PRESS

LONDON

WILLIAM HEINEMANN LTD

MCMLIX

Printed in Great Britain

CONTENTS

ERRATA

Vol. I

Page 234, line 4: *for* ἀνακλᾳ *read* ἀνακλᾷ
" 274, last line but one: *for* ὀιστοῖς *read* οἰστοῖς, and again on
" 308 end of ch. 16

Vol. II

" 43, note *d*: *for* Issus aud *read* Issus and
" 102, note 6: delete full stop after '*corrupt*'
" 107, line 4 from bottom: *for* nowise *read* no wise
" 153, middle: *for* Maltese *read* Melitean
" 197, line 12: *for* hidden it may be in *read* hidden, it may be, in
" 290, note 2: *for* συν *read* σὺν
" 299, line 7: *for* mastich *read* mastic
" 300, note 1: *for* γᾶρ *read* γὰρ
" 371, *Add footnote*: '[b] Menis became King *c.* 3400 B.C. and united the Northern and Southern Kingdoms of Egypt.'

AELIAN

ON THE CHARACTERISTICS OF ANIMALS

SUMMARY

SUMMARY

SUMMARY

Book XVI

SUMMARY

Book XVII

BOOK XII

ΑΙΛΙΑΝΟΥ

ΠΕΡΙ ΖΩΩΝ ΙΔΙΟΤΗΤΟΣ

ΙΒ

1. Μυρέων τῶν ἐν Λυκίᾳ κόλπος ἐστί, καὶ ἔχει πηγήν, καὶ ἐνταῦθα νεὼς Ἀπόλλωνός ἐστι, καὶ ὁ τοῦδε τοῦ θεοῦ ἱερεὺς κρέα μόσχεια διασπείρει τῶν τῷ θεῷ τεθυμένων, ὀρφώ [1] τε οἱ ἰχθύες ἀθρόοι προσνέουσι, καὶ τῶν κρεῶν ἐσθίουσιν οἷα δήπου καλούμενοι δαιτυμόνες. καὶ χαίρουσιν οἱ θύσαντες, καὶ τὴν τούτων δαῖτα πιστεύουσιν εἶναί σφισιν ὅτταν ἀγαθήν, καὶ λέγουσιν ἵλεων εἶναι τὸν θεόν, διότι [2] οἱ ἰχθύες ἐνεπλήσθησαν τῶν κρεῶν. εἰ δὲ ταῖς οὐραῖς αὐτὰ ἐς τὴν γῆν ἐκβάλοιεν ὥσπερ οὖν ἀτιμάσαντες καὶ μυσαρὰ κρίναντες, τοῦτο δὴ τοῦ θεοῦ μῆνις εἶναι πεπίστευται. γνωρίζουσι δὲ καὶ τὴν τοῦ ἱερέως φωνὴν οἱ ἰχθύες, καὶ ὑπακούσαντες [3] μὲν εὐφραίνουσι δι' οὓς [4] κέκληνται, τοὐναντίον δὲ δράσαντες λυποῦσιν.

[1] ὀρφοίω, ὀρφοί. [2] δι' ὅν

AELIAN

ON THE CHARACTERISTICS OF ANIMALS

BOOK XII

1. There is a bay at Myra in Lycia and it has a spring and there is a shrine of Apollo there, and the priest of this god scatters the flesh of calves that have been sacrificed to the god, and Sea-perch [a] come swimming up in shoals and eat the flesh, as though they were guests invited to the feast. And the sacrificers are delighted, for they believe that this feasting of the fishes is a good omen for them, and they say that the god is propitious because the fish gorged themselves upon the flesh. If however the fish cast the food ashore with their tails as though they despised it and regarded it as tainted, this is believed to signify the wrath of the god. And the fish recognise the priest's voice, and if they obey his summons they gladden those on whose behalf they have been summoned; in the opposite event they cause them grief.

Sacred Fish at Myra

[a] Evidently not the 'Great Sea-perch' (5. 18), but Thompson declines to identify it.

---

[3] ἐπακούσαντες. [4] τούτους δι' οὕς.

2. Κατὰ τὴν πάλαι Βαμβύκην (καλεῖται δὲ νῦν Ἱεράπολις, Σελεύκου ὀνομάσαντος τοῦτο αὐτήν) ἰχθύες εἰσὶν ἱεροί, καὶ κατ᾽ ἴλας νήχονται καὶ ἔχουσιν ἡγεμόνας, καὶ τῶν ἐμβαλλομένων αὐτοῖς τροφῶν προεσθίουσιν οὗτοί γε. φυλάττουσι δὲ καὶ τὴν πρὸς ἀλλήλους φιλίαν μάλιστα ἰχθύων, καὶ ἔστιν ἀεὶ ἔνσπονδα αὐτοῖς, ἤτοι τῆς θεοῦ τὴν ὁμόνοιαν καταπνεούσης, ἢ διότι τῶν ἐμβαλλομένων τροφῶν ἐμπιπλάμενοι οὕτως τῆς ἀλλήλων βορᾶς ἄγευστοί τε καὶ ἀμαθεῖς [1] διαμένουσιν.

3. Λέγουσιν Αἰγύπτιοι, καὶ ἐμὲ μὲν ἥκιστα πείθουσι, λέγουσι δ᾽ οὖν ἄρνα καὶ ὀκτάπουν καὶ δίκερκον κατὰ τὸν Βόκχοριν τὸν ᾀδόμενον ἐκεῖνον γενέσθαι, καὶ ῥῆξαι φωνήν. καὶ δύο κεφαλὰς ᾄδουσι τῆς ἀρνός, καὶ τετράκερω γενέσθαι φασὶ τὴν αὐτήν. Ὁμήρῳ μὲν οὖν φωνὴν Ξάνθῳ τῷ ἵππῳ δόντι συγγνώμην νέμειν ἄξιον,[2] ποιητὴς γάρ· καὶ Ἀλκμὰν δὲ μιμούμενος ἐν τοῖς τοιούτοις Ὅμηρον οὐκ ἂν φέροιτο αἰτίαν, ἔχει γὰρ ἀξιόχρεων ἐς αἰδῶ [3] τὴν πρωτίστην τόλμαν· Αἰγυπτίοις δὲ τοιαῦτα κομπάζουσι προσέχειν πῶς οἷόν τε; εἴρηται δ᾽ οὖν,[4] εἰ καὶ μυθώδη, τὰ τῆσδε τῆς ἀρνὸς ἴδια.

4. Καὶ ἐκεῖνο δὲ ὑπὲρ τῶν ἱεράκων ἀκούσας οἶδα. πρὸ τοῦ τὸν Νεῖλον ἐπιπολάζειν τῇ Αἰγύπτῳ καὶ ἐς τὰς ἀρούρας ἀνέρχεσθαι, ἀφιᾶσι τῶν πτερῶν [5] τὰ ἤδη γηρῶντα ὥσπερ οὖν τῶν φυτῶν οἱ κλάδοι ⟨τὰ⟩ [6] φύλλα τὰ ξηρά, καὶ ἀναφύουσι

[1] ἀμαθεῖς εἰκότως. [2] ἄξια. [3] αἰδῶ *corrupt*, *H.*
[4] γοῦν. [5] *Ges*: πτερύγων. [6] ⟨τά⟩ *add. H.*

2. In the ancient Bambyce [a] (it is now called Hierapolis since Seleucus gave it this name) there are sacred fish which swim in companies and have leaders; these are the first to eat of the food which is thrown in to them. More than all other fish do they maintain friendly relations with one another and are always at peace, either because the goddess [b] inspires them with unanimity, or because being satisfied with the food that is thrown in to them, they therefore abstain from eating one another and know nothing of it.

Sacred Fish at Hierapolis

3. The Egyptians assert (though they are far from convincing me), they assert, I say, that in the days of the far-famed Bocchoris a Lamb was born with eight feet and two tails, and that it spoke. They say also that this Lamb had two heads and four horns. It is right to forgive Homer who bestows speech upon Xanthus the horse [*Il.* 19. 404], for Homer is a poet. And Alcman could not be censured for imitating Homer in such matters, for the first venture of Homer is a plea sufficient to justify forgiveness. But how can one pay any regard to Egyptians who exaggerate like this? However, fabulous though they be, I have related the peculiarities of this lamb.

A monstrous Lamb

4. Here is another fact touching Hawks that I remember to have heard. Before the Nile inundates Egypt and comes up over the ploughlands Hawks shed their old feathers just as the branches of trees shed their withered leaves, and grow new and

The Hawk: various species

[a] On the E border of Syria some 12 mi. from the Euphrates. Renamed by Seleucus Nicator (*c.* 358 –280 B.C.) in honour of the goddess Astarte.
[b] Atargatis, Astarte.

πτίλα νεαρὰ καὶ ὡραῖα οἱ ἱέρακες ὡς τὴν ἄνθην τὰ δένδρα. γένη δὲ ἱεράκων πλείονα ἄρα ἦν, καὶ ἔοικεν ὑπαινίττεσθαι καὶ Ἀριστοφάνης τοῦτο. φησὶ γοῦν

> ἀλλ' ἐπέμψαμεν
> τρισχιλίους ἱέρακας ἱπποτοξότας.
> χωρεῖ δὲ πᾶς τις ὄνυχας ἠγκυλωμένος
> κερχνῂς τριόρχης γὺψ κύμινδις αἰετός.

νενέμηνται δὲ καὶ ἀπεκρίθησαν θεοῖς πολλοῖς. ὁ μὲν περδικοθήρας καὶ ὠκύπτερος Ἀπόλλωνός ἐστι θεράπων φασί, φήνην δὲ καὶ ἅρπην Ἀθηνᾷ προσνέμουσιν, Ἑρμοῦ δὲ τὸν φασσοφόντην ἄθυρμα εἶναί φασιν, Ἥρας δὲ τὸν τανυσίπτερον, καὶ τὸν τριόρχην οὕτω καλούμενον Ἀρτέμιδος. μητρὶ δὲ θεῶν τὸν μέρμνον . . .[1], καὶ ἄλλον[2] ἄλλῳ θεῷ. γένη δὴ[3] ἱεράκων ἐστὶ πάμπολλα.

5. Αἰγύπτιοι μὲν οὖν σέβοντές τε καὶ ἐκθεοῦντες γένη ζῴων διάφορα γέλωτα ὀφλισκάνουσι παρά γε τοῖς πολλοῖς· Θηβαῖοι[4] δὲ σέβουσιν Ἕλληνες ὄντες ὡς ἀκούω γαλῆν, καὶ λέγουσί γε Ἡρακλέους αὐτὴν γενέσθαι τροφόν, ἢ τροφὸν μὲν οὐδαμῶς, καθημένης δὲ ἐπ' ὠδῖσι τῆς Ἀλκμήνης καὶ τεκεῖν οὐ δυναμένης, τὴν δὲ παραδραμεῖν καὶ τοὺς τῶν ὠδίνων λῦσαι δεσμούς, καὶ προελθεῖν τὸν Ἡρακλέα καὶ ἕρπειν ἤδη. καὶ οἱ τὴν Ἁμαξιτὸν τῆς[5] Τρῳάδος κατοικοῦντες μῦν σέβουσιν· ἔνθεν τοι

[1] *Lacuna* : ⟨ἀνάπτουσι⟩ *ex. gr. H.* [2] ἄλλον δέ. [3] δέ. [4] καὶ Θ. δέ. [5] Ἁ. δὲ τῆς.

beautiful plumage as trees do foliage. It seems that there are in fact several species of Hawks, and Aristophanes appears to hint as much. At any rate he says [*Av.* 1179]

> 'But we have despatched three thousand Hawks, mounted archers. And each one moves forward with talons crooked—kestrel, buzzard, vulture, night-hawk,[a] eagle.'

They are allotted separately to many gods. The partridge-catcher,[b] they say, and the ocypterus[c] are servants of Apollo; the lämmergeier and the shearwater they assign to Athena; the dove-killer is said to be the darling of Hermes, the wide-wing, of Hera, and the buzzard, as it is called, of Artemis. To the Mother of the Gods ⟨they assign⟩ the mermnus, and to one god one bird, to another another. There are in fact a great many kinds of Hawks.

The Marten and Alcmena

5. The Egyptians incur the derision at any rate of most people for worshipping and deifying various kinds of animals. But the inhabitants of Thebes, although Greeks, worship a marten, so I hear, and allege that it was the nurse of Heracles, or if it was not the nurse, yet when Alcmena was in labour and unable to bring her child to birth, the marten ran by her and loosed the bonds of her womb, so that Heracles was delivered and at once began to crawl. And those who live in Hamaxitus in the Troad worship a Mouse, and that is why,

The Mouse worshipped in the Troad

[a] Or 'Hawk-owl.'
[b] Perh. 'Sparrow-hawk,' Gossen § 182.
[c] Perh. 'Lesser Hen-harrier,' *ib.*

καὶ τὸν Ἀπόλλω τὸν παρ' αὐτοῖς τιμώμενον Σμίνθιον καλοῦσί φασιν. ἔτι γὰρ καὶ τοὺς Αἰολέας καὶ τοὺς Τρῶας τὸν μῦν προσαγορεύειν σμίνθον, ὥσπερ οὖν καὶ Αἰσχύλος ἐν τῷ Σισύφῳ

ἀλλ' ἀρουραῖος τίς ἐστι σμίνθος ὧδ' ὑπερφυής;[1]

καὶ τρέφονται μὲν ἐν τῷ Σμινθείῳ[2] μύες τιθασοὶ δημοσίας τροφὰς λαμβάνοντες, ὑπὸ δὲ τῷ βωμῷ φωλεύουσι[3] λευκοί, καὶ παρὰ τῷ τρίποδι τοῦ Ἀπόλλωνος ἕστηκε μῦς. μυθολόγημα δὲ ὑπὲρ τῆσδε τῆς θρησκείας καὶ ἐκεῖνο προσακήκοα. τῶν Αἰολέων καὶ τῶν Τρώων τὰ λήια πολλὰς μυῶν μυριάδας ἐπελθούσας ἄωρα[4] ὑποκείρειν καὶ ἀτελῆ τὰ θέρη τοῖς σπείρασιν ἀποφαίνειν. οὐκοῦν τὸν ἐν Δελφοῖς θεὸν πυνθανομένων εἰπεῖν ὅτι δεῖ θύειν Ἀπόλλωνι Σμινθεῖ, τοὺς δὲ πεισθέντας ἀπαλλαγῆναι τῆς ἐκ τῶν μυῶν ἐπιβουλῆς καὶ τὸν πυρὸν αὐτοῖς ἐς τὸν[5] νενομισμένον ἄμητον ἀφικνεῖσθαι. ἐπιλέγουσι δὲ ἄρα τούτοις καὶ ἐκεῖνα. ἐς ἀποικίαν Κρητῶν οἱ σταλέντες οἴκοθεν ἔκ τινος τύχης καταλαβούσης αὐτοὺς ἐδεήθησαν τοῦ Πυθίου φῆναί τινα αὐτοῖς χῶρον ἀγαθὸν καὶ ἐς τὸν συνοικισμὸν λυσιτελῆ. ἐκπίπτει δὴ λόγιον, ἔνθα ἂν αὐτοῖς οἱ γηγενεῖς πολεμήσωσιν, ἐνταῦθα καταμεῖναι καὶ ἀναστῆσαι πόλιν. οὐκοῦν ἥκουσι μὲν ἐς τὴν Ἁμαξιτὸν τήνδε καὶ στρατοπεδεύουσιν ὥστε ἀναπαύσασθαι, μυῶν δὲ ἄφατόν τι πλῆθος ἐφερπύσαν τά τε ὄχανα αὐτοῖς τῶν ἀσπίδων διέτραγε καὶ τὰς τῶν τόξων νευρὰς διέφαγεν· οἱ

[1] *Hermann*: ἀρουραῖός τις . . . ὑπερφυής MSS, *H.*

[2] εἰς τοῦς Σμινθίους MSS, ἐν τῷ Σμινθίου *Ges*, Σμίνθεως *Radermacher.*

according to them, they give the name of *Sminthian* to Apollo whom they worship, for the Aeolians and the people of the Troad still call a mouse *sminthus*, just as Aeschylus too in his *Sisyphus* [*fr.* 227 N] writes

> ' Nay, but what sminthus of the fields is so monstrous ? '

And in the temple of Smintheus tame Mice are kept and fed at the public expense, and beneath the altar white Mice have their nests, and by the tripod of Apollo there stands a Mouse. And I have also heard the following mythical tale about this cult. Mice came in tens of thousands and cut off before they ripened the crops of the Aeolians and Trojans, rendering the harvest barren for the sowers. Accordingly the god at Delphi said when they enquired of him, that they must sacrifice to Apollo Smintheus; they obeyed and freed themselves from the conspiracy of Mice, and their wheat attained the normal harvest. And they add the following story. Some Cretans who owing to a disaster that befell them were sent out to found a colony, besought the Pythian Apollo to tell them of some good place where it would be advantageous to found a city. There issued from the oracle this answer: in the place where the earth-born made war upon them, there they should settle and raise a city. So they came to this place Hamaxitus and pitched their camp in order to rest; but a countless swarm of Mice crept stealthily upon them, gnawed through their shield-straps and ate through their bow-strings. So they guessed that these were the

---

[3] καὶ φωλεύουσι. [4] *Ges*: ἀώρους.
[5] *Schn*: ἐς τόνδε τόν.

δὲ ἄρα συνέβαλον τούτους ἐκείνους εἶναι τοὺς γηγενεῖς, καὶ μέντοι καὶ ἐς ἀπορίαν ἥκοντες τῶν ἀμυντηρίων τόνδε τὸν χῶρον οἰκίζουσι, καὶ Ἀπόλλωνος ἱδρύονται νεὼν Σμινθίου. ἡ μὲν οὖν τῶν μυῶν μνήμη προήγαγεν ἡμᾶς ἐς θεολογίαν τινά, χείρους δὲ αὐτῶν οὐ γεγόναμεν καὶ τοιαῦτα προσακούσαντες.

6. Ἦσαν δὲ ἄρα δελφῖνες καὶ νεκρῶν μνήμονες καὶ τῶν συννόμων ἀπελθόντων τοῦ βίου οὐδαμῶς προδόται. τὸν γοῦν ἑαυτῶν τεθνεῶτα ὑποδύντες εἶτα μέντοι κομίζουσι φοράδην ἐς τὴν γῆν τοῖς ἀνθρώποις πιστεύοντες θάψαι, καὶ Ἀριστοτέλης μαρτυρεῖ τούτῳ· ἕπεται δὲ πλῆθος ἕτερον οἱονεὶ τιμῶντες ἢ καὶ νὴ Δία ὑπερμαχοῦντες, μή ποτε ἄλλο κῆτος ἐπιδράμῃ καὶ τὸν νεκρὸν ἁρπάσαν εἶτα καταδαίσηται. ὅσοι μὲν οὖν εἰσιν ἔνδικοι καὶ τῆς μουσικῆς ἐπαΐοντες, τῆς τῶν δελφίνων φιλομουσίας αἰδοῖ θάπτουσιν αὐτούς· οἱ δὲ ἀπό τε Μουσῶν φασιν ἀπό τε Χαρίτων ἀκηδῶς αὐτῶν ἔχουσι. καὶ δότε συγγνώμην, ὦ δελφῖνες φίλοι, τῇ τῶν ἀνθρώπων ἀγριότητι, εἴγε καὶ Ἀθηναῖοι Φωκίωνα τὸν χρηστὸν ἔρριψαν ἄταφον. καὶ Ὀλυμπιὰς δὲ ἔκειτο γυμνὴ ἡ τεκοῦσα τὸν τοῦ Διός, ὡς ἐκόμπαζέ τε αὐτὴ καὶ ἐκεῖνος ἔλεγε. καὶ τὸν Ῥωμαῖον Πομπήιον τὸν Μέγαν ἐπίκλην ἀποκτείναντες Αἰγύπτιοι τοσαῦτα ἐργασάμενον καὶ νίκας νικήσαντα ἄγαν σεμνὰς καὶ θριαμβεύσαντα τρὶς καὶ τὸν τοῦ φονέως πατέρα σώσαντα καὶ ἐς

'earth-born' referred to, and, besides, having now no means of getting weapons of defence, they settled in this spot and built a temple to Apollo Smintheus. Well, this mention of Mice has led us to touch upon a matter of theology; however we are none the worse for having listened even to such tales as this.

The Dolphin and its dead

6. It seems that Dolphins are mindful even of their dead and by no means abandon their fellows when they have departed this life. At any rate they get underneath their dead companion and then carry him along to the shore, confident that men will bury him, and Aristotle bears witness to this [*HA* 631 a 18]. And another company of Dolphins follow them by way of doing honour to, or even actually fighting to protect, the dead body, for fear lest some other great fish should rush up, seize it, and then devour it. All just men who appreciate music bury dead Dolphins out of respect for their love of music. But those to whom, as they say, the Muses and the Graces are alien care nothing for Dolphins. And so, beloved Dolphins, you must pardon the savage nature of man, since even the people of Athens cast out the excellent Phocion [a] unburied. And even Olympias lay unburied, although she was the mother of the son of Zeus,[b] as she herself boasted and as he asserted. And the Egyptians after killing the Roman Pompey, surnamed 'the Great,' who had achieved so much, who had had such distinguished victories and had celebrated three triumphs, who

[a] Phocion, distinguished Athenian general and statesman, 4th cent. B.C., opposed Demosthenes in advocating peace with Philip of Macedon. Later was wrongly suspected of treachery and put to death, 318 B.C.

[b] Alexander the Great.

τὴν Αἰγυπτίων βασιλείαν ἐπαναγαγόντα εἴασαν ἐρριμμένον, ἄμοιρον τῆς κεφαλῆς, πλησίον τῆς θαλάττης καὶ ἐκεῖνον, ὡς ὑμᾶς πολλάκις ἐῶσι.[1] τὸ γάρ τοι ζῷον τὸ πάμβορον τοῦτο οὐδὲ ὑμῶν φείδεται, ἀλλὰ καὶ ὑμᾶς ταρίχους ἐργάσασθαι τολμῶσι, καὶ σφᾶς αὐτοὺς λελήθασι ταῖς Μούσαις ταῖς Διὸς θυγατράσι ταῦτα ἀπὸ θυμοῦ δρῶντες.

7. Λέοντας μὲν ἐν Αἰγύπτῳ σέβουσι, καὶ ἐξ αὐτῶν κέκληται πόλις· καὶ τὰ ἴδιά γε τῶν ἐκεῖ λεόντων εἰπεῖν ἄξιον. ἔχουσι νεὼς καὶ διατριβὰς εὖ μάλα ἀφθόνους, καὶ κρέα βοῶν αὐτοῖς ἐστιν ὁσημέραι, καὶ διασπαρακτὰ κεῖται γυμνὰ ὀστῷν καὶ ἰνῶν, καὶ ἐσθιόντων ἐπᾴδουσιν Αἰγυπτίᾳ φωνῇ. ἡ δὲ ὑπόθεσις τῆς ᾠδῆς, 'μὴ βασκήνητέ τινα τῶν ὁρώντων,' καὶ ἔοικεν ὡς ἂν εἴποις ἀντὶ περιάπτων τὸ ᾆσμα. ἐκθεοῦνται δὲ ἄρα παρ' αὐτοῖς πολλοί, καὶ ἀντιπρόσωποί γε δίαιται ἀνειμέναι αὐτοῖς εἰσι. καὶ αἱ μὲν πρὸς τὴν ἕω θυρίδες, αἱ δὲ πρὸς τὴν ἑσπέραν ἀνεῳγμέναι κεχαρισμενωτέραν αὐτοῖς τὴν δίαιταν ἀποφαίνουσιν. ἔστι δὲ αὐτοῖς καὶ γυμνάσια ὑγιείας[2] χάριν, καὶ πλησίον παλαῖστραι, ὁ δὲ ἀντίπαλος μόσχος τῶν εὐτραφῶν.[3] καὶ πρὸς τοῦτον γυμνασάμενος, ὃν[4] αὐτὸν καθέλῃ (δρᾷ δὲ βραδέως ὑπ' ἀργίας αὐτὸ καὶ ἀθηρίας), ἐμφορεῖταί τε καὶ ὑποστρέφει ἐς τὸ αὔλιον τὸ ἴδιον. διάπυρον δέ ἐστι τὸ ζῷον

[1] ἐῶσι. ἐγὼ δὲ εἶδον καὶ τέττιγας εἴραντάς τινας καὶ πιπράσκοντας ἐπὶ δεῖπνον καὶ μάλα γε ἐδείπνει.

[2] *Schn* : ὑγείας.

[3] *Jac* : τῶν εὐτραφῶν μόσχος.

[4] *Jac* : ἵνα.

had saved the life of his murderer's father [a] and had re-established him on the throne of Egypt, left him cast out, a headless corpse, by the sea, just as men often leave you. For this all-devouring creature man does not even spare you, but goes so far as to pickle you, and is unconscious that his action is hateful to the Muses, the daughters of Zeus.

7. In Egypt they worship Lions, and there is a city called after them.[b] It is worth recording the peculiarities of the Lions there. They have temples and very many spaces in which to roam; the flesh of oxen is supplied to them daily and it lies, stripped of bones and sinews, scattered here and there, and the Lions eat to the accompaniment of song in the Egyptian language. And the theme of the song is 'Do not bewitch any of the beholders'; this singing appears, as you might say, to be a substitute for amulets. Many of the Lions are deified in Egypt, and there are chambers face to face consecrated to their use. The windows of some open to the east, others to the west, making life more pleasant for them. And to preserve their health they have places for exercise, and wrestling-grounds near by, and their adversary is a well-nourished calf. And if, after practising his skill against the calf, the Lion brings it down (this takes time for he is lazy and unused to hunting), he eats his fill and goes back to his own stall.

The Lion in Egypt

[a] Ptolemy XII, 'Auletes,' took refuge in Rome from his rebellious subjects, where he was befriended by Pompey who aided his restoration, 55 B.C. His son Ptolemy XIII succeeded him (51), and it was at the instigation of his council that Pompey was murdered on landing in Egypt (48).

[b] Leontopolis, in the Delta of Egypt.

ἰσχυρῶς, καὶ ἐντεῦθεν καὶ Ἡφαίστῳ ἀνῆψαν αὐτὸ Αἰγύπτιοι· τὸ δὲ ἔξωθεν πῦρ δυσωπεῖται καὶ φεύγει πλήθει τοῦ ἔνδοθέν φασιν. ἐπειδὴ δὲ ἄγαν πυρῶδές ἐστι,[1] οἶκον Ἡλίου φασὶν εἶναι· καὶ ὅταν γε ᾖ ἑαυτοῦ θερμότατος καὶ θερειότατος ὁ ἥλιος, λέοντι αὐτὸν πελάζειν[2] φασί. προσέτι γε μὴν καὶ οἱ τὴν μεγάλην οἰκοῦντες Ἡλίου πόλιν ἐν τοῖς τοῦ θεοῦ προπυλαίοις τούσδε τρέφουσι τοὺς λέοντας, θειοτέρας τινὸς μοίρας ὡς Αἰγύπτιοί φασι μετειληχότας. καὶ γάρ τοι καὶ ὄναρ οἷσπερ οὖν ὁ θεός ἐστιν ἵλεως ἐπιστάντες προθεσπίζουσί τινα, καὶ τοὺς ἐπίορκον ὀμόσαντας οὐκ ἐς ἀναβολὰς ἀλλὰ ἤδη δικαιοῦσι, τοῦ θεοῦ τὴν ὀργὴν τὴν δικαίαν αὐτοῖς καταπνέοντος. λέγει δὲ καὶ Ἐμπεδοκλῆς τὴν ἀρίστην εἶναι μετοίκησιν τὴν τοῦ ἀνθρώπου, εἰ μὲν ἐς ζῷον ἡ λῆξις αὐτὸν[3] μεταγάγοι, λέοντα γίνεσθαι· εἰ δὲ ἐς φυτόν, δάφνην. ἃ δὲ Ἐμπεδοκλῆς λέγει, ταῦτά ἐστιν·

ἐν θήρεσσι[4] λέοντες ὀρειλεχέες χαμαιεῦναι
γίνονται, δάφναι δ' ἐνὶ δένδρεσιν ἠυκόμοισιν.

εἰ δὲ δεῖ καὶ τῆς τῶν Αἰγυπτίων σοφίας ἐς φύσιν ἐκτρεπόντων καὶ τὰ τοιαῦτα ὥραν τίθεσθαι (δεῖ δέ), τὰ μὲν πρόσθια τοῦδε τοῦ ζῴου πυρὶ ἀποκρίνουσιν, ὕδατί ⟨γε⟩[5] μὴν τὰ κατόπιν. καὶ τὴν σφίγγα μέντοι τὴν διφυῆ Αἰγύπτιοί τε χειρουργοὶ γλύφοντες καὶ Θηβαῖοι μῦθοι κομπάζοντες δίμορφον ἡμῖν πειρῶνται δεικνύναι, σεμνύνοντες τῇ τε τοῦ παρθενωποῦ καὶ τῇ τοῦ λεοντοειδοῦς σώματος

[1] ἐστι καὶ αὐτόν.
[2] πελάζειν τῷ οὐρανίῳ.
[3] *Ges*: αὐτήν.
[4] *Schol. Aphthon.*: θηρσὶ δέ.
[5] ⟨γε⟩ *add. Reiske.*

The Lion is a very fiery animal, and this is why the Egyptians connect him with Hephaestus, but, they say, he dislikes and shuns the fire from without because of the great fire within himself. And since he is of a very fiery nature, they say the Lion [a] is the house of the Sun, and when the sun is at its hottest and at the height of summer, they say it is approaching the Lion. Moreover the inhabitants of the great city of Heliopolis keep these Lions in the entrance to the temples of the god as sharing (so the Egyptians say) to some extent the lot of the gods. And further, they appear in dreams to those whom the god regards with favour and utter prophecies, and those who have committed perjury they punish not after some delay but immediately, for the god inspires them with a righteous indignation. And Empedocles maintains that if his lot translates a man into an animal, then it is best for him to transmigrate into a lion; if into a plant, then into a sweet-bay. Empedocles' words are [*fr.* 127, Diels *Vorsok.*[6] 1. 362]

> 'Among wild beasts they become lions that couch upon the mountains and sleep on the earth, and among trees with fair foliage sweet-bay-trees.'

But if we are (as we ought) to take into consideration the wisdom of the Egyptians who refer such manifestations to natural causes, they assign the foreparts of this animal to fire, and the hinder parts to water. Again, Egyptian artificers in their sculpture, and the vainglorious legends of Thebes attempt to represent the Sphinx, with her two-fold nature, as of two-fold shape, making her awe-inspiring by

The Sphinx

[a] The sign *Leo* in the zodiac.

κράσει αὐτήν. τοῦτό τοι καὶ Εὐριπίδης ὑπαινίττεται λέγων

> οὐρὰν δ' ὑπίλλασ' ὑπὸ λεοντόπουν βάσιν
> καθίζετο.

καὶ μέντοι καὶ τὸν Νεμεαῖον λέοντα τῆς σελήνης ἐκπεσεῖν φασι. λέγει γοῦν [1] καὶ τὰ Ἐπιμενίδου ἔπη

> καὶ γὰρ ἐγὼ γένος εἰμὶ Σελήνης ἠυκόμοιο,
> ἣ δεινὸν φρίξασ' ἀπεσείσατο θῆρα λέοντα
> ἐν Νεμέᾳ, ἀνάγουσ' αὐτὸν διὰ πότνιαν Ἥραν.

καὶ ταῦτα μὲν ἐς τοὺς μύθους ἀποκρίνωμεν,[2] τά γε μὴν λεόντων ἴδια καὶ ἀνωτέρω καὶ νῦν ⟨δὲ⟩ [3] ἀποχρώντως εἴρηται.

8. Ζῷόν ἐστιν ὁ πυραύστης, ὅπερ οὖν χαίρει μὲν τῇ λαμπηδόνι τοῦ πυρὸς καὶ προσπέτεται τοῖς λύχνοις ἐνακμάζουσιν,[4] ἐμπεσὼν δὲ ὑπὸ ῥύμης [5] εἶτα μέντοι καταπέφλεκται. μέμνηται δὲ αὐτοῦ καὶ Αἰσχύλος ὁ τῆς τραγῳδίας ποιητὴς λέγων

> δέδοικα μωρὸν κάρτα πυραύστου μόρον.

9. Ὁ δὲ κίγκλος ζῷόν ἐστι πτηνὸν ἀσθενὲς τὰ [6] κατόπιν, καὶ διὰ τοῦτό φασι μὴ ἰδίᾳ μηδὲ καθ' ἑαυτὸν δυνάμενον αὐτὸν νεοττιὰν [7] συμπλέξαι, ἐν ταῖς ἄλλων δὲ τίκτειν. ἔνθεν ⟨τοι⟩ [8] καὶ τοὺς πτωχοὺς κίγκλους ἐκάλουν αἱ τῶν ἀγροίκων

[1] *Reiske*: οὖν.
[2] *Reiske*: ἀπεκρίναμεν.
[3] ⟨δέ⟩ *add. H.*
[4] ἐνακμαζούσῃ τῇ φλογί.
[5] *Ges*: ῥώμης.
[6] τό.
[7] νεοττείαν *most* MSS.
[8] ⟨τοι⟩ *add. H.*

fusing the body of a maiden with that of a lion. And Euripides suggests this when he says [*fr.* 540 N.]

'And drawing her tail in beneath her lion's feet she sat down.'

The Nemean Lion

And moreover they say that the Lion of Nemea fell from the moon. At any rate Epimenides also has these words [*fr.* 2, Diels *Vorsok.*[6] 1. 32]:

'For I am sprung from the fair-tressed Moon, who in a fearful shudder shook off the savage lion in Nemea, and brought him forth at the bidding of Queen Hera.'

Let us however relegate these matters to the region of myth; but the peculiarities of Lions have been sufficiently dealt with both earlier on and in the present chapter.

The Wax-moth

8. The Wax-moth is a creature that delights in the brilliance of fire and flies to lamps burning brightly, but falls into them owing to its momentum and is burned to death. And Aeschylus the Tragic poet mentions it in these words [*fr.* 288 N]:

'I greatly dread the foolish fate of the wax-moth.'

The Wagtail

9. The Wagtail[a] is a winged creature weak in its hinder parts, and that is why (they say) it is incapable of building a nest of its own accord or for itself, but lays its eggs in the nests of other birds. Hence in the proverbs of country folk poor men are

[a] So Thompson renders; but L-S[9] 'dabchick, *Podiceps ruficollis*.'

παροιμίαι. κινεῖ δὲ τὰ οὐραῖα πτερά, ὥσπερ οὖν ὁ παρὰ τῷ Ἀρχιλόχῳ κηρύλος. μέμνηται δὲ καὶ τοῦ ὄρνιθος τοῦδε Ἀριστοφάνης ἐν τῷ Ἀμφιαράῳ λέγων

ὀσφὺν δ' ἐξ ἄκρων διακίγκλισον ἠύτε κίγκλος [1]
ἀνδρὸς πρεσβύτου, τελέειν δ' ἀγαθὴν ἐπαοιδήν.

καὶ ἐν τῷ Γήρᾳ

λορδοῦ κιγκλοβάταν ῥυθμόν.

καὶ Αὐτοκράτης [2] ἐν Τυμπανισταῖς

οἷα παίζουσιν φίλαι
παρθένοι Λυδῶν κόραι
κοῦφα πηδῶσαι πόδας,[3]
κἀνακρούουσαι χεροῖν,
Ἐφεσίαν παρ' Ἄρτεμιν
καλλίσταν, καὶ τοῖν ἰσχίοιν
τὸ μὲν κάτω τὸ δ' αὖ
εἰς ἄνω ἐξαίρουσαι,[4]
οἷα κίγκλος ἅλλεται.

10. Οἱ μύες ἀποθνήσκοντες καθ' ἑαυτοὺς καὶ ἐκ μηδεμιᾶς ἐπιβουλῆς ἀπορρεόντων αὐτοῖς τῶν μελῶν κατὰ μικρὰ ἀπέρχονται τοῦ βίου. ἔνθεν ⟨τοι⟩ [5] καὶ ἡ παροιμία λέγει κατὰ μυὸς ὄλεθρον, μέμνηται δὲ αὐτῆς Μένανδρος ἐν τῇ Θαΐδι. τρυγόνος δὲ λαλίστερον ἔλεγον· ἡ γάρ τοι τρυγὼν καὶ διὰ τοῦ στόματος μὲν ἀπαύστως φθέγγεται, ἤδη δὲ καὶ ἐκ τῶν κατόπιν μερῶν ὥς φασι πάμπλειστα. μέμνηται δὲ καὶ ταύτης τῆς παροιμίας ἐν τῷ Πλοκίῳ ὁ αὐτός. καὶ Δημήτριος

called 'wagtails.' The bird moves its tail-feathers, like the ceryl in the passage of Archilochus [*fr.* 49 D]. And Aristophanes also mentions this bird in his *Amphiaraus* [*fr.* 29 K] thus:

'Give the old man's loins a thorough shaking, as the Wagtail does, and work a powerful spell.'

And in his *Geras* [*fr.* 140 K]:

'Rhythmic wagtail-gait of a belly-arching fellow.'

And Autocrates in his *Tympanistae* [*fr.* 1 K]:

'As sweet maidens, daughters of Lydia, sport and lightly leap and clap their hands in the temple of Artemis the Fair at Ephesus, now sinking down upon their haunches and again springing up, like the hopping wagtail.'

Two proverbs: (*a*) the Mouse

10 (i). When Mice die a natural death and not through any design upon them, their limbs dissolve and little by little they depart this life. That, you see, is the origin of the saying 'Like a mouse's death,' and Menander mentions it in his *Thaïs* [*fr.* 219 K]. And men commonly say 'More talkative than a turtle-dove,' because the turtle-dove not only never stops uttering through its mouth, but they do say that it utters a great deal through its hinder parts also. And the same writer mentions this proverb in his *Necklace* [*fr.* 416 K]. And Demetrius in

(*b*) the Turtle-dove

[1] *Mein*: κίγκλου. [2] *Ges*: αὐτοκρατήσας.
[3] *Fiorillo*: κόμαν MSS *H.* [4] *Thompson*: -ουσα MSS, *edd.*
[5] ⟨τοι⟩ *add. H.*

ἐν τῇ Σικελίᾳ τῷ δράματι μέμνηται ὅτι καὶ τῇ πυγῇ λαλοῦσιν αἱ τρυγόνες.

Λέγουσι δὲ τοὺς μύας λαγνιστάτους εἶναι, καὶ μάρτυρά γε Κρατῖνον ἐπάγονται εἰπόντα ἐν ταῖς Δραπετίσι

φέρε νῦν σοι
ἐξ αἰθρίας καταπυγοσύνην μυὸς ἀστράψω Ξενοφῶντος.

καὶ ἔτι μᾶλλον τὸν θῆλυν ἔλεγον ἐς τὰ ἀφροδίσια εἶναι λυττητικόν. καὶ πάλιν παρὰ Ἐπικράτει ἐν τῷ Χορῷ [1]

τελέως δὲ μ' ὑπῆλθεν ἡ κατάρατος μαστροπὸς
ἐπομνύουσα τὰν Κόραν τὰν Ἄρτεμιν
τὰν Φερρέφατταν [2] ὡς δάμαλις, ὡς παρθένος,
ὡς πῶλος ἀδμής. ἣ δ' ἄρ' ἦν μυωνία,

ἐς ὑπερβολὴν δὲ λαγνιστάτην αὐτὴν εἰπεῖν ἠθέλησε 'μυωνίαν ὅλην' ὀνομάσας. καὶ Φιλήμων

μῦς λευκός, ὅταν αὐτήν τις (ἀλλ' αἰσχύνομαι
λέγειν), κέκραγε τηλικοῦτον εὐθὺς ἡ
κατάρατος,[3] ὥστ' οὐκ ἔστι πολλάκις λαθεῖν.

11. Σέβουσι δὲ Αἰγύπτιοι καὶ μέλανα ταῦρον, καὶ καλοῦσιν Ὄνουφιν αὐτόν. καὶ τὸ ὄνομα τοῦ χώρου ἔνθα τρέφεται Αἰγύπτιοι λεγέτωσαν ἡμῖν λόγοι· τραχὺ γάρ. ἀντίαι ⟨δὲ⟩ [4] αὐτῷ τρίχες ἤπερ οὖν τοῖς ἄλλοις εἰσίν· ἴδια γάρ τοι καὶ τοῦδε

[1] Χορῷ, δρᾶμα δέ ἐστι τῷ Ἐπικράτει τοῦτο.
[2] *Mein*: φερσέφατταν.
[3] *Bentley*: κατάρατος μαστροπός.
[4] ⟨δέ⟩ *add. H.*

his play *Sicelia* [*fr.* 3 K] mentions that turtle-doves chatter through their rump as well.

(ii). They say that Mice are exceedingly salacious, and they cite Cratinus as a witness, when he says in his *Drapetides* (Runaway slave-girls) [*fr.* 53 K]:

The Mouse, its character

'Look you, from a clear sky will I blast with lightning the debauchery of that mouse Xenophon.'

And they say that the female mouse is even more madly amorous. And again from the *Chorus* of Epicrates [*fr.* 9 K] they cite these words:

'The accursed go-between fooled me completely, swearing by the Maiden, by Artemis, by Persephone,[a] that the wench was a heifer, a virgin, an untamed filly—and all the time she was an absolute mousehole.'

By calling her an 'absolute mousehole' he meant to say that she was beyond measure lecherous. And Philemon says [*fr.* 126 K]:

'A white mouse, when someone tries to—but I am ashamed to say the word, the confounded woman at once lets out such a yell, that it is often impossible to avoid attracting attention.'

11. The Egyptians also worship a black bull which they call Onuphis. And the name of the place where it is reared let the Egyptian narratives tell us, for it is a hard name. Its hair grows the opposite way to that on other bulls; that is another of its

Onuphis, the sacred bull

[a] The go-between is humorously depicted as not knowing that 'the Maiden' and 'Persephone' are one and the same person.

ταῦτα. μέγιστος δὲ ἦν ἄρα βοῶν οὗτος καὶ ὑπὲρ τοὺς Χάονας, οὕσπερ οὖν καὶ λαρινοὺς καλοῦσι Θεσπρωτοί τε καὶ Ἠπειρῶται τῆς σπορᾶς τῆς τῶν Γηρυόνου βοῶν γενεαλογοῦντες αὐτούς. καὶ σιτεῖταί γε Ὄνουφις πόαν Μηδικὴν οὗτος.

12. Ὀξύτατος δὲ ἦν ἄρα καὶ ἁλτικώτατος ἰχθύων ὁ δελφίς, ἀλλὰ καὶ τῶν χερσαίων ἁπάντων. ὑπερπηδᾷ γοῦν καὶ ναῦν, ὡς Ἀριστοτέλης λέγει, καὶ τήν γε αἰτίαν πειρᾶται προστιθέναι, καὶ ἔστιν αὕτη. συνέχει τὸ πνεῦμα, ὥσπερ οὖν καὶ οἱ ὕφυδροι κολυμβηταί· καὶ γάρ τοι καὶ ἐκεῖνοι ἐντείναντες [1] ἔνδον τὸ πνεῦμα, ὥσπερ οὖν νευράν, εἶτα τὸ σῶμα ὡς βέλος ἀφιᾶσι. τὸ δὲ θλιβόμενόν φησιν ἔνδον ὠθεῖ τε καὶ ἐξακοντίζει αὐτούς.

13. Ἡ δὲ φῦσά ἐστιν ἰχθὺς Αἰγύπτιος θαυμάσαι ἄξιος. οἶδε γὰρ ὥς φασιν ὁπότε ἡ σελήνη λήγει, οἶδε δὲ αὐτῆς καὶ τὴν αὔξησιν. καὶ οὖν καὶ τὸ ἧπαρ αὐτοῦ συναύξεται [2] τῇ θεῷ ἢ συμφθίνει, καὶ πῇ μὲν εὐτραφές ἐστι, πῇ δὲ λεπτότερον.[3]

14. Ὁ δὲ γλάνις [4] ἐστὶ μὲν περὶ τὸν Μαίανδρον καὶ τὸν Λύκον τοὺς Ἀσιανοὺς ποταμούς, τῆς δὲ Εὐρώπης περὶ τὸν Στρυμόνα, καὶ σιλούρῳ μὲν τὸ εἶδος ὅμοιός ἐστι. πέφυκε δὲ φιλοτεκνότατος ἰχθύων οὗτος. ὅταν γοῦν ἡ θήλεια ἀποκυήσῃ, ἡ

[1] καὶ οὗτοι καὶ ἐκεῖνοι συντείναντες. [2] συναύξει.
[3] εὐτραφής . . . λεπτότερος. [4] *Schn*: λάγνις.

---

[a] Coastal district in the N of Epirus.

peculiarities. It is larger, it seems, than all other bulls, even than those of Chaonia [a] which the inhabitants of Thesprotia and Epirus call 'fatted,' tracing their descent from the oxen of Geryones.[b] This Onuphis is fed upon lucerne.

12. It seems that the Dolphin is swifter and can leap higher than all other fish, in fact than all land animals also. At any rate it leaps even over a vessel, as Aristotle says [*HA* 631 a 22]; and he attempts to assign a cause for this, which is as follows. It holds its breath as divers do when under water. For, you know, divers straining the breath in their bodies, let it go like a bowstring, and with it their bodies like an arrow; and, says Aristotle, the breath compressed inside them thrusts and shoots them upwards.

The Dolphin

13. The *Physa* [c] is an Egyptian fish that fills one with astonishment, for it knows, they say, when the Moon is waning and when it is waxing. Moreover its liver grows or dwindles as that goddess does: at one time it is well-nourished, at another it is more shrunken.

The 'Physa' fish

14. The Catfish is found in the Maeander and the Lycus, the rivers of Asia Minor, and in the Strymon in Europe, and resembles the European sheat-fish. It is of all fishes the most devoted to its offspring. At any rate the female after parturition ceases to pay attention to her children, like a woman who has

The Catfish

[b] A monster possessing three heads (or bodies) and living in Spain. The capture of his oxen was the tenth Labour of Heracles.

[c] Not certainly identified; perh. the Globe-fish.

μὲν ἀφεῖται τῆς ὑπὲρ τῶν τέκνων φροντίδος, οἷα δήπου λεχώ, ὁ δὲ ἄρρην τῇ φρουρᾷ τῇ τῶν βρεφῶν ἑαυτὸν ἐπιτάξας παραμένει, πᾶν ἀναστέλλων τὸ ἐπιβουλεῦον. ἱκανὸς δέ ἐστι καὶ ἄγκιστρον καταπιεῖν, ὡς Ἀριστοτέλης φησίν.

15. Βάτραχος ὕδρον μισεῖ καὶ δέδοικεν ἰσχυρῶς. οὐκοῦν τῇ βοῇ τῇ πολλῇ πειρᾶται ἀντεκπλήττειν αὐτὸν καὶ ἀντιφοβεῖν. κροκοδίλου δὲ κακουργία [1] ἐς ἀνθρώπου τε θήραν καὶ ζῴου ἑτέρου,[2] τὴν ἀτραπὸν δι᾽ ἧς οἶδε κατιόντας ἐς ποταμὸν ἢ ἐφ᾽ ὑδρείαν ἢ ἵππου [3] ἀρδείαν ἢ καμήλου ἢ καὶ νὴ Δία ὥστε ἐπιβῆναι πλοίου, ταύτην [4] τοι νύκτωρ πολλῷ τῷ ὕδατι καταρραίνει, καὶ ἐμπλήσας τὸ στόμα ἐγχεῖ κατὰ τῆς ἀτραποῦ πολλάκις, ὀλισθηρὰν αὐτὴν ἐργάσασθαι θέλων καὶ εὐκολωτέραν ἑαυτῷ τὴν ἄγραν ἀποφαίνων· τὰ γάρ τοι [5] κατολισθάνοντα οὐ κρατεῖ τῆς ἐπιβάθρας, ἀλλ᾽ ἐκεῖνα μὲν κατηνέχθη, ὁ δὲ ὑπεπήδησεν ὑπολαβὼν καὶ δειπνεῖ. ὀλίγα δὲ κροκοδίλων πέρι ἐρῶ καὶ νῦν. οὐ πρὸς πᾶν τὸ τῶν τροχίλων γένος ἐστὶ τῷδε τῷ θηρίῳ ἔνσπονδα (πολλὰ δὲ αὐτῶν γένη καὶ ὀνόματα, τραχέα δὲ καὶ ἀκοῦσαι ἀντίτυπα, καὶ διὰ τοῦτο ἐῶ αὐτά) μόνον δὲ τὸν καλούμενον κλαδαρόρυγχον ἑταῖρον καὶ φίλον ἔχει· δύναται γὰρ οὗτος ἀλύπως ἐκλέγειν αὐτῷ τὰς βδέλλας.

[1] κακουργία καὶ ἐκείνη MSS, πανουργία *Radermacher.*

[2] ἑτέρου ἐτράπη MSS, ἐτράπη *del. edd.*; *H marks a lacuna, Radermacher places a comma, after* ἑτέρου.

[3] ἵππου τινός.

[4] *Pauw* : ταύτῃ.

[5] *Perh. a subst. is missing, H.*

newly given birth, whereas the male takes charge of the young things, stays by them, and wards off every attempt upon them. And he is quite capable, according to Aristotle [*HA* 621 b 2], of swallowing [a] a fish-hook.

Frog and Water-snake

15. The Frog abhors and greatly dreads the water-snake. Accordingly, in return it tries to terrify and scare the water-snake by its loud croaking. The malice of the Crocodile in its pursuit of men and other animals ⟨is shown by the following example⟩. When it knows the path by which men come down to a river either to draw water or to water a horse or a camel or even to embark on a vessel, it floods the track with a quantity of water by night and filling its mouth, pours the contents on the path again and again, meaning to make it slippery and to render the capture easier for itself. For when ⟨men or animals⟩ slip they do not retain their hold on the gang-plank but fall off, whereupon the Crocodile, leaping up, seizes and makes a meal of them. I have still to mention a few facts touching Crocodiles. This animal is not well-disposed to every species of Egyptian plover (and there are many species, with names harsh and repulsive to the ear, and so I omit them); it is only the Clapperbill,[b] as it is called, that it treats as companion and friend, for this bird is able to pick off the leeches without coming to harm.

The Crocodile

and the Clapperbill

[a] Ar. says *συνδάκνων διαφθείρει τὰ ἄγκιστρα.*

[b] Another name for the *τροχίλος*, the Egyptian plover. See above, 3. 11; 8. 25.

16. Λέγει Δημόκριτος πολύγονα εἶναι ὗν καὶ κύνα, καὶ τὴν αἰτίαν προστίθησι λέγων, ὅτι πολλὰς ἔχει τὰς μήτρας καὶ τοὺς τόπους τοὺς δεκτικοὺς τοῦ σπέρματος. ὁ τοίνυν θορὸς οὐκ ἐκ μιᾶς ὁρμῆς ἁπάσας αὐτὰς ἐκπληροῖ, ἀλλὰ δίς τε καὶ τρὶς ταῦτα τὰ ζῷα ἐπιθόρνυται, ἵνα ἡ συνέχεια πληρώσῃ τὰ τοῦ γόνου δεκτικά. ἡμιόνους δὲ λέγει μὴ τίκτειν· μὴ γὰρ ἔχειν ὁμοίας μήτρας τοῖς ἄλλοις ζῴοις, ἑτερομόρφους δέ, ἥκιστα δυναμένας γονὴν δέξασθαι· μὴ γὰρ εἶναι φύσεως ποίημα τὴν ἡμίονον, ἀλλὰ ἐπινοίας ἀνθρωπίνης καὶ τόλμης ὡς ἂν εἴποις μοιχιδίου [1] ἐπιτέχνημα τοῦτο καὶ κλέμμα. δοκεῖ δέ μοι, ἦ δ᾽ ὅς, ὄνου ἵππον βιασαμένου [2] κατὰ τύχην κυῆσαι, μαθητὰς δὲ ἀνθρώπους τῆς βίας ταύτης γεγενημένους εἶτα μέντοι προελθεῖν ἐπὶ τὴν τῆς γονῆς αὐτῶν συνήθειαν. καὶ μάλιστά γε τοὺς τῶν Λιβύων ὄνους μεγίστους ὄντας ἐπιβαίνειν ταῖς ἵπποις οὐ κομώσαις ἀλλὰ κεκαρμέναις· ἔχουσα γὰρ τὴν ἑαυτῆς ἀγλαΐαν τὴν διὰ τῆς κόμης οὐκ ἂν ὑπομείνειε [3] τὸν τοιόνδε γαμέτην οἱ σοφοὶ τοὺς τούτων γάμους φασίν.

17. Ἐν τοῖς νοτίοις μᾶλλον ἐκπίπτειν τὰ ἔμβρυα Δημόκριτος λέγει ἢ ἐν τοῖς βορείοις, καὶ εἰκότως· χαυνοῦσθαι γὰρ ὑπὸ τοῦ νότου τὰ σώματα ταῖς κυούσαις καὶ διίστασθαι. ἅτε τοίνυν τοῦ σκήνους διακεχυμένου καὶ οὐχ ἡρμοσμένου ἀλεαίνεσθαι [4] τὰ κυόμενα καὶ θερμαινόμενα δεῦρο καὶ ἐκεῖσε διολισθάνειν καὶ ἐκπίπτειν ῥᾷον· εἰ δὲ εἴη πάγος καὶ βορρᾶς καταπνέοι, συμπέπηγε μὲν τὸ ἔμβρυον,

[1] *Reiske*: μοιχίδιον.

[2] *Diels*: ὄνος . . . βιασάμενος MSS, βιάσασθαι *H*, κυῆσαι *del. H*.

16. Democritus states that the Pig and the Dog bring forth many at a birth, and he assigns the cause to the fact that they have many wombs and many places for the reception of semen. Now the seed does not fill them all at a single ejaculation, but these animals copulate twice or three times in order that the continuance of the act may fill the receptacles of the seed. Mules however, he says, do not give birth, for they have not got wombs like other animals but of a different formation and quite incapable of receiving seed; for the mule is not the product of nature but a surreptitious contrivance of the ingenuity and, so to say, adulterous daring of man. And I fancy, said Democritus, that a mare became pregnant from being by chance violated by an ass, and that men were its pupils in this deed of violence, and presently accustomed themselves to the use of the offspring. And it is especially the asses of Libya which, being very big, mount mares that have no manes, having been clipped. For those who know about the coupling of horses say that a mare in possession of the glory of her mane would never tolerate such a mate.

Democritus on the fecundity of certain animals

The Libyan Ass and mares

17. Democritus says that the foetus is dropped more easily in southern countries than in northern; and this is natural because the south wind makes the bodies of pregnant females relax and expand. So as the shelter has been loosened and is no longer close-fitting, the embryo grows warm and the heat causes it to slip this way and that and to drop out with greater ease. If however there is a frost and the north wind is blowing, the embryo is congealed and

Democritus on the effects of climate on the animal foetus

[3] ὑπομείνῃ. [4] πλανᾶσθαι καί.

δυσκίνητόν δέ ἐστι καὶ οὐ ταράττεται ὡς ὑπὸ κλύδωνος, ἅτε δὲ ἄκλυστον καὶ ἐν γαλήνῃ ὂν ἔρρωταί τε καὶ ἔστι σύντονον καὶ διαρκεῖ πρὸς τὸν κατὰ φύσιν χρόνον τῆς ζωογονίας. οὐκοῦν ἐν κρυμῷ μέν φησιν ὁ Ἀβδηρίτης συμμένει, ἐν ἀλέᾳ δὲ ὡς τὰ πολλὰ ἐκπτύεται. ἀνάγκην δὲ εἶναι λέγει τῆς θέρμης πλεοναζούσης διίστασθαι καὶ τὰς φλέβας καὶ τὰ ἄρθρα.

18. Αἰτίαν δὲ ὁ αὐτὸς λέγει τοῖς ἐλάφοις τῆς τῶν κεράτων ἀναφύσεως ἐκείνην εἶναι. ἡ γαστὴρ αὐτοῖς ὥς ἐστι θερμοτάτη ὁμολογεῖ, καὶ τὰς φλέβας δὲ αὐτῶν τὰς διὰ τοῦ σώματος πεφυκυίας παντὸς ἀραιοτάτας λέγει, καὶ τὸ ὀστοῦν τὸ κατειληφὸς τὸν ἐγκέφαλον λεπτότατον εἶναι καὶ ὑμενῶδες καὶ ἀραιόν, φλέβας τε ἐντεῦθεν [καὶ][1] ἐς ἄκραν τὴν κεφαλὴν ὑπανίσχειν παχυτάτας. τὴν γοῦν τροφὴν καὶ ταύτης γε τὸ γονιμώτατον ὤκιστα ἀναδίδοσθαι. καὶ ἡ μὲν πιμελὴ αὐτοῖς ἔξωθέν φησι περιχεῖται, ἡ δὲ ἰσχὺς τῆς τροφῆς ἐς τὴν κεφαλὴν διὰ τῶν φλεβῶν ἀναθόρνυται. ἔνθεν οὖν τὰ κέρατα ἐκφύεσθαι διὰ πολλῆς ἐπαρδόμενα τῆς ἰκμάδος. συνεχὴς οὖν οὖσα ἐπιρρέουσά τε ἐξωθεῖ τὰ πρότερα. καὶ τὸ μὲν ὑπερίσχον ὑγρὸν ἔξω τοῦ σώματος σκληρὸν γίνεται, πηγνύντος αὐτὸ καὶ κερατοῦντος τοῦ ἀέρος, τὸ δὲ ἔνδον ἔτι μεμυκὸς ἁπαλόν ἐστι. καὶ τὸ μὲν σκληρύνεται ὑπὸ τῆς ἔξωθεν ψύξεως, τὸ δὲ ἁπαλὸν μένει ὑπὸ τῆς ἔνδον ἀλέας. οὐκοῦν ἡ ἐπίφυσις τοῦ νέου κέρατος τὸ πρεσβύτερον ὡς ἀλλότριον ἐξωθεῖ,

[1] *καί del. H.*

is not easily moved, and is not rocked as it were by a wave, but as though it were in a waveless calm, remains firm and taut and endures until the time ordained by nature for its birth. And so in cold, according to the philosopher of Abdera, the foetus remains in its place, but in warmth it is generally ejected. For when the heat is excessive, he says that the veins and sex-organs are bound to expand.

18. And the same writer says that the reason why Deer grow horns is as follows. He agrees that their stomach is extremely hot, and that the veins throughout their entire body are extremely fine, while the bone containing the brain is extremely thin, like a membrane, and loose in texture, and the veins that rise from it to the crown of the head are extremely thick. The food at all events, or at any rate the most productive part of it, is distributed through the body at great speed: the fatty portion of it, he says, envelops their body on the outside, while the solid portion mounts through the veins to the brain. And this is how horns, being moistened with plentiful juices, come to sprout. The continuous flow therefore extrudes the earlier horns. And the moisture which rises and emerges from the body solidifies, the air congealing and hardening it into horns, while that which is still enclosed in the body is soft. The one portion is rendered solid by the external cold; the other remains soft owing to the internal heat. Accordingly the added growth of the new horn extrudes the older as alien, because what is within chafes and tries to push it upwards, swelling and throbbing as though it were in haste to be born and to emerge, for the juice, you see, burst-

Democritus on the horns of Deer

θλίβοντος τοῦ ἔνδοθεν καὶ ἀνωθεῖν τοῦτο ἐθέλοντος καὶ οἰδάνοντος [1] καὶ σφύζοντος, ὥσπερ οὖν ἐπειγομένου τεχθῆναι καὶ προελθεῖν. ἡ γάρ τοι ἰκμὰς ῥηγνυμένη [2] καὶ ὑπανατέλλουσα ἀτρεμεῖν ἀδύνατός ἐστι, γίνεται δὲ ἄρα [3] καὶ αὐτὴ σκληρὰ καὶ ἐπωθεῖται τοῖς προτέροις. καὶ τὰ μὲν πλείω ἐκθλίβεται ὑπὸ τῆς ἰσχύος τῆς ἔνδον, ἤδη δέ τινα καὶ κλάδοις περισχεθέντα καὶ ἐμποδίζοντα ἐς τὸν ὠκὺν δρόμον ὑπὸ ῥύμης [4] τὸ θηρίον ὠθούμενον ἀπήραξε. καὶ τὰ μὲν ἐξώλισθε, τὰ δὲ ἕτοιμα ἐκκύπτειν ἡ φύσις προάγει.

19. Οἱ τομίαι βόες (Δημόκριτος λέγει), σκολιὰ καὶ λεπτὰ καὶ μακρὰ φύεται τὰ κέρατα αὐτοῖς, τοῖς δὲ ἐνόρχοις παχέα τὰ πρὸς τῇ ῥίζῃ καὶ ὀρθὰ καὶ ἐς [5] μῆκος προήκοντα ἧττον. καὶ πλατυμετώπους εἶναι λέγει τούτους τῶν ἑτέρων πολλῷ μᾶλλον· τῶν γὰρ φλεβῶν πολλῶν ἐνταῦθα οὐσῶν, εὐρύνεσθαι τὰ ὀστᾶ ὑπ' αὐτῶν. καὶ ἡ ἔκφυσις δὲ τῶν κεράτων παχυτέρα οὖσα ἐς πλάτος τὸ αὐτὸ τῷ ζῴῳ μέρος προάγει καὶ ἐκείνη· οἱ δὲ τομίαι μικρὸν ἔχοντες τὸν κύκλον τῆς ἕδρας τῆς τῶν κεράτων πλατύνονται ἧττόν φησιν.

20. Οἱ δὲ ἄκερῳ ταῦροι τὸ τενθρηνιῶδες [6] (οὕτω δὲ ὀνομάζει Δημόκριτος, εἴη δ' ἂν τὸ σηραγγῶδες λέγων) [7] ἐπὶ τοῦ βρέγματος οὐκ ἔχοντες ἀντιτύπου τοῦ παντὸς ὄντος ὀστοῦ καὶ τὰς συρροίας τῶν χυμῶν οὐ δεχομένου, γυμνοί τε καὶ ἄμοιροι γίνονται τῶν ἀμυντηρίων. καὶ αἱ φλέβες δὲ αἱ

[1] *Pauw*: ὀδυνῶντος. [2] *Triller*: πηγνυμένη MSS, *H*.
[3] δὲ ἄρα] γάρ. [4] *Ges*: ῥώμης.

ing out and mounting upwards from below cannot remain stationary, but it too solidifies and is impelled against the parts above it. And the older horns are in most cases forced out by the strength of that which is within, although in some cases the animal, forced ahead by its own momentum, has broken off horns that have got entangled in branches and hinder it from running swiftly. These then drop off, but the new horns which are ready to peep out are pushed forward by nature.

Democritus on the growth of horns in Oxen

19. Castrated Oxen, says Democritus, grow curved, thin, and long horns; whereas those of uncastrated Oxen are thick at the base, straight, and of shorter length. And he says that these have a much wider forehead than the others, for as there are many veins in that part, the bones are in consequence broader. And the growth of the horns being thicker makes that part of the animal broader, whereas castrated Oxen in which the circumference at the base of the horns is but small, have a narrower forehead, says he.

Democritus on hornless Bulls

20. But hornless Bulls, not possessing the 'honeycombed' part of the forehead (so Democritus styles it; his meaning would be 'porous'), since the entire bone is solid and does not permit the conflux of the body's juices, are unprotected and destitute of the means of self-defence. And since the veins in this

---

[5] πρός.

[6] *Schn*: θρηνῶδες.

[7] εἴη δ' ἄν . . . λέγων *transposed by Warmington*, οὐκ ἔχοντες (εἴη . . . λέγων) MSS, *H*.

κατὰ τοῦ ὀστοῦ τοῦδε ἀτροφώτεραι οὖσαι, λεπτότεραί τε καὶ ἀσθενέστεραι γίνονται. ἀνάγκη δὲ καὶ ξηρότερον τὸν αὐχένα τῶν ἀκεράτων εἶναι· λεπτότεραι γὰρ καὶ αἱ τούτου φλέβες. ταύτῃ τοι καὶ ἐρρωμέναι ἧττον. ὅσαι δὲ Ἀράβιοι βόες θήλειαι μέν εἰσι τὸ γένος, εὐφυεῖς δὲ τὰ κέρατα, ταύταις [1] ἥ γε πολλὴ ἐπίρροια τῶν χυμῶν φησι τροφὴ τῆς εὐγενοῦς βλάστης τοῖς κέρασίν ἐστιν. ἄκερῳ δὲ καὶ αὗται ὅσαι τὸ δεκτικὸν τῆς ἰκμάδος ὀστοῦν στερεώτερόν τε ἔχουσι καὶ δέχεσθαι τοὺς χυμοὺς ἥκιστον. καὶ συνελόντι εἰπεῖν αὔξης ἡ ἐπιρροὴ αἰτία τοῖς κέρασι· ταύτην δὲ ἄρα ἐποχετεύουσι φλέβες πλεῖσταί τε καὶ παχύταται καὶ ὑγρὸν κύουσαι ὅσον καὶ δύνανται στέγειν.

21. Ἴδιον δὲ τῶν ζῴων καὶ ἡ φιλανθρωπία. ἀετὸς γοῦν ἔθρεψε βρέφος. καὶ εἰπεῖν τὸν πάντα λόγον ἐθέλω, ὡς ἂν γένηται [2] μάρτυς ὧν προεθέμην. Βαβυλωνίων βασιλεύοντος Σευηχόρου Χαλδαῖοι λέγουσι τὸν γενόμενον ἐκ τῆς ἐκείνου θυγατρὸς τὴν βασιλείαν ἀφαιρήσεσθαι τὸν πάππον.[3] τοῦτο ἐκεῖνος πέφρικε, καὶ ἵνα εἴπω τι καὶ ὑποπαίσας Ἀκρίσιος γίνεται ἐς τὴν παῖδα· ἐφρούρει γὰρ πικρότατα. λάθρᾳ δὲ ἡ παῖς (ἦν γὰρ τοῦ Βαβυλωνίου σοφώτερον τὸ χρεὼν) τίκτει [4] ὑποπλησθεῖσα ἔκ τινος ἀνδρὸς ἀφανοῦς. τοῦτο οὖν οἱ φυλάττοντες δέει τοῦ βασιλέως ἔρριψαν ἐκ τῆς ἀκροπόλεως· ἦν γὰρ ἐνταῦθα ἀφειργμένη ἡ προειρημένη. οὐκοῦν ἀετὸς [5] τὴν ἔτι τοῦ παιδὸς καταφορὰν ὀξύτατα ἰδών, πρὶν ἢ τῇ γῇ προσαραχθῆναι τὸ βρέφος,

[1] καὶ ταύταις. [2] γένοιτο.

[3] πάππον. καὶ Χαλδαίων μὲν ἦν τὸ εἰρημένον θέσπισμα.

bone are somewhat under-nourished, they grow thinner and feebler. The neck too is of necessity drier in hornless Bulls, for the veins in it also are thinner. And that is why the veins are not so strong. But all the Arabian cows that have finely developed horns, have them (he says) because the copious influx of animal juices promotes the splendid growth of the horns. But even Arabian cows are hornless when they have the frontal bone that receives the moist secretions too solid and unreceptive of the animal juices. In a word, this influx is the cause of growth in horns, and the flow is introduced where the veins are most numerous, thickest, and as full of moisture as they can hold.

Eagle saves the baby Gilgamos

21. A love of man is another characteristic of animals. At any rate an Eagle fostered a baby. And I want to tell the whole story so that I may have evidence of my proposition. When Seuechorus was king of Babylon the Chaldeans foretold that the son born of his daughter would wrest the kingdom from his grandfather. This made him afraid and (if I may be allowed the small jest) he played Acrisius [a] to his daughter: he put the strictest of watches upon her. For all that, since fate was cleverer than the king of Babylon, the girl became a mother, being pregnant by some obscure man. So the guards from fear of the King hurled the infant from the citadel, for that was where the aforesaid

[a] King Acrisius for the same reason immured his daughter Danae in a brazen tower, where she was visited by Zeus in a shower of gold and gave birth to Perseus.

---

[4] *Perh.* ἄρρεν *has fallen out after* τίκτει *H.* [5] ὁ ἀετός.

ὑπῆλθεν αὐτὸ καὶ τὰ νῶτα ὑπέβαλε, καὶ κομίζει ἐς κῆπόν τινα, καὶ τίθησι πεφεισμένως εὖ μάλα. ὁ τοίνυν τοῦ χώρου μελεδωνὸς τὸ καλὸν παιδίον θεασάμενος ἐρᾷ αὐτοῦ καὶ τρέφει· καὶ καλεῖται Γίλγαμος, καὶ βασιλεύει Βαβυλωνίων. εἰ δέ τῳ δοκεῖ μῦθος τοῦτο, σύμφημι πειρώμενος ἐς ἰσχὺν κατεγνωκέναι αὐτόν· Ἀχαιμένη ⟨γε⟩ [1] μὴν τὸν Πέρσην, ἀφ' οὗ καὶ κάτεισιν ἡ τῶν Περσῶν εὐγένεια, ἀετοῦ τρόφιμον ἀκούω γενέσθαι.

22. Ἐν δὲ Κρήτῃ Ῥοκκαίας οὕτως Ἀρτέμιδος καλεῖται νεώς. ἐνταῦθα οἱ κύνες λυττῶσιν ἰσχυρῶς. ἐς ταύτην οὖν ὅταν τὴν νόσον ἐμπέσωσιν, εἶτα μέντοι ἑαυτοὺς ἐκ τῆς ἄκρας ἐπὶ τὴν κεφαλὴν ὠθοῦσιν ἐς τὴν θάλατταν.

23. Ἐν τῇ Ἐλυμαίᾳ χώρᾳ νεώς ἐστιν Ἀναΐτιδος,[2] καί εἰσιν ἐνταυθοῖ τιθασοὶ λέοντες, καὶ τοὺς ἐς τὸν νεὼν παριόντας ἀσπάζονταί τε καὶ σαίνουσι. καὶ εἰ καλοίης ἐσθίων, οἱ δὲ ὡς κλητοὶ δαιτυμόνες ἔρχονται, καὶ ὅσα ἂν ὀρέξῃς λαβόντες εἶτα ἀπίασι σωφρόνως τε καὶ κεκοσμημένως.

24. Ἐν τῇ θαλάττῃ τῇ Ἐρυθρᾷ ἰχθὺς γίνεταί φασι, καὶ ὄνομα αὐτῷ ὑγρὸς φοῖνιξ, καὶ γραμμὰς

[1] ⟨γε⟩ *add. H.* [2] *Valesius*: Ἀδώνιδος.

---

[a] The legendary (or semi-legendary) hero of the Gilgamesh Epic. See M. Jastrow, *Religion of Babylonia and Assyria*, pp. 469, 524.

[b] Rhocca, a settlement a little way S of Methymna at the western end of Crete.

girl was imprisoned. Now an Eagle which saw with its piercing eye the child while still falling, before it was dashed to the earth, flew beneath it, flung its back under it, and conveyed it to some garden and set it down with the utmost care. But when the keeper of the place saw the pretty baby he fell in love with it and nursed it; and it was called Gilgamos [a] and became king of Babylon.

If anyone regards this as a legend, I, after testing it to the best of my ability, concur in the verdict. I have heard however that Achaemenes the Persian, from whom the Persian aristocracy are descended, was nursed by an Eagle.

Dogs at Rhocca

22. In Crete there is a temple to Artemis Rhoccaea,[b] as she is called. The dogs there go raving mad. So when they are afflicted with this disease they hurl themselves head foremost from the promontory into the sea.

Tame Lions in Elam

23. In the country of Elam [c] there is a shrine to Anaïtis [d] and there are tame lions there which welcome and fawn upon those on their way to the shrine. And if you call them while you are eating they come like guests invited to a meal, and after taking whatever you offer, they depart in a modest and becoming manner.

The Water-Phoenix

24. In the Red Sea, so they say, there is a fish, and its name is the 'Water-Phoenix.' It has black

[c] A part of Susiana, at the N end of the Persian Gulf.

[d] Perhaps a Babylonian goddess, identified by the Greeks sometimes with Athena, at others with Aphrodite, most commonly with Artemis.

ἔχει μελαίνας, καὶ μεταξὺ τούτων κυαναῖς[1] σταγόσι κατέστικται.

25. Τῷ δὲ σαύρῳ τῷ ἐκεῖθι τὸ μὲν μῆκος τῷ κατὰ τὴν ἡμετέραν γινομένῳ θάλατταν ἴσον ἐστί, ῥάβδοι δὲ αὐτὸν περιέρχονται χρυσῷ προσεικασμέναι ἀπὸ τῶν βραγχίων ἐς τὴν οὐρὰν καθήκουσαι, μέση δὲ αὐτὰς διατέμνει[2] ἀργύρῳ προσεικασμένη. τὸ στόμα δὲ αὐτῷ κέχηνε, καὶ ἡ κάτω γένυς ἐς τὴν ἄνω νεύειν πέφυκε[3]· πρασίνους δὲ ἔχει τοὺς ὀφθαλμούς, βλέφαρα δὲ αὐτοὺς περιέρχεται χρυσοειδῆ. ἔστι δὲ καὶ ὁ χάραξ καλούμενος[4] τῆς αὐτῆς θαλάττης θρέμμα. ἔχει δὲ πτερύγια, καὶ χρυσῷ προσείκασται ὅσα γε ἰδεῖν τὰ παρ᾽ ἑκάτερα, καὶ νωτιαῖα ὅσα καὶ ταῦτα ἔχει χρυσοειδῆ. κατωτέρω δὲ ἄρα εἰσὶ πορφυραῖ ζῶναι τὴν χρόαν, χρυσοειδὲς δὲ καὶ τὸ οὐραῖόν μοι νόει τοῦ αὐτοῦ, πορφυραῖ δὲ ἄρα στιγμαὶ[5] τοὺς ὀφθαλμοὺς αὐτῷ μέσους ἐς κάλλος γράφουσιν. ὁ δὲ τοξότης ἐν τῇ αὐτῇ θαλάττῃ γινόμενος ἐχίνῳ ὅμοιός ἐστι τὸ εἶδος, κέντρα δὲ ἔχει στερεὰ καὶ μακρά.

26. Αἱ δὲ ὕστριχες αἱ Λιβυκαὶ κεντοῦσί τε[6] τοὺς ἁπτομένους πικρῶς καὶ μέντοι καὶ ὀδύνας ἐνεργάζονται χαλεπάς.[7] καὶ τεθνεώτων δὲ πονηρὰ τὰ ἐκ τῶν ἀκανθῶν νύγματα ἀπαντᾷ, ὥς φασιν.

27. Ἔστι δὲ ἐν τῇ θαλάττῃ τῇ Ἐρυθρᾷ καὶ πίθηκος, οὐκ ἰχθύς, ἀλλὰ σελαχῶδες ζῷον,[8] οὐ

[1] κυανέαις.
[2] ὑποπέφυκε.
[3] *Jac*: διατέμνει χρυσῇ.
[4] ὁ καλούμενος.
[5] ἀραστεγκιαί V, ἄρα γε σκιαί *other* MSS.

stripes, and between them it is speckled with dark blue dots.

25. The Horse-mackerel in the Red Sea is the same length as that which occurs in our sea: its body is encircled with stripes like gold which extend from the gills to the tail, and a silvery stripe parts them in two. Its mouth is open and the lower jaw projects beyond the upper; its eyes are green and are surrounded by lids of a golden colour.

The Horse-mackerel

The fish called *Charax* is another product of the same sea. It has fins, and the lateral ones are like gold in appearance, and so are all its dorsal fins. On the lower part of its body are rings of purple, but the tail, believe me, is golden, while purple dots colour beautifully the centre of its eyes.

The 'Charax'

The Archer,[a] which occurs in the same sea, resembles the sea-urchin in appearance and has hard, long prickles.

The 'Archer fish'

26. The Porcupines of Libya administer a sharp prick to those who touch them and even cause severe pains. Even when dead their bristles can give a nasty stab, so they say.

The Porcupine

27. There is also a Monkey[b] in the Red Sea; it is not a fish but a cartilaginous creature, and not

The Red Sea 'Monkey'

[a] The Globe- or Porcupine-fish.

[b] Thompson (*Gk. fishes*, s.v. πίθηκος) takes this to be 'a fanciful description of *Malthe*, a . . . relation of the . . . Fishing-frog.'

[6] *Reiske*: γε. [7] χαλεπὰς τὰ κέντρα.
[8] ζῷον οἱονεὶ ἄλεπον.

μέγα δὲ οὐδὲ τοῦτο. ἔοικέ γε μὴν τῷ χερσαίῳ ὁ θαλάττιος τὴν χρόαν, καὶ τὸ πρόσωπον δὲ πιθηκῶδές οἵ ἐστι. προβέβληται δὲ τοῦ λοιποῦ σώματος ἔλυτρον, οὐκ ἰχθυῶδες, ἀλλὰ ὥς γε τὸ τῆς χελώνης εἶναι. ὑπόσιμος δὲ καὶ οὗτος, οἷα δήπου καὶ ὁ χερσαῖος. τὸ δ' ἄλλο σῶμα πλατὺς κατὰ σχῆμα τὸ τῆς νάρκης, ὡς εἰπεῖν ὄρνιν εἶναι τὰς πτέρυγας ἁπλώσαντα· καὶ νηχόμενός γε ἔοικε πετομένῳ. παραλλάττει δὲ τοῦ χερσαίου [1] καὶ ταύτῃ. κατάστικτός ἐστι, πυρροὶ δέ εἰσίν οἱ κατὰ τοῦ ἰνίου πλατεῖς,[2] ὡς βράγχια. τὸ δὲ στόμα ἐπ' ἄκρῳ [3] τῷ προσώπῳ ἔχει μακρόν, συμφυῶς [4] τῇ τοῦ χερσαίου πλάσει καὶ κατὰ τοῦτο ὁ ἰχθὺς εἰκασμένος.

28. Ἡ ἀηδὼν διὰ τοῦ θέρους καὶ τὴν χρόαν ἐκτρέπει ἐς εἶδος ἕτερον, καὶ μεταβάλλει τὸ φώνημα· οὐ γὰρ ᾄδει πολυήχως καὶ ποικίλως, ἑτέρως δὲ ἤπερ οὖν διὰ τοῦ ἦρος. κόσσυφος δὲ θέρους μὲν ᾄδει, χειμῶνος δὲ παταγεῖ καὶ τεταραγμένον φθέγγεται, καὶ τὴν χρόαν ὡς στολὴν μεταμφιεσάμενος [5] ἀπὸ τοῦ πρόσθεν μέλανος ὑπόξανθός ἐστιν. ἥ γε μὴν κίχλη χειμῶνός ἐστι ψαροτέρα ἰδεῖν, θέρους δὲ τὸν αὐχένα ποικίλον ἐπιδείκνυσι. καὶ ἰχθῦς δὲ τὴν χρόαν μεταβλητικοὶ οἵδε, κίχλαι τε καὶ κόσσυφοι καὶ φυκίδες τε καὶ μαινίδες. οἱ δὲ θῶες, ὡς Ἀριστοτέλης λέγει, διὰ μὲν τοῦ θέρους εἰσὶ ψιλοί, δασεῖς δὲ διὰ τοῦ χειμῶνος.

29. Ἐν Βουβάστῳ δὲ τῇ Αἰγυπτίᾳ λίμνη ἐστί, καὶ τρέφει σιλούρων πάμπολυ πλῆθος, καὶ χει-

large at that. And this sea-monkey resembles the land-monkey in colour, and its face is ape-like. But the rest of its body is protected by a sheath, not like a fish but resembling that of a tortoise. It is also somewhat flat-nosed, as the land-monkey is. But the rest of its body is a flat shape like the torpedo, so that one might say that it was a bird with outspread wings; at any rate when swimming it looks like a bird in flight. But it differs from the land-monkey in this way: it is speckled, and the flat parts on the nape of the neck are red, and so are the gills. It has a large mouth at the extremity of its face, and in this respect also the fish bears a natural resemblance to the shape of the land-monkey.

Change of colour in birds and fishes

28. During the summer the Nightingale assumes a different colour and alters its note, for its song is not resonant and varied but different from its song in spring. The blackbird sings in summertime, but in winter it utters a chattering and confused sound, and changing its colour like a garment, from being black appears light brown. And the thrush in winter appears somewhat speckled, whereas in summer it displays a mottled neck. The following fish too change their colour, various wrasses (*ciclae*, *cossyphi*, and *phycides*), and sprats. And jackals, according to Aristotle [*HA* 630 a 15], are hairless throughout the summer but in winter have thick coats.

The Nile Perch

29. At Bubastus in Egypt there is a pool and it fosters an immense multitude of Nile Perch, and

[1] τῷ χερσαίῳ. [2] πλατεῖς *a substantive is missing.*
[3] οὐκ ἐπ' ἄκρῳ. [4] συμφυῶς δέ.
[5] μεταμφιασάμενος.

ροήθεις εἰσὶν οὗτοί γε καὶ ἰχθύων πραότατοι. καὶ ἐμβάλλουσιν αὐτοῖς ἄρτων τρύφη, οἱ δὲ ἀνασκιρτῶσι καὶ πηδῶντες ἄλλος πρὸ ἄλλου τὰς ἐμβαλλομένας τροφὰς ἐκλέγουσι. γίνεται δὲ ἄρα ὁ ἰχθὺς ὅδε καὶ ἐν ποταμοῖς, ὥσπερ οὖν ἐν τῷ Κύδνῳ τῷ Κιλικίῳ· βραχὺς δὲ οὗτός ἐστι τὸ μέγεθος. τὸ δὲ αἴτιον, οὐ τρέφει τοῦτον ἀφθόνως διειδὲς νᾶμα καὶ καθαρὸν καὶ προσέτι καὶ ψυχρόν (τοιοῦτος δὲ ὁ Κύδνος ἐστί), τεθολωμένῳ δὲ καὶ ἰλύος μεστῷ φιληδεῖ μᾶλλον καὶ ἐνταῦθα πιαίνεται. Πύραμος δὲ καὶ Σάρος τρέφουσι τούτων ἁδροτέρους, καὶ οὗτοι δὲ Κίλικές εἰσιν. εἶεν δ' ἂν οἱ αὐτοὶ τρόφιμοι καὶ Ὀρόντου τοῦ Σύρων, καὶ μέντοι καὶ Πτολεμαῖος [1] ποταμὸς μεγίστους τρέφει, καὶ λίμνη δὲ ἡ Ἀπαμεῖτις.

30. Χειροήθεις δὲ ἰχθῦς καὶ ὑπακούοντες τῇ κλήσει καὶ τροφὰς ἀσμένως δεχόμενοι πολλαχόθι καὶ εἰσὶ καὶ τρέφονται, ὥσπερ οὖν καὶ ἐν Ἠπείρῳ ἐν † ἐστῶτι † [2] μὲν τῇ πόλει, ἣν ἐκάλουν πάλαι Στεφανήπολιν, ἐν τῷ νεῷ τῆς Τύχης ἐν ταῖς παρ' ἑκάτερα ἀνιόντων δεξαμεναῖς, καὶ ἐν Ἑλώρῳ δὲ τῆς Σικελίας, ὅπερ ἦν πάλαι Συρακοσίων φρούριον, καὶ ἐν τῷ ἱερῷ δὲ τοῦ Λαβρανδέως Διὸς ἐν κρήνῃ [3] διειδοῦς νάματος, καὶ ἔχουσιν ὁρμίσκους

[1] *Πτολεμαίων.* [2] *Corrupt*, ἐν Κασσώπῃ? *H* (1858).
[3] *Jac*: ἐστι κρήνη.

[a] This is A.'s name for a canal, begun in the 14th cent. B.C. and intended to afford a passage for ships from the Mediterranean to the Red Sea. It linked the Nile with the Bitter Lakes, turned S, and again linked them with the Red Sea.

these are tame and the gentlest of fish. People throw in morsels of bread to them, and they leap up, each trying to jump quicker than the other, and pick out the food that is being thrown in. This fish is also found in rivers, for instance in the Cydnus in Cilicia; but there it is small. And the reason is that a stream which is clear, pure, and cold besides (for such is the Cydnus) does not afford it plentiful nourishment, for the fish prefers turbid water full of mud, and fattens on it. But the Pyramus and the Sarus breed larger kinds; these also are rivers of Cilicia. And it must be the same fish that are bred in the Syrian Orontes, but the largest of all are bred in the river Ptolemaeus [a] and in the lake of Apamea.[b]

30. Tame fishes which answer to a call and gladly accept food are to be found and are kept in many places, in Epirus for instance, at the town . . .[c] formerly called Stephanepolis, in the temple of Fortune in the cisterns on either side of the ascent; at Helorus too in Sicily which was once a Syracusan fortress; and at the shrine of Zeus of Labranda [d] in a spring of transparent water. And there fish have golden necklaces and earrings also of gold. The

Tame fish of various lands

After silting up it was cleared by order of Darius. It had to be dug again in the time of the Ptolemies, but by the 8th cent. A.D. had ceased to be navigable. See Hdt. 2. 158, Diod. Sic. 1. 33, Strabo 17. 1. 25.

[b] Apamea was an important town in the Valley of the Orontes. Schol. on Opp. *Cyn.* 2. 120 gives the name of the lake as Meliboea.

[c] Cassope, suggested by H., was a town in Epirus, a few mi. N of the Ambracian gulf; but it is not known to have been called Stephanepolis, nor is any town of this name recorded elsewhere.

[d] Labranda and Mylasa, towns in Caria.

χρυσοῦς καὶ ἐλλόβια, χρυσᾶ μέντοι καὶ ταῦτα. ἀφέστηκε δὲ ὁ νεὼς τοῦ Διὸς τοῦδε τῆς Μυλασέων πόλεως σταδίους ἑβδομήκοντα. τὸ δὲ ἄγαλμα [1] ξίφος παρήρτηται, καὶ τιμᾶται καλούμενος Κάριός τε καὶ Στράτιος· πρῶτοι γὰρ οἱ Κᾶρες ἀγορὰν πολέμου ἐπενόησαν, καὶ ἐστρατεύσαντο ἀργυρίου, ὄχανά τε ταῖς ἀσπίσι προσήρτησαν, καὶ λόφους ἐνέπηξαν τοῖς κράνεσιν. ἐκλήθησαν δὲ τὸ ὄνομα τοῦτο ἀπὸ Καρὸς τοῦ Κρήτης καὶ Διός· Ζεὺς δὲ Λαβρανδεὺς ὕσας λάβρῳ καὶ πολλῷ τὴν ἐπωνυμίαν τήνδε ἠνέγκατο.[2] καὶ ἐν Χίῳ δὲ ἐν τῷ καλουμένῳ Γερόντων λιμένι τιθασῶν ἰχθύων πλῆθός ἐστιν, οὕσπερ οὖν ἐς παραμυθίαν τοῦ γήρως τοῖς πρεσβυτάτοις οἱ Χῖοι τρέφουσι. καὶ ἐν τῇ γῇ δὲ τῇ τῶν ποταμῶν τοῦ τε Εὐφράτου καὶ τοῦ Τίγρητος μέσῃ πηγὴ ὑμνεῖται καὶ ἐς βυθὸν [3] διειδὴς εἶναι καὶ ἐκβάλλειν ὕδωρ ἰδεῖν λευκόν, καὶ γίνεται ποταμὸς τὸ ἐκπῖπτον ὄνομα Ἀβόρρας.[4] ἐπᾴδουσί τε τῷ ὀνόματι οἱ ἐπιχώριοι καὶ λόγον ἱερόν, καὶ ἔστιν ὁ λόγος, ἡ Ἥρα μετὰ τοὺς γάμους τοῦ Διὸς ἐνταῦθα ἀπελούσατο, ὥς φασι Σύροι,[5] καὶ ἐς νῦν ὁ χῶρος εὐωδίαν ἀναπνεῖ, καὶ πᾶς ὁ ἀὴρ κύκλῳ ταύτῃ κίρναται. καὶ ἐνταῦθα σκιρτῶσιν ἰχθύων πράων ἀγέλαι.

31. Τὰ δὲ ἴδια τῶν ζῴων εἰδέναι οὐδὲ θεοὶ ὑπερορῶσιν. ἀκούω γοῦν Εὐρυσθένην καὶ Προκλέα τοὺς ⟨Ἀριστοδήμου τοῦ⟩ [6] Ἀριστομάχου τοῦ

[1] *Jahn*: εἰς τὸ ἄγαλμα δέ.
[2] ἐνέγκατο ἔχειν.
[3] βυθὸν κάτω.
[4] Ἀβύρρας *Schn*: βούρρας.

shrine of this Zeus is 70 *stades*[a] distant from the city of Mylasa. A sword is attached to the side of the statue, and the god is worshipped under the name of 'Zeus of Caria' and 'God of War,' for the Carians were the first to think of making a trade of war and to serve as soldiers for pay, to fit arm-straps to their shields, and to fix plumes on their helmets. And they were called 'Carians' after Car the son of Creta and Zeus, and Zeus received the title of *Labrandeus* because he sent down furious (*labros*) and heavy rainstorms. And in Chios in what is called 'The Old Men's Harbour' there are multitudes of tame fish, which the inhabitants of Chios keep to solace the declining years of the very aged. And in the country that lies between the Euphrates and the Tigris there is a spring which is celebrated as being transparent to the bottom and as sending forth bright, clear water, which as it brims over becomes the river Aborras.[b] And the people of the country attach a sacred story to the name, which is as follows. After her marriage with Zeus Hera bathed herself there, so the Syrians say, and to this day the spot exhales a fragrance, and all the air round about is permeated with it. And there tame fishes gambol in shoals.

31. Even the gods do not disdain to take cognisance of the characteristics of animals. At any rate I learn that Eurysthenes and Procleus, the sons of

The sons of Aristodemus and the Delphic oracle

[a] About $7\frac{1}{2}$ miles.

[b] The Aborras (or Chaborras, the form preferred by some) is a large river with many tributaries, and itself becomes a tributary of the Euphrates.

---

[5] *οἱ Σύροι.*

[6] Ἀριστοδήμου τοῦ *add. Sylburg.*

Κλεόδα [1] τοῦ Ὕλλου τοῦ Ἡρακλέους παῖδας βουλομένους ἄγεσθαι γυναῖκας ἐλθόντας ἐς Δελφοὺς τὸν θεὸν ἐρέσθαι τίνι ἂν κηδεύσαντες Ἑλλήνων ἢ βαρβάρων εἶτα μέντοι καλῶς καὶ εὐβούλως γῆμαι δόξαιεν, τὸν δὲ θεὸν αὐτοῖς ἀποκρίνασθαι ἐπανιέναι μὲν ἐς Λακεδαίμονα, ὑποστρέφειν δὲ κατὰ τὴν ὁδὸν ταύτην, καθ᾽ ἣν καὶ ἀφίκοντο. ἐν ᾗ δ᾽ ἂν αὐτοῖς χώρᾳ τὸ ἀγριώτατον ἀπαντήσῃ ζῷον φέρον τὸ πραότατον, ἐνταῦθά τοι ἁρμόσασθαι γάμους· οὕτω γὰρ αὐτοῖς ἔσεσθαι λῷον. καὶ οἱ μὲν ἐπείθοντο, γίνονται δὲ [2] κατὰ τὴν Κλεωναίων χώραν, ἐντυγχάνει δὲ αὐτοῖς λύκος φέρων ἄρνα ἔκ τινος ποίμνης [3] αὐτὸν συνηρπακώς. συνέβαλον οὖν ἐκεῖνοι λέγειν ταῦτα τὰ ζῷα τὸν χρησμόν, καὶ ἡρμόσαντο τὰς Θερσάνδρου τοῦ Κλεωνύμου θυγατέρας δοκίμου ἀνδρός. εἰ δὲ οἱ θεοὶ ἴσασι τὸ ἡμερώτατον ζῷον καὶ τὸ ἀγριώτατον, οὐδὲ ἡμῖν ἐκμελὲς τὰς φύσεις αὐτῶν εἰδέναι.

32. Ἡ Ἰνδῶν γῆ φέρει [4] πολλὰ καὶ ποικίλα. καὶ τὰ μὲν εὐδαίμονός ἐστι καὶ θαυμαστῆς μαρτύρια φορᾶς, τὰ δὲ οὐκ ἀξιόζηλα [5] οὐδὲ οἷα ἐπαινεῖν ἢ ποθεῖν ἄξια. καὶ ὑπὲρ μὲν τῶν λυσιτελῶν ἢ ἁβρῶν τε καὶ πολυτελῶν τὰ μὲν εἶπον, τὰ δὲ εἰρήσεται σὺν τοῖς θεοῖς· τὸ δὲ νῦν ἔχον ὅπως ὠδῖνα ὄφεων ἡ γῆ ἐπιδείκνυται [6] εἰπεῖν ὥρμημαι. πολλοὺς τίκτει καὶ διαφόρους, καὶ † τὸ λειφθὲν τοῖς ἀπείροις ἄπειρον.† [7] οὗτοι οὖν [8] οἱ ὄφεις καὶ

[1] *Lobeck* : Κλεάδα *or* Κλεόδου.
[2] οἱ δὲ γίνονται *or* γ. δή.
[3] *Ges* : ποιμένος.
[4] φέρει μέν.

Aristodemus, son of Aristomachus, son of Cleodas, son of Hyllus the son of Heracles, wishing to wed, went to Delphi to ask the god with whom, whether Greek or barbarian, they should ally themselves in order to appear as having made a prosperous and wise marriage. And the god answered: Go back to Sparta, returning by the way you came, and wherever the fiercest animal carrying the gentlest meets you, there plight your troth; for that will be better for you. So they obeyed and arrived in the territory of Cleonae[a] where a wolf met them carrying a lamb which it had snatched from a flock. So they reckoned that the oracle meant these animals, and they took the daughters of Thersander, son of Cleonymus, a man of good repute, to wife.

Now if the gods know what animal is the gentlest and what the fiercest, it is not unfitting that we too should know their natures.

32. The land of India bears a great number and variety of creatures. And some are evidence of its beneficent and wonderful fertility, others are not to be envied nor such as one can commend or desire. Something about those that are profitable or are luxuries of great price I have already said; more shall be, please god, said hereafter. But for the present I intend to describe how the earth shows the pain with which it bears snakes. Many and various

The Snakes of India

[a] Town some 7 or 8 mi. SW of Corinth.

---

[5] ἀξιόζηλα αὐτῆς. [6] ἀποδείκνυται.

[7] τὸ λειφθὲν . . . ἄπειρον *corrupt. Perh.* ἄπιστον *Gow*, τὸ λ. τοι ἀπειράκις ἄπειρον *Post.*

[8] οὖν ἄρα.

ἀνθρώπους καὶ τὰ ἄλλα ζῷα ἀδικοῦσι. τίκτει δὲ ἡ αὐτὴ γῆ καὶ πόας τῶν δηγμάτων ἀμυντηρίους, ἔχουσί τε αὐτῶν τὴν ἐμπειρίαν τε καὶ σοφίαν οἱ ἐπιχώριοι, καὶ ποῖον φάρμακον ὄφεως τίνος ἀντίπαλόν ἐστι κατεγνώκασι, καὶ ἀμύνουσιν ὡς ὅτι τάχιστα ἀλλήλοις, ἐπιτεμέσθαι πειρώμενοι τὴν τοῦ ἰοῦ κατὰ τοῦ σώματος ἐπινομὴν ὀξυτάτην τε οὖσαν καὶ ὠκίστην. καὶ ταῦτα μὲν αὐτοῖς ἐς ἐπικουρίαν τὴν ἀναγκαίαν καὶ μάλα εὐπόρως ἀνίησιν ἡ χώρα καὶ ἀφθόνως· ὄφις δὲ ὃς ἂν ἀποκτείνῃ ἄνθρωπον, ὡς Ἰνδοὶ λέγουσιν (καὶ μάρτυρας ἐπάγονται Λιβύων πολλοὺς καὶ τοὺς περὶ Θήβας οἰκοῦντας Αἰγυπτίων), οὐκέτι καταδῦναι καὶ ἐσερπύσαι ἐς τὴν ἑαυτοῦ οἰκίαν ἔχει, τῆς γῆς αὐτὸν μὴ δεχομένης, ἀλλ᾽ ἐκβαλλούσης τῶν οἰκείων ὡς ἂν εἴποις φυγάδα κόλπων. ἀλήτης δ᾽ ἐντεῦθεν καὶ πλάνης περιέρχεται, καὶ ταλαιπωρεῖται ὑπαίθριος καὶ διὰ τοῦ θέρους καὶ διὰ τοῦ χειμῶνος, καὶ οὔτε ἔτι σύννομος αὐτῷ πρόσεισιν, οὔτε οἱ ἐξ αὐτοῦ γεννώμενοι γνωρίζουσι ⟨τὸν⟩ [1] πατέρα. τιμωρία μὲν δὴ καὶ τοῖς ἀλόγοις ἐπ᾽ ἀνδροφονίᾳ παρὰ τῆς φύσεως τοιάδε ἐδείχθη, [προνοίᾳ τοῦ θείου,] [2] κατά γε τὴν μνείαν τὴν ἐμήν· εἴρηται δὲ ἐς παίδευσιν τῷ συνιέντι.

33. Φυλάττειν δὲ ἄρα κύνες χηνῶν ἀχρειότεροι, καὶ τοῦτο κατεφώρασαν Ῥωμαῖοι. ἐπολέμουν γοῦν αὐτοῖς οἱ Κελτοί, καὶ πάνυ καρτερῶς ὠσάμενοι [3] αὐτοὺς ἐν αὐτῇ τῇ πόλει ἦσαν, καὶ ᾕρητό γε

[1] ⟨τόν⟩ *add. H.*

[2] [προνοίᾳ τοῦ θ.] *gloss, H*: *cp.* 9. 30 *fin.*

[3] καὶ ὠσάμενοί γε.

are the snakes it bears . . .[a] Now these snakes are injurious to man and all other animals. But the same land produces herbs that counteract their bites, and the natives have experience and knowledge of them, and have observed which drug is an antidote to which snake, and come to one another's aid with all possible speed in their effort to arrest the very violent and rapid spread of the poison throughout the body. And the country produces these drugs in generous abundance to help when needed. But any snake that kills a man, so the Indians say (and they cite numerous witnesses from Libya and the inhabitants of Egyptian Thebes), can no longer descend and creep into its own home: the earth declines to receive it, but casts it out like an exile from its own bosom. Thenceforward it moves around, a vagabond and wanderer, living in distress beneath the open sky throughout summer and winter; none of its mates goes near it any more, nor do those which it has begotten recognise their sire. Such is the punishment for manslaughter which Nature has shown to befall even dumb animals [it is by divine providence], as my memory tells me. This is said for the instruction of persons of understanding.

The Geese of the Capitol

33. Dogs are less useful at keeping watch than geese, as the Romans discovered. At any rate the Celts were at war with them, and had thrust them back with overwhelming force and were in the city

[a] Reading ἄπιστον, tentatively suggested by Gow, we might render 'and what is omitted would be incredible to the uninformed'; or following Post, 'and what is omitted is of course absolutely infinite.'

αὐτῶν ἡ Ῥώμη πλὴν τοῦ λόφου τοῦ Καπετωλίου· ἦν γὰρ αὐτοῖς οὐκ ἐπιβατὸς ἐκ τοῦ ῥᾴστου. τὰ μὲν οὖν δοκοῦντα δέξασθαι οἷά τε χωρία τοὺς ἐπιόντας σὺν ἐπιβουλῇ, ἐπέφρακτο [1] μέντοι ταῦτα. ἦν δὲ ὁ χρόνος, καθ' ὃν Μάρκος Μάλλιος ὑπατεύων τὸν λόφον τὸν προειρημένον ἐγχειρισθέντα οἱ διεφύλαττεν. οὗτός τοι καὶ τὸν υἱὸν ἀριστεύσαντα μὲν ἀνέδησε στεφάνῳ, ὅτι δὲ ἐκ τῆς ἑαυτοῦ μετῆλθε τάξεως, ἀπέκτεινεν. ἐπεὶ δὲ οἱ Κελτοὶ πανταχόθεν ἄβατα ἐθεώρουν εἶναί σφισι, τῆς νυκτὸς τὸ ἄκρατον [2] ἔκριναν ἐλλοχήσαντες εἶτα ἐπιθέσθαι καθεύδουσι βαθύτατα, ἔσεσθαι δὲ ἐπιβατὰ ἑαυτοῖς ἤλπισαν κατά τε [3] τὸ ἀφύλακτον καὶ ἔνθα ἐρημία [4] ἦν, τῶν Ῥωμαίων πεπιστευκότων μὴ ἂν ἐντεῦθεν ἐπιθέσθαι [5] τοὺς Γαλάτας. καὶ μέντοι καὶ ἐκ τούτων ἀκλεέστατα ἐλήφθη ἂν καὶ αὐτὸς καὶ ἡ ἄκρα τοῦ Διός, εἰ μὴ χῆνες παρόντες ἔτυχον· οἱ μὲν γὰρ κύνες πρὸς τὴν ῥιφεῖσαν τροφὴν κατεσιώπησαν, ἴδιον δὲ ἄρα χηνῶν πρὸς τὰ ῥιπτούμενα ἐς ἐδωδήν σφισι βοᾶν καὶ μὴ ἀτρεμεῖν. οὐκοῦν ἀνέστησάν τε τὸν Μάλλιον ἀνακλάγξαντες καὶ τὴν περικειμένην φυλακήν. ταῦτά τοι τίνουσι δίκας οἱ κύνες παρὰ Ῥωμαίοις καὶ νῦν ἀνὰ πᾶν ἔτος προδοσίας ἀρχαίας μνήμῃ, τιμᾶται δὲ χὴν τεταγμέναις ἡμέραις, καὶ ἐν φορείῳ πρόεισιν εὖ μάλα πομπικῶς.

34. Καὶ ταῦτα μέντοι [6] ὑπὲρ ζῴων εἰπεῖν οὐκ ἔστιν ἀπὸ μούσης. Σκύθαι ξύλων ἀπορίᾳ ἅτινα ἂν καταθύσωσι τοῖς αὐτῶν ὀστοῖς ἕψουσι. Φρύγες δὲ ἐὰν παρ' αὐτοῖς τις ἀροτῆρα ἀποκτείνῃ βοῦν,

[1] πέφρακτο. [2] ἀόρατον. [3] γε.

itself; indeed they had captured Rome, except for the hill of the Capitol, for that was not easy for them to scale. For all the spots which seemed open to assault by stratagem had been prepared for defence. It was the time at which Marcus Manlius, the consul, was guarding the aforesaid height as entrusted to him. (It was he, you remember, who garlanded his son for his gallant conduct, but put him to death for deserting his post.) But when the Celts observed that the place was inaccessible to them on every side, they decided to wait for the dead of night and then fall upon the Romans when fast asleep; and they hoped to scale the rock where it was unguarded and unprotected, since the Romans were confident that the Gauls would not attack from that quarter. And as a result Manlius himself and the Citadel of Jupiter would have been captured with the utmost ignominy, had not some geese chanced to be there. For dogs fall silent when food is thrown to them, but it is a peculiarity of geese to cackle and make a din when things are thrown to them to eat. And so with their cries they roused Manlius and the guards sleeping around him. This is the reason why up to the present day dogs at Rome annually pay the penalty of death in memory of their ancient treachery, but on stated days a goose is honoured by being borne along on a litter in great state.

34. It would not be out of place to mention these further facts touching animals. The Scythians for want of fire-wood cook with the bones of any animal that they sacrifice. Among the Phrygians any man

Various customs relating to animals

[4] *Klein* : ἠρεμία MSS, *H.* [5] ἐπιθήσεσθαι. [6] μέν.

ἡ ζημία θάνατος αὐτῷ. Σαγαραῖοι δὲ τῇ Ἀθηνᾷ καμήλων ἀγῶνα ὅσα ἔτη σὺν αἰδοῖ τῇ τῆς θεοῦ ἐπιτελοῦσι, γίνονται δὲ ἄρα παρ' αὐτοῖς αὗται δρομικώταταί τε ἅμα καὶ ὤκισται. Σαρακόροι δὲ οὔτε ἀχθοφόρους οὔτε ἀλοῦντας ἔχουσι τοὺς ὄνους ἀλλὰ πολεμιστάς, καὶ ἐπ' αὐτῶν γε τοὺς ἐνοπλίους κινδύνους ὑπομένουσιν, ὥσπερ οὖν οἱ Ἕλληνες ἐπὶ τῶν ἵππων. ὅστις δὲ ἄρα ⟨τῶν⟩[1] παρ' αὐτοῖς ὄνων ογκωδέστερος εἶναι δοκεῖ, τοῦτον τῷ Ἄρει προσάγουσιν ἱερόν. λέγει δὲ Κλέαρχος ὁ ἐκ τοῦ περιπάτου μόνους Πελοποννησίων Ἀργείους ὄφιν μὴ ἀποκτείνειν· ἐν δὲ ταῖς ἡμέραις, ἃς καλοῦσιν ἀρνηίδας οἱ αὐτοί, ἐὰν κύων ἐς τὴν ἀγορὰν παραβάλῃ, ἀναιροῦσιν αὐτόν. ἐν Θετταλίᾳ δὲ ὁ μέλλων γαμεῖν θύων τὰ γαμοδαίσια[2] ἵππον ἐσάγει πολεμιστὴν τὸν χαλινὸν περικείμενον καὶ τὴν ἐνόπλιον σκευὴν καὶ ἐκείνην πᾶσαν· εἶτα ὅταν ἀπὸ τῆς ἱερουργίας γένηται καὶ σπείσῃ, τῇ νύμφῃ τὸν ἵππον ἀπὸ τοῦ ῥυτῆρος ἀγαγὼν παραδίδωσι. τί δὲ νοεῖ τοῦτο Θετταλοὶ λεγέτωσαν. Τενέδιοι δὲ τῷ[3] ἀνθρωπορραίστῃ[4] Διονύσῳ τρέφουσι κύουσαν βοῦν, τεκοῦσαν δὲ ἄρα αὐτὴν οἷα δήπου λεχὼ θεραπεύουσι. τὸ δὲ ἀρτιγενὲς βρέφος καταθύουσιν ὑποδήσαντες κοθόρνους. ὅ γε μὴν πατάξας αὐτὸ τῷ πελέκει λίθοις βάλλεται δημοσίᾳ,[5] καὶ ἔστε ἐπὶ τὴν θάλατταν

[1] ⟨τῶν⟩ *add. H.*
[2] *Ges* : γαμοδέσια.
[3] τῷ πάλαι.
[4] *Unger* : ἂν ἀρίστην.
[5] τῇ ὁσίᾳ.

who kills a ploughing ox is punished with death. The Sagaraeans [a] every year hold camel races in honour of the goddess Athena, and their camels are good at racing and very swift. The Saracori keep asses, not to carry burdens nor to grind corn but to ride in war, and mounted on them they brave the dangers of battle, just as the Greeks do on horseback. And any ass of theirs that appears to be more given to braying than others they offer as a sacrifice to the God of War. Clearchus, the Peripatetic philosopher, states that the inhabitants of Argos are the only people in the Peloponnese who refuse to kill a snake. And these same people, if a dog comes near the market-place on the days which they call *Arneïd*, kill it. In Thessaly a man about to marry, when offering the wedding sacrifice, brings in a war-horse bitted and even fully equipped with all its gear; then when he has completed the sacrifice and poured the libation, he leads the horse by the rein and hands it to his bride. The significance of this the Thessalians must explain. The people of Tenedos keep a cow that is in calf for Dionysus the Man-slayer, and as soon as it has calved they tend it as though it were a woman in child-bed. But they put buskins on the newly born calf and then sacrifice it. But the man who dealt it the blow with the axe is pelted with stones by the populace and flees until he reaches

[a] If these are to be identified with Strabo's *Sacarauli* (Ptolemy's *Sacaraucae*) they were a tribe living on the E side of the Caspian. If the word means 'dweller by the River Sagaris' they were a Sarmatian tribe between the Caspian and the sea of Azov. Herodotus (1. 125; 7. 85) mentions *Sagartians* among the nomads of Persia.—The Saracori seem to be otherwise unknown.

φεύγει. Ἐρετριεῖς δὲ τῇ ἐν Ἀμαρύνθῳ [1] Ἀρτέμιδι κολοβὰ θύουσιν.

35. Πέπυσμαι δὲ πρὸς τοῖς ἤδη μοι προειρημένοις κύνας γενέσθαι φιλοδεσπότους Ξανθίππου τοῦ Ἀρίφρονος.[2] μετοικιζομένων γὰρ τῶν Ἀθηναίων ἐς τὰς ναῦς, ἡνίκα τοῦ χρόνου ὁ Πέρσης τὸν μέγαν πόλεμον ἐπὶ τὴν Ἑλλάδα ἐξῆψε, καὶ ἔλεγον οἱ χρησμοὶ λῷον εἶναι τοῖς Ἀθηναίοις τὴν μὲν πατρίδα ἀπολιπεῖν, ἐπιβῆναι δὲ τῶν τριήρων, οὐδὲ οἱ κύνες τοῦ προειρημένου ἀπελείφθησαν, ἀλλὰ συμμετῳκίσαντο [3] τῷ Ξανθίππῳ, καὶ διανηξάμενοι ἐς τὴν Σαλαμῖνα ἀπέσβησαν.[4] λέγετον δὲ ἄρα ταῦτα Ἀριστοτέλης καὶ Φιλόχορος.

36. Τὸ ὕδωρ ὁ Κρᾶθις λευκῆς χρόας ποιητικὸν μεθίησι.[5] τὰ γοῦν πρόβατα πιόντα αὐτοῦ καὶ οἱ βόες καὶ πᾶσα ἡ τετράπους ἀγέλη, καθά φησι Θεόφραστος, λευκὰ ἐκ μελάνων γίνεται [6] ἢ πυρρῶν. καὶ ἐν Εὐβοίᾳ δὲ οἱ βόες λευκοὶ τίκτονται σχεδὸν πάντες, ἔνθεν τοι καὶ ἀργιβόειον [7] ἐκάλουν οἱ ποιηταὶ τὴν Εὔβοιαν.

37. Οἰνοχόου βασιλικοῦ (καὶ ἦν ὁ βασιλεὺς Νικομήδης ὁ Βιθυνῶν) ἀλεκτρυὼν ἠράσθη Κένταυ-

[1] *Gron*: ἐν μυρίνθῳ. [2] *Schn*: Ἀρίφρου.
[3] *Valck*: συμμετῴκισαν. [4] *Jac*: ἀπέβησαν.
[5] μεθίησι ποταμὸς ὤν. [6] ἐγίνετο.
[7] ἀργίβοιον *Lobeck, H.*

---

[a] Village on the W coast of Euboea, between 2 and 3 mi. from Eretria.

the sea. The people of Eretria sacrifice maimed animals to Artemis at Amarynthus.[a]

The Dogs of Xanthippus

35. I have learnt in addition to what I have already said that the dogs of Xanthippus,[b] son of Ariphron, were devoted to their master, for when the people of Athens were emigrating on to their ships at the time when the Persians lit the flames of their great war against Greece, and the oracles declared that it was better for the Athenians to abandon their country and to embark upon their triremes, not even the dogs of Xanthippus were left behind, but emigrated along with him, and after swimming across to Salamis died. The story is narrated by Aristotle[c] and Philochorus.

The River Crathis

36. The river Crathis[d] has water that turns things white. At all events sheep and cattle and every four-footed herd that drink of it, according to the account given by Theophrastus,[e] from being black or red turn white. And in Euboea almost all oxen are born white, hence poets used to call Euboea 'white-kined.'[f]

Birds in love with human beings

37. A cockerel of the name of Centaurus fell in love with the cup-bearer of a king (the king was

[b] Father of Pericles, commanded the Athenian fleet in the Persian war.

[c] The story does not appear in any extant writing of Aristotle; *fr.* 354 (Rose, p. 420). Plutarch (*Them.* 10) says there was but *one* dog, and it died, exhausted by its long swim.

[d] In Bruttian territory.

[e] Not in any extant work.

[f] But the word ἀργιβόειος is known only from this passage.

ρος ὄνομα, καὶ λέγει Φίλων τοῦτο. ἠράσθη δὲ ἄρα καὶ κολοιὸς ὡραίου παιδός. καὶ μελίττας δέ τινας ἐρωτικὰς εἶναι πέπυσμαι, εἰ καὶ αἱ πλείους σωφρονοῦσιν.

38. Τὴν Σφίγγα ὑπόπτερον γράφουσί τε καὶ πλάττουσι πᾶν ὅσον περὶ χειρουργίαν σπουδαῖον καὶ πεπονημένον. ἀκούω δὲ καὶ ἐν Κλαζομεναῖς σῦν γενέσθαι πτηνόν, ἥπερ οὖν ἐλυμαίνετο τὴν χώραν τὴν Κλαζομενίαν· καὶ λέγει τοῦτο Ἀρτέμων ἐν τοῖς Ὥροις [1] τοῖς Κλαζομενίων. ἔνθεν τοι καὶ χῶρος ἐκεῖ κέκληται ὑὸς πτερωτῆς ὀνομαζόμενός τε καὶ ᾀδόμενος. τοῦτο δὲ εἴ τῳ δοκεῖ μῦθος εἶναι, δοκείτω, ἐμὲ δ' οὖν περὶ ζῴου λεχθὲν καὶ μὴ λαθὸν οὐκ ἐλύπησεν εἰρημένον.

39. Ἁλίᾳ τῇ Συβάρεως παριούσῃ [2] ἐς ἄλσος Ἀρτέμιδος (ἦν δὲ ἐν Φρυγίᾳ τὸ ἄλσος) δράκων ἐπεφάνη θεῖος, μέγιστος τὴν ὄψιν, καὶ ὡμίλησεν αὐτῇ. καὶ ἐντεῦθεν οἱ καλούμενοι Ὀφιογενεῖς τῆς σπορᾶς τῆς πρώτης ὑπῆρξαν.

40. Τιμῶσι δὲ ἄρα Δελφοὶ μὲν λύκον, Σάμιοι δὲ πρόβατον, Ἀμπρακιῶταί γε μὴν τὸ ζῷον τὴν λέαιναν· τὰ δὲ αἴτια τῆς ἑκάστου τιμῆς εἰπεῖν οὐκ ἔστιν ἔξω τῆσδε τῆς σπουδῆς. Δελφοῖς μὲν χρυσίον ἱερὸν σεσυλημένον καὶ ἐν τῷ Παρνασῷ [3]

[1] *Cobet*: Ὅροις MSS, *H*. [2] *Ges*: περιούσης.
[3] Παρνασσῷ.

---

[a] Nicomedes was the name of three Bithynian kings. Athenaeus (13. 606B) gives the name of the cup-bearer as Secundus.

Nicomedes[a] of Bithynia); Philo tells the story. And a jackdaw also fell in love with a handsome boy. I learn also that some bees are amorous, although the majority are more restrained.

38. Every painter and every sculptor who devotes himself and has been trained to the practice of his art figures the Sphinx as winged. And I have heard that on Clazomenae[b] there was a sow with wings, and it ravaged the territory of Clazomenae. And Artemon records this in his *Annals of Clazomenae*. That is why there is a spot named and celebrated as 'The Place of the Winged Sow,'[c] and it is famous. But if anyone regards this as a myth, let him do so; for my part I am not sorry to have mentioned what has been related and what has not escaped my notice touching an animal.

A winged Sow

39. Halia, the daughter of Sybaris, was entering a grove of Artemis (the grove was in Phrygia) when a divine serpent appeared to her—it was of immense size—and lay with her. And from this union sprang the *Ophiogeneis* (snake-born) of the first generation.

The Snake-born

40. At Delphi they pay honour to a wolf, in Samos to a sheep, in Ambracia to a lioness; and it is not irrelevant to our present study to set out the reasons for this honour in each case. At Delphi it was a wolf that tracked down some sacred gold that

Honours paid to Animals

[b] Island some 20 mi. W from Smyrna.

[c] The fore-part of a winged boar is represented on some of the coins of Clazomenae, see *Brit. Mus. Cat. of Coins; Ionia*, pl. iii. 18, pl. vii. 2.

κατορωρυγμένον ἀνίχνευσε λύκος,[1] Σαμίοις δὲ καὶ αὐτοῖς τοιοῦτο χρυσίον κλαπὲν πρόβατον ἀνεῦρε, καὶ ἐντεῦθεν Μανδρόβουλος ὁ Σάμιος τῇ Ἥρᾳ πρόβατον ἀνάθημα ἀνῆψε· καὶ τὸ μὲν Πολέμων λέγει τὸ πρότερον, τὸ δὲ Ἀριστοτέλης τὸ δεύτερον. Ἀμπρακιῶται δέ, ἐπεὶ τὸν τύραννον αὐτῶν Φαΰλον διεσπάσατο λέαινα, τιμῶσι τὸ ζῷον αἴτιον αὐτοῖς ἐλευθερίας γεγενημένον. Μιλτιάδης δὲ τὰς ἵππους τὰς τρὶς Ὀλύμπια ἀνελομένας ἔθαψεν ἐν Κεραμεικῷ, καὶ Εὐαγόρας δὲ ὁ Λάκων καὶ ἐκεῖνος Ὀλυμπιονίκας ἵππους ἔθαψε μεγαλοπρεπῶς.

41. Ὁ Γάγγης ὁ παρὰ τοῖς Ἰνδοῖς ῥέων ὑπαρχόμενος μὲν ἐκ τῶν πηγῶν βαθύς ἐστιν ἐς ὀργυιὰς εἴκοσι, πλατὺς δὲ ἐς ὀγδοήκοντα σταδίους· ἔτι γὰρ αὐθιγενεῖ τῷ ὕδατι πρόεισι καὶ ἀμιγεῖ πρὸς ἕτερον· προϊὼν δὲ τῶν ἄλλων ἐς αὐτὸν ἐμπιπτόντων καὶ ἀνακοινουμένων οἱ τὸ ὕδωρ ἐς βάθος μὲν ἥκει καὶ ἑξήκοντα ὀργυιῶν, πλατύνεται δὲ καὶ ὑπερεκχεῖται ἐς σταδίους τετρακοσίους. καὶ ἔχει νήσους Λέσβου τε καὶ Κύρνου μείζονας, καὶ τρέφει κήτη, καὶ ἐκ τῆς τούτων πιμελῆς ἄλειφα ἐργάζονται. εἰσὶ δὲ ἐν αὐτῷ καὶ χελῶναι, καὶ αὐταῖς τὸ χελώνιον πιθάκνης καὶ εἴκοσιν ἀμφορέας δεχομένης οὐ μεῖόν ἐστι. κροκοδίλων δὲ παιδεύει διπλᾶ γένη. καὶ τὰ μὲν αὐτῶν ἥκιστα βλάπτει, τὰ δὲ παμβορώτατα

[1] ὁ λύκος.

---

[a] A mythical character whose name passed into a proverb. He was said to have dedicated to Hera a golden ram one year, a silver the next, a bronze the third, thereafter nothing.

had been pillaged and buried on Parnassus. So too for the Samians it was a sheep that discovered some stolen gold; for that reason Mandrobulus of Samos [a] dedicated a sheep to Hera. The first story is recorded by Polemon, the second by Aristotle.[b] And the people of Ambracia since the day when a lioness tore their tyrant Phaÿlus [c] to pieces, do honour to this animal as the instrument of their liberation. And Miltiades buried in Cerameicus the mares which had won three Olympic victories; Evagoras the Spartan also gave his horses which had won at Olympia a magnificent funeral.

The Ganges and its Turtles and Crocodiles

41. At its rising from wells the Ganges, the river of India, is 20 fathoms deep and 80 *stades* [d] wide, for it is still flowing with its own native waters unmixed with any other. But as it flows on and other rivers fall into it and join their water with it, it reaches a depth of 60 fathoms, and widens and overflows to an extent of four hundred *stades* [e]. And it contains islands larger than Lesbos and Cyrnus,[f] and breeds monstrous fishes, and from their fat men manufacture oil. There are also in the river turtles whose shell is as large as a jar holding as much as 20 *amphorae*.[g] And it fosters two kinds of crocodiles. Some of them are perfectly harmless, but others eat

Hence the saying ἐπὶ τὰ Μανδροβούλου χωρεῖ τὸ πρᾶγμα, 'things get steadily worse.' See Leutsch, *Paroem. Gr.* 2. 114.

[b] Not in any extant work; *fr.* 525 (Rose, p. 520).

[c] Antoninus Liberalis (4) gives the name as Phalaecus; his date is unknown.

[d] Nearly 9 miles.

[e] Just over 44 miles.

[f] The Greek name for Corsica.

[g] The ἀμφορεύς contained nearly 9 gallons. This turtle may be the *Trionyx gangeticus*.

σαρκῶν ἐσθίει καὶ ἀφειδέστατα, καὶ ἔχουσιν ἐπ' ἄκρου τοῦ ῥύγχους ἐξοχὴν ὡς κέρας. τούτοις τοι καὶ πρὸς τὰς τῶν κακούργων τιμωρίας ὑπηρέταις χρῶνται· τοὺς γὰρ ἐπὶ τοῖς μεγίστοις τῶν ἀδικημάτων ἑαλωκότας ῥίπτουσιν αὐτοῖς, καὶ δημίου δέονται ἥκιστα.

42. Δέλεαρ δὲ καθιᾶσιν οἱ σοφοὶ τὰ θαλάττια τοῖς μὲν σκάροις, ὡς Λεωνίδης φησί, κορίαννα καὶ καρτά, καὶ ἔστιν εὔθηρα ταῦτα καὶ ἑλεῖν ῥᾷστα· προσνεῖ γὰρ αὐτοῖς ὁ σκάρος γοητευόμενος ὥσπερ ἡδύσμασι. φύλλα δὲ τευτλίων αἱρεῖ τὰς τρίγλας· χαίρει γὰρ τῷδε τῷ λαχάνῳ τὸ ζῷον, καὶ δι' αὐτοῦ ἁλίσκεταί τε καὶ δουλοῦται ῥᾷστα.[1]

43. Ἐνύδρου δὲ θήρας διαφοραὶ τέτταρες, φασί, δικτυεία [2] ⟨καὶ⟩ [3] κόντωσις καὶ κυρτεία καὶ ἀγκιστρεία προσέτι. καὶ ἡ μὲν δικτυεία πλουτοφόρος, καὶ ἔοικεν ἁλισκομένῳ στρατοπέδῳ καὶ αἱρουμένοις αἰχμαλώτοις τισί, καὶ δεῖται χορηγίας ποικίλης, οἷον σπάρτου καὶ λίνου λευκοῦ καὶ μέλανος ἄλλου καὶ κυπείρου καὶ φελλῶν μολίβου τε καὶ πίτυος καὶ ἱμάντων καὶ ῥοῦ καὶ λίθου καὶ βύβλου καὶ κεράτων καὶ νεὼς ἐξήρους ἄξονός τε καὶ σκυταλίδων καὶ κοττάνης καὶ τυμπάνου καὶ σιδήρου καὶ ξύλων καὶ πίττης. ἐμπίπτει δὲ γένη τε ἰχθύων διάφορα καὶ ἀγέλαι ποικίλαι τε καὶ πολλαί. ἡ δὲ κόντωσίς [4] ἐστι μὲν τῶν ἄλλων

[1] *Ges*: ἕκαστα.
[2] *Schn*: δικτυία.
[3] ⟨καί⟩ *add. H.*
[4] *Schn*: διακόντωσις.

[a] The *Gavialis gangeticus* is said to be harmless and to have a 'horn' at the end of its snout; the other, flesh-eating kind

flesh with the utmost voracity and ruthlessness, and on the end of their snout they have an excrescence like a horn.[a] These the people employ as agents for punishing criminals, for those who are detected in the most flagrant acts are thrown to the crocodiles, and there is no need of a public executioner.

The Parrot Wrasse

42. Those who are skilled in sea-fishing let down as bait for Parrot Wrasses coriander and chopped leeks, so says Leonidas; and these herbs are successful as bait and afford an easy capture. For the Parrot Wrasse, as though bewitched by spices, swims up to them. And the leaves of beet capture the Red Mullet, for the fish delights in this vegetable, and with its aid the fish is caught and enslaved with the utmost ease.

The Red Mullet

Four methods of fishing: (a) with a net

43. There are, they say, four different methods of fishing, viz with nets, with a pole, with a weel, and with a hook. Netting fish brings wealth, and may be compared to the capture of a camp and the taking of prisoners; it requires a variety of gear, for instance rope, fishing-line white and black, cord made from galingale, corks, lead, pine timber, thongs, sumach, a stone, papyrus, horns, a six-oared ship, a windlass with handles, a *cottane*,[b] a drum, iron, timber, and pitch. And there fall into the nets fish of different kinds, varied droves in their multitude.

(b) with a pole

Fishing with a pole is the most manly form and

is the *Crocodilus palustris*. Gossen would therefore transpose καὶ ἔχουσιν . . . ὡς κέρας after ἥκιστα βλάπτει. See *RE* 11. 1947, Gadow, *Amphibia and Reptiles*, 452 (Camb. Nat. Hist. 8).

[b] κοττάνη is so far unexplained; it may be conjectured to have been some piece of machinery.

ἀνδρειοτάτη, καὶ δεῖται θηρατοῦ ῥωμαλεωτάτου. παρεῖναι δὲ χρὴ κάμακα ὀρθὴν ἐλατίνην [1] καὶ σχοινία σπάρτινα πυρεῖά τε πεύκης τῆς λιπαρωτάτης· ⟨δεῖται δὲ⟩ [2] καὶ νεὼς μικρᾶς ⟨καὶ⟩ [3] ἐρετῶν συντόνων καὶ βραχίονας ἀγαθῶν. ἡ δὲ κυρτεία δολερωτάτη θήρα καὶ ἐπιβουλοτάτη δεινῶς ἐστι, καὶ ἐλευθέροις πρέπειν [4] δοκεῖ ἥκιστα. δεῖται δὲ ὁλοσχοίνων τε ἀβρόχων καὶ λύγου καὶ χερμάδος καὶ εὐναίων [5] καὶ φύκους θαλαττίου σχοίνων τε καὶ κυπαρίττου κόμης καὶ φελλῶν καὶ ξύλων καὶ δελέατος καὶ νεὼς μικρᾶς. ἡ δὲ ἀγκιστρεία σοφωτάτη ἐστὶ καὶ τοῖς ἐλευθέροις πρεπωδεστάτη.[6] δεῖται δὲ ἄρα [7] ἱππείων τριχῶν, τὰς χρόας καὶ λευκὰς [8] καὶ μελαίνας καὶ πυρρὰς καὶ μεσαιπολίους· τῶν δὲ βαπτομένων ἐγκρίνουσι τὰς γλαυκὰς καὶ [9] τὰς ἁλιπορφύρους· αἱ γὰρ ἄλλαι πᾶσαι πονηραί, φασίν. χρῶνται δὲ καὶ τῶν ἀγρίων συῶν ταῖς θριξὶ ταῖς ὀρθαῖς καὶ τερμίνθῳ [10] δέ, καὶ χαλκῷ πλείστῳ καὶ μολίβῳ καὶ σπαρτίναις καὶ πτεροῖς, μάλιστα μὲν λευκοῖς καὶ μέλασι [11] καὶ ποικίλοις. χρῶνταί γε μὴν οἱ ἁλιεῖς καὶ φοινικοῖς ἐρίοις καὶ ἁλουργέσι καὶ φελλοῖς καὶ ξύλοις· καὶ σιδήρου καὶ ἄλλων δέονται, ἐν δὲ τοῖς καὶ καλάμων εὐφυῶν καὶ ἀβρόχων καὶ ὁλοσχοίνων βεβρεγμένων καὶ νάρθηκος ἐξεσμένου καὶ ῥάβδου κρανείας καὶ χιμαίρας κεράτων καὶ

[1] ἔλαιον. [2] ⟨δεῖται δέ⟩ *add. Schn.* [3] ⟨καί⟩ *add. Jac.*
[4] *Ges* : πρέπει. [5] εὐναίων καὶ λίθον.
[6] *Ges* : σοφώτατον . . . πρεπωδέστατον. [7] δὲ ἄρα] γάρ.
[8] τριχῶν. ⟨ὧν⟩ τ. χ. ⟨εἶναι χρῆ⟩ λευκάς *Bernhardy.*
[9] γλαυκὰς καί] λευκὰς ἤ. [10] *Schn* : τερίνθῳ.
[11] ἢ μέλασι.

needs a hunter of very great strength. He must have a straight pole of pine-wood, ropes of esparto, and firesticks of thoroughly sappy pine. He also needs a small boat and vigorous oarsmen with strong arms.

(c) with a weel

Fishing with a weel is a pursuit that calls for much craft and deep design, and seems highly unbecoming to free men. The essentials are club-rushes unsoaked, withies, a large stone, anchors, sea-weed, leaves of rushes and cypress, corks, pieces of wood, a bait, and a small skiff.

(d) with a rod and line

Fishing with a hook is the most accomplished form and the most suitable for free men. One needs horse-hair,[a] white, black, red, and grey in colour. If the hairs are dyed, men select only those coloured blue-grey and sea-purple; for all the rest, they say, are bad. Men also use the straight bristles of wild boars and flax [b] also, and a quantity of bronze and lead, cords of esparto, feathers,[c] especially white, black, and particoloured. And anglers also use crimson and sea-purple wool, corks, and pieces of wood. Iron and other materials are needed; among them reeds of straight growth and unsoaked, club-rushes that have been soaked, stalks of fennel rubbed smooth, a fishing-rod of cornel-wood, the horns and hide of a goat.[d] Some fish are caught by one device, others by another, and the

[a] For fishing-line; see 15. 10.

[b] τέρμινθος: 'a flax-like plant from which the Athenians made fishing lines' (L-S[9]).

[c] The purpose of *feathers* and *wool* is not explicitly stated until we reach 15. 1, where fishing with an artificial fly is first mentioned. See also 15. 10.

[d] Used in fishing for Sargues, 1. 23.

δέρματος. ἄλλος δὲ ἄλλῳ τούτων ἰχθὺς αἱρεῖται, καὶ τάς γε θήρας ἤδη εἶπον αὐτῶν.

44. Λόγω δὲ ἄρα τῶδε Ἰνδὸς καὶ Λίβυς τὸ γένος διαφόρω· ἐρεῖ δὲ ὁ μὲν Ἰνδὸς τὰ ἐπιχώρια, ὁ δὲ Λίβυς ὅσα οἶδε καὶ ἐκεῖνος· ἃ δ' οὖν ᾄδετον ἄμφω τὼ λόγω ἐστὶν ἐκεῖνα. ἐν Ἰνδοῖς ἐὰν ἁλῷ τέλειος ἐλέφας, ἡμερωθῆναι χαλεπός ἐστι, καὶ τὴν ἐλευθερίαν ποθῶν φονᾷ. ἐὰν δὲ αὐτὸν καὶ δεσμοῖς διαλάβῃς, ἔτι καὶ μᾶλλον ἐς [1] θυμὸν ἐξάπτεται, καὶ δοῦλος εἶναι καὶ δεσμώτης [2] οὐχ ὑπομένει. ἀλλ' οἱ Ἰνδοὶ καὶ ταῖς τροφαῖς κολακεύουσιν αὐτόν, καὶ ποικίλοις καὶ ἐφολκοῖς δελέασι πραΰνειν πειρῶνται, παρατιθέντες ὅσα πληροῖ τὴν γαστέρα καὶ θέλγει [3] τὸν θυμόν. ὁ δὲ ἄχθεται αὐτοῖς καὶ ὑπερορᾷ. τί οὖν ἐκεῖνοι κατασοφίζονται [4]; μοῦσαν αὐτοῖς προσάγουσιν ἐπιχώριον, καὶ κατᾴδουσιν αὐτοὺς ὀργάνῳ τινὶ καὶ τούτῳ συνήθει· καλεῖται δὲ σκινδαψὸς τὸ ὄργανον. ὁ δὲ ὑπέχει τὰ ὦτα καὶ θέλγεται, καὶ ἡ μὲν ὀργὴ πραΰνεται, ὁ δὲ θυμὸς ὑποστέλλεταί τε καὶ στόρνυται, κατὰ μικρὰ δὲ καὶ ἐς τὴν τροφὴν ὁρᾷ. εἶτα ἀφεῖται μὲν τῶν δεσμῶν, μένει δὲ τῇ μούσῃ δεδεμένος, καὶ δειπνεῖ προθύμως ἁβρὸς δαιτυμὼν [5]· πόθῳ γὰρ τοῦ μέλους οὐκ ἂν ἔτι ἀποσταίη. Λιβύων δὲ ἵπποι (δεῖ γὰρ ἀκοῦσαι καὶ τὸν λόγον τὸν ἕτερον), ἐς τοσοῦτον αὐτὰς αἱρεῖ ἡ αὔλησις. πραΰνονταί τε καὶ ἡμεροῦνται, καὶ ὑπολήγουσι μὲν τοῦ ὑβρίζειν τε καὶ σκιρτᾶν, ἕπονται δὲ τῷ νομεῖ ὅποι [6] ἂν αὐτὰς τὸ μέλος

[1] ἐς τόν.
[2] δεσπότης.
[3] ὡς πληροῦν . . . θέλγειν.
[4] κατασοφίζονται καὶ δρῶσι.
[5] δαιτυμὼν καταδεδεμένος.
[6] ὅπου.

various methods of catching them I have already described.

44. These two accounts from India and Libya show a difference. The Indian shall relate the practice in his country, and the Libyan shall relate what he knows. So their two accounts are as follows.

Music and the Elephant

In India if a full-grown Elephant is captured he is hard to tame and his craving for freedom makes him thirst for blood, and if you make him fast with ropes his anger is inflamed all the more and he will not stand being a slave and a prisoner. But the Indians blandish him with food and try to mollify him with a variety of attractive baits, offering him what will fill his stomach and assuage his passion. Yet he is displeased with them and takes no notice of them. So what device do the Indians adopt to meet this? They introduce native music and charm the Elephants with a musical instrument that is in common use; it is called *scindapsus*.[a] And the Elephant lends an ear and is pacified; his rage is softened, and his passion is subdued and allayed, and little by little he begins to notice his food. Then he is freed from his bonds but remains captivated by the music, and eats his food with the eagerness of a man faring sumptuously: for in his love for the music he will no longer run away.

and the Libyan Mare

But the mares of Libya (for we must listen to the second account as well) are equally captivated by the sound of the pipe. They become gentle and tame and cease to prance and be skittish, and follow the herdsman wherever the music leads them; and

[a] A four-stringed musical instrument.

ἀπάγῃ, ἐπιστάντος δὲ καὶ ἐκεῖναι ἐφίστανται· ἐὰν δὲ ἐπανατείνῃ [1] τὸ αὔλημα, λείβεται δάκρυα ὑφ' ἡδονῆς αὐταῖς. οἱ μὲν οὖν βουκόλοι τῶν ἵππων ῥοδοδάφνης κλάδον κοιλάναντες καὶ αὐλὸν ἐργασάμενοι καὶ ἐς αὐτὸν ἐμπνέοντες εἶτα οὕτω ⟨τῶν⟩ [2] προειρημένων καταυλοῦσι. λέγει δὲ Εὐριπίδης καὶ ποιμνίτας τινὰς ὑμεναίους· ἔστι δὲ ἄρα τοῦτο αὔλημα, ὅπερ οὖν τὰς μὲν ἵππους τὰς θηλείας ἐς ἔρωτα ἐμβάλλει καὶ οἶστρον ἀφροδίσιον, τοὺς δὲ ἄρρενας μίγνυσθαι αὐταῖς ἐκμαίνει. τελοῦνται μὲν ⟨δὴ⟩ [3] ἱππικοὶ γάμοι τὸν τρόπον τοῦτον, καὶ ἔοικεν ὑμέναιον ᾄδειν τὸ αὔλημα.

45. Τὸ τῶν δελφίνων φῦλον ὡς εἰσι φιλῳδοί τε καὶ φίλαυλοι, τεκμηριῶσαι ἱκανὸς καὶ Ἀρίων ὁ Μηθυμναῖος ἔκ τε τοῦ ἀγάλματος τοῦ ἐπὶ Ταινάρῳ καὶ τοῦ ἐπ' [4] αὐτῷ γραφέντος ἐπιγράμματος. ἔστι δὲ τὸ ἐπίγραμμα

ἀθανάτων πομπαῖσιν Ἀρίονα Κυκλέος [5] υἱὸν
ἐκ Σικελοῦ πελάγους σῶσεν ὄχημα τόδε.

ὕμνον δὲ χαριστήριον τῷ Ποσειδῶνι, μάρτυρα τῆς τῶν δελφίνων φιλομουσίας, οἱονεὶ καὶ τούτοις ζωάγρια ἐκτίνων ὁ Ἀρίων ἔγραψε. καὶ ἔστιν ὁ ὕμνος οὗτος·

Ὕψιστε θεῶν,
πόντιε, χρυσοτρίαινε Πόσειδον,
γαιάοχ' [6] ἐγκύμον' ⟨ἀν'⟩ ἅλμαν· [7]
βράγχιοι [8] περὶ δὲ σὲ πλωτοὶ
θῆρες χορεύουσι κύκλῳ,
κούφοισι ποδῶν ῥίμμασιν

if he stands still, so do they. But if he plays his pipe with greater vigour, tears of pleasure stream from their eyes. Now the herdsmen of the mares hollow a stick of rose-laurel, fashion it into a pipe, and blow into it, and thereby charm the aforesaid animals. And Euripides speaks of some 'marriage songs of shepherds' [*Alc.* 577]; this is the pipe-music which throws mares into an amorous frenzy and makes horses mad with desire to couple. This in fact is how the mating of horses is brought about, and the pipe-music seems to provide a marriage song.

45. Sufficient proof that Dolphins love song and the music of pipes is supplied by Arion of Methymna in his statue on Taenarum and the inscription written upon it. The inscription runs

Arion and the Dolphins

'Sent by the immortals this mount saved Arion son of Cycleus from the Sicilian main.'

And Arion wrote a hymn of thanks to Poseidon that bears witness to the Dolphins' love of music and is a kind of payment of the reward due to them also for having saved his life.

This is the hymn.

'Highest of the gods, lord of the sea, Poseidon of the golden trident, earth-shaker in the swelling brine, around thee the finny monsters in a ring

---

[1] *Jac*: παρατείνῃ.
[2] ⟨τῶν⟩ *add. Jac.*
[3] ⟨δή⟩ *add. H.*
[4] ὑπ'.
[5] *Salmasius*: Κύκλονος.
[6] *Bergk*: γαιήοχ' MSS, *H.*
[7] *Hermann*: ἐγκυμονάλμαν.
[8] βραγχίοις *Hermann, H.*

ἐλάφρ᾿ ἀναπαλλόμενοι, σιμοὶ
φριξαύχενες ὠκυδρόμοι
σκύλακες, φιλόμουσοι
δελφῖνες, ἔναλα θρέμματα
κουρᾶν Νηρεΐδων θεᾶν,
ἃς ἐγείνατ᾿ Ἀμφιτρίτα·
οἵ μ᾿ εἰς Πέλοπος γᾶν ἐπὶ Ταιναρίαν ἀκτὰν
ἐπορεύσαν [1] πλαζόμενον Σικελῷ ἐνὶ πόντῳ,
κυρτοῖσι νώτοις ὀχέοντες,[2]
ἄλοκα Νηρεΐας πλακὸς
τέμνοντες, ἀστιβῆ πόρον, φῶτες δόλιοι
ὥς μ᾿ ἀφ᾿ [3] ἁλιπλόου γλαφυρᾶς νεὼς
εἰς οἶδμ᾿ ἁλιπόρφυρον λίμνας ἔριψαν.[4]

Arion and the Dolphins

ἴδιον μὲν δήπου δελφίνων πρὸς τοῖς ἄνω λεχθεῖσι καὶ τὸ φιλόμουσον.

46. Λόγος που διαρρεῖ Τυρρηνὸς ὁ λέγων τοὺς ὗς τοὺς ἀγρίους καὶ τὰς παρ᾿ αὐτοῖς ἐλάφους ὑπὸ [5] δικτύων μὲν καὶ κυνῶν ἁλίσκεσθαι, ᾗπερ οὖν θήρας νόμος, συναγωνιζομένης δὲ αὐτοῖς τῆς μουσικῆς καὶ μᾶλλον. πῶς δέ, ἤδη ἐρῶ.[6] τὰ μὲν δίκτυα περιβάλλουσι καὶ τὰ λοιπὰ θήρατρα, ὅσα ἐλλοχᾷ τὰ ζῷα· ἕστηκε δὲ ἀνὴρ αὐλῶν τεχνίτης, καὶ ὡς ὅτι μάλιστα πειρᾶται τοῦ μέλους ὑποχαλᾶν, καὶ ὅ τι ποτέ ἐστι τῆς μούσης σύντονον ἐᾷ, πᾶν δὲ ὅ τι γλύκιστον αὐλῳδίας τοῦτο ᾄδει. . .[7] ἡσυχία τε καὶ ἠρεμία ῥᾳδίως διαπορθμεύει, καὶ ἐς τὰς ἄκρας καὶ ἐς τοὺς αὐλῶνας καὶ ἐς τὰ δάση καὶ ἐς ἁπάσας συνελόντι εἰπεῖν τὰς τῶν

[1] *Brunck*: ἐπορεύσατε MSS, *H*, *v.l.* -το.
[2] *Brunck*: χορεύοντες.
[3] *Brunck*: με ἀπό.

swim and dance, with nimble flingings of their feet leaping lightly, snub-nosed hounds with bristling neck, swift runners, music-loving dolphins, sea-nurslings of the Nereïd maids divine, whom Amphitrite bore, even they that carried me, a wanderer on the Sicilian main, to the headland of Taenarum in Pelops' land, mounting me upon their humped backs as they clove the furrow of Nereus' plain, a path untrodden, when deceitful men had cast me from their sea-faring hollow ship into the purple swell of ocean.[a]

So to the characteristics of dolphins mentioned earlier on I think we may add a love of music.

46. There is an Etruscan story current which says that the wild boars and the stags in that country are caught by using nets and hounds, as is the usual manner of hunting, but that music plays a part, and even the larger part, in the struggle. And how this happens I will now relate. They set the nets and other hunting gear that ensnare the animals in a circle, and a man proficient on the pipes stands there and tries his utmost to play a rather soft tune, avoiding any shriller note, but playing the sweetest melodies possible. The quiet and the stillness easily carry ⟨the sound⟩ abroad; and the music streams up to the heights and into ravines and thickets—in a word into every lair and resting-place of these

Music as a means of capturing Animals

[a] The poem is apocryphal and is the work of some writer of dithyrambs perhaps of the late 5th cent. B.C. See H. W. Smyth, *Gk. melic poets*, pp. 15, 205.

[4] *Hermann*: ῥίψαν.
[5] καὶ ὑπό.
[6] λέγω.
[7] *Lacuna.*

θηρίων κοίτας καὶ εὐνὰς τὸ μέλος ἐσρεῖ. καὶ τὰ μὲν πρῶτα παριόντος ἐς τὰ ὦτα αὐτοῖς τοῦ ἤχου ἐκπέπληγε [1] καί που καὶ δείματος ὑποπίμπλαται, εἶτα ἄκρατος καὶ ἄμαχος [2] αὐτὰ ἡδονὴ τῆς μούσης περιλαμβάνει, καὶ κηλούμενα λήθην ἔχει καὶ ἐκγόνων [3] καὶ οἰκιῶν.[4] καίτοι φιλεῖ τὰ θηρία μὴ ἀπὸ τῶν συντρόφων χωρίων πλανᾶσθαι. τὰ δ᾽ οὖν Τυρρηνὰ κατ᾽ ὀλίγον ὥσπερ ὑπό τινος ἴυγγος ἀναπειθούσης ἕλκεται,[5] καὶ καταγοητεύοντος τοῦ μέλους ἀφικνεῖται καὶ ἐμπίπτει ταῖς πάγαις τῇ μούσῃ κεχειρωμένα.

47. Ἀνθίαι δὲ βαλλόμενοι ὅταν ἁλῶσιν οἴκτιστόν εἰσι θεαμάτων, καὶ ἀποθνήσκοντες ἑαυτοὺς ἐοίκασι θρηνεῖν καὶ τρόπον τινὰ ἱκετεύειν, ὥσπερ οὖν ἄνθρωποι λησταῖς ἐντυχόντες ἀνοικτίστοις τε καὶ φονικωτάτοις. οἱ μὲν γὰρ αὐτῶν ἀποδιδράσκειν πειρώμενοι εἶτα τοῖς δικτύοις ἐμπαλάσσονται,[6] ὑπεράλλεσθαι δὲ αὐτοὺς πειρωμένους τὸν λόχον εἶτα μέντοι καταλαμβάνει αἰχμή· οἱ δὲ ἀποδιδράσκοντες τόνδε τὸν θάνατον ἐς τὴν τέως πολεμίαν ἰχθύσι γῆν ἐξεπήδησαν, τὸ τέλος τοῦ βίου τὸ χωρὶς τοῦ ξίφους προῃρημένοι καὶ μάλα ἀσμένως.

[1] ἐκπέπληγε καὶ διὰ τὸ ἄηθες.
[2] ἀκρατῶς καὶ ἀμάχως.
[3] ἐγγόνων.
[4] οἰκιῶν καὶ χώρων.
[5] *Reiske*: ἕλκονται.
[6] *Schn*: ἐμπλάσσονται.

animals. Now at first when the sound penetrates to their ears it strikes them with terror and fills them with dread, and then an unalloyed and irresistible delight in the music takes hold of them, and they are so beguiled as to forget about their offspring and their homes. And yet wild beasts do not care to wander away from their native haunts. But little by little these creatures in Etruria are attracted as though by some persuasive spell, and beneath the wizardry of the music they come and fall into the snares, overpowered by the melody.

The 'Anthias' fish

47. The Anthias, if wounded while it is being captured, is a most pitiful sight, and as it dies seems to be mourning for itself and to be somehow imploring, like men who have fallen among pitiless and most bloodthirsty brigands. For some of these fish in their attempt to escape get entangled in the nets, and as they try to leap out of the ambush are caught by the harpoon. Others which contrive to escape this death, spring out on to the shore, hitherto the fishes' enemy, preferring, and gladly so, death without the aid of the sword.

BOOK XIII

IΓ

1. Ἀετὸν ἀκούω Γορδίῳ τὴν τοῦ παιδὸς αὐτοῦ Μίδου[1] βασιλείαν ὑποσημῆναι, ἡνίκα ἀροῦντι τῷ Γορδίῳ ἐπιπτάς, εἶτα μέντοι κατὰ τοῦ ζυγοῦ καθίσας συνδιημέρευσεν, οὐδὲ προαπέστη πρὶν ἢ γενομένης ἑσπέρας καὶ ἐκεῖνος κατέλυσε τὴν ἄροσιν ἐπιστάντος τοῦ βουλυτοῦ. Γέλωνος δὲ τοῦ Συρακοσίου παιδὸς ὄντος λύκος μέγιστος ἐσπηδήσας ἐς τὸ διδασκαλεῖον ἐξήρπασε τῶν χειρῶν τοῖς ὀδοῦσι τὴν δέλτον, καὶ ὁ Γέλων ἐξαναστὰς τοῦ θάκου ἐδίωκεν αὐτόν, τὸ μὲν θηρίον μὴ καταπτήξας, περιεχόμενος δὲ τῆς δέλτου ἰσχυρῶς. ἐπεὶ δὲ ἔξω τοῦ διδασκαλείου ἐγένετο, τὸ μὲν κατηνέχθη καὶ τοὺς παῖδας αὐτῷ διδασκάλῳ κατέβαλε, θείᾳ δὲ προμηθείᾳ ὁ Γέλων περιῆν[2] μόνος. καὶ τό γε παράδοξον, οὐκ ἀπέκτεινεν ἄνθρωπον ἀλλ' ἔσωσε λύκος, οὐκ ἀτιμασάντων τῶν θεῶν οὐδὲ διὰ τῶν ἀλόγων τῷ μὲν τὴν βασιλείαν προδηλῶσαι, τὸν δὲ τοῦ μέλλοντος κινδύνου σῶσαι. ἴδιον δὴ τῶν ζῴων καὶ τὸ θεοφιλές.

2. Οἱ Κᾶρες αἱροῦσι τοὺς σαργοὺς τὸν τρόπον τοῦτον. νότου καταπνέοντος ἡσυχῇ καὶ προσβάλλοντος αὔρας μαλακωτέρας καὶ τοῦ κύματος στορεσθέντος καὶ πράως ταῖς ψάμμοις ἐπηχοῦντος,

[1] *Ges*: Μήδου. [2] περιῄει.

BOOK XIII

1. I have heard that an eagle intimated to Gordius that his son Midas[a] would be king when, as he was ploughing, it flew over Gordius, and then settling upon the yoke, remained with him all day long and did not depart before he finished his ploughing at eventide when the hour for unyoking was at hand.

Gordius and an Eagle

And when Gelon[b] of Syracuse was a boy an immense wolf sprang into the schoolroom and with its teeth snatched his writing-tablet from his hands. And Gelon rose from his seat and gave chase, not being afraid of the beast but clinging valiantly to his writing-tablet. And when he got outside the schoolroom it fell and crushed the boys along with the master. It was by divine providence that Gelon was the only one to escape. And the strange thing is that the wolf did not kill a man but saved his life, for the gods did not disdain to foreshow a kingdom to one even by means of a dumb animal, and to save the other from danger that threatened.

Gelon and a Wolf

So it is characteristic of animals to be beloved of the gods.

2. This is how the people of Caria catch Sargues. When the south wind is blowing gently and sending softer breezes and when the waves are at rest and chime lightly upon the sands, then the fisherman has

The Sargue

[a] Mythical King of Phrygia.

[b] Gelon, *c.* 540–478 B.C., became Tyrant of S. in 485.

τηνικαῦτα ὁ θηρατὴς καλάμου μὲν οὐ δεῖται οὐδὲ ἕν, λαβὼν δὲ ἀρκεύθου ῥάβδον πάνυ σφόδρα ἐρρωμένης, ἀπ' [1] ἄκρας αὐτῆς ἐξάπτει σειράν, καὶ περιπείρει [2] τῷ ἀγκίστρῳ λυκόστομον [3] ὄντα ἡμιτάριχον, καὶ καθίησιν ἐς τὴν θάλατταν. καὶ κάθηται μὲν ἐπὶ τῇ πρύμνῃ τῆς πορθμίδος καὶ τὸν δόλον ὑποκινεῖ, ὑπερέττει δέ οἱ παῖς [4] ἡσυχῇ, προμαθὼν τῆς ἐλάσεως τὸ σχολαῖον ἐπίτηδες, καὶ ὡς ἐπὶ τὴν γῆν προάγει τὸ σκάφος. πολλοὶ δὲ οἱ σαργοὶ περισκιρτῶσιν ἐκ τῶν συντρόφων φωλεῶν ἀναθορόντες, ἀθροίζονται δὲ ἐπὶ τὸ ἄγκιστρον· ἄγει γὰρ αὐτοὺς οἱονεὶ ἴυγγι ὁ πάλαι μὲν τεθνηκὼς ἐς τὸ ἑλεῖν δὲ σκευασθεὶς [5] ἰχθύς. εἶτα πλησίον τῆς γῆς γενόμενοι ῥᾳδίως ἁλίσκονται, τῇ λιχνείᾳ τῆς γαστρὸς δεδεμένοι.

3. Διατριβαὶ δὲ ἰχθύων πολλαί, καὶ γίνονται οἱ μὲν ἐν ταῖς πέτραις, οἱ δὲ ἐν ταῖς ψάμμοις, ἄλλοι δὲ ἐν ταῖς πόαις. καὶ γάρ τοι καὶ πόαι θαλάττιαί εἰσι, καὶ αἱ μὲν αὐτῶν καλοῦνται βρύα, αἱ δὲ ἄμπελοι, καὶ σταφυλαί τινες, καὶ φύκια ἄλλα· ἦν δὲ ἄρα θαλαττίας [6] πόας καὶ κράμβη ὄνομα, καὶ μνία καλεῖταί τινα ἐν αὐταῖς καὶ τρίχες. τροφὴ δὲ ἄρα τούτων ἄλλῳ ἄλλη [7] ἦν, καὶ οὐκ ἂν πάσαιτο ἑτέρας ὁ εἰθισμένος τῇ συντρόφῳ καὶ ὁμοεθνεῖ, ὡς ἂν εἴποι τις.

4. Ἀκούσειας δ' ἂν ἁλιέων καὶ ἰχθύων τινὰ καλλιώνυμον οὕτω λεγόντων. καὶ ὑπὲρ αὐτοῦ

[1] ἐπ'. [2] *Reiske*: περί.
[3] *Ges*: κυκλόστομον. [4] καὶ παῖς.

no need of his reed, but taking a rod of very tough juniper he fastens a cord on the end and spits a half-pickled anchovy on the hook and lets it down into the sea. And he sits in the prow of the skiff and dangles the lure, while his boy rows gently, having purposely been instructed beforehand in the art of leisurely propulsion, and makes the skiff move in the direction of the shore. And the Sargues dart up in their numbers from their native lairs and gambol around and collect about the hook. For the fish, long dead indeed but prepared for catching, draws them as it were with a spell. Presently when they are close to the shore they are easily caught, being made prisoners through their belly's greed.

Fishes, their haunts and their food

3. The haunts of fishes are numerous: some are found among rocks, others in sand, others again among vegetation, for you must know there is vegetation even in the sea, and some is called 'oyster-green,' some 'vines,' certain kinds 'grapes,' and others 'grass-wrack.' And it seems that the name 'cabbage' also is attached to marine vegetation, and some kinds are called 'seaweed' and some 'hair.' And some fish feed on one kind, others on another, and a fish that is accustomed to the food on which it has been reared and to which it is, so to say, akin would never touch any other kind.

The Star-gazer fish

4. You may hear fishermen speak also of a fish they call *Callionymus* (Star-gazer). And concerning

---

[5] *διασπασθείς*. [6] *Ges*: *θαλαττίου*.
[7] *Gron*: *ἄλλο*.

Ἀριστοτέλης λέγει ὅτι ἄρα ἐπὶ τοῦ λοβοῦ τοῦ δεξιοῦ καθημένην [1] ἔχει χολὴν πολλήν, τὸ δὲ ἧπαρ αὐτῷ [2] κατὰ τὴν λαιὰν φορεῖται πλευράν. καὶ μαρτυρεῖ τούτοις καὶ ὁ Μένανδρος ἐν τῇ Μεσσηνίᾳ οἶμαι λέγων

> τίθημ' ἔχειν χολήν σε καλλιωνύμου
> πλείω,

καὶ Ἀνάξιππος ἐν Ἐπιδικαζομένῳ

> ἐάν με κινῇς καὶ ποιήσῃς τὴν χολὴν
> ἅπασαν ὥσπερ καλλιωνύμου ζέσαι,
> ὄψει διαφέροντ' οὐδὲ ἓν ξιφίου κυνός.

εἰσὶ μὲν οὖν οἳ καί φασιν αὐτὸν ἐδώδιμον, οἱ δὲ πλείους ἀντιλέγουσιν αὐτοῖς. οὐ ῥᾳδίως δὲ αὐτοῦ μνημονεύουσιν ἐν ταῖς † ὑπὲρ τῶν ἰχθύων πανθοινίαις, ὧν τι καὶ ὄφελός ἐστι ποιηταὶ θέμενοι [3] σπουδὴν ἐς μνήμην ἔνθεσμον,† [4] Ἐπίχαρμος μὲν ἐν Ἥβας [5] Γάμῳ καὶ Γᾷ καὶ Θαλάσσᾳ καὶ προσέτι ⟨καὶ⟩ [6] Μώσαις, [7] Μνησίμαχος δὲ ἐν τῷ Ἰσθμιονίκῃ.

5. Βάτραχος δὲ θαλάττιος τίκτει κατὰ τοὺς ὄρνιθας ᾠὸν καὶ οὗτος. οὐ ζῳογονεῖ γὰρ ἐν

[1] *Ges*: καθειμένην.
[2] αὐτῷ δὲ τὸ ἧ.
[3] ποιητῶν θεμένων.
[4] ὑπὲρ τῶν . . . ἔνθεσμον *corrupt*.
[5] *Cas*: Ἥρας.
[6] ⟨καὶ⟩ *add. H.*
[7] *Hemst*: Μούσαις.

[a] Ar. only says that its gall-bladder is close to the liver and very large in relation to the size of the fish. See *fr.* 286 (Rose, p. 307).

it Aristotle says [*HA* 506 b 10] [a] that it has a considerable quantity of gall stored close to the right-hand lobe of the liver, and that its liver is situated on its left side. And Menander bears witness to these statements when he says in his *Messenian woman* [*fr.* 31 K], I think,

'I will make you have more gall than a Star-gazer';

and Anaxippus in his *Epidicazomenus* [*fr.* 2K]:

'If you rouse me and make all my gall boil like a Star-gazer's, you will find that I differ no whit from a sword-fish.'

There are those who assert that it is edible; most people however assert the contrary. But you will not easily discover any mention of the Star-gazer in any description of fish-banquets, although poets have been at pains to record every fish of any value; they are [b] Epicharmus in his *Hebe's Wedding* [Kaibel *CGF* p. 98], his *Land and Sea* [*ib.* 94], and also his *Muses* [*ib.* 98], and Mnesimachus in his *Isthmian Victor* [*fr.* 5K].

5. The Fishing-frog [c] also lays an egg, as birds do, for it is not viviparous, because its new-born young

The Fishing-frog

[b] The passage is corrupt and the translation gives what may be the general sense.

[c] More commonly called 'Angler'; see above, 9. 24. It has a huge, broad, flat head but a very thin body. Of the three filaments projecting from its head the front one alone is movable and tipped with a lappet: this is the 'lure' (δέλεαρ) of 9. 24. The 'account of its reproduction and of its egg . . . is quite untrue' (Thompson). See *Enc. Brit.* (11th ed.), art. 'Angler.'

ἑαυτῷ· κεφαλὴν γὰρ ἔχει καὶ τὰ ἀρτιγενῆ μεγάλην τε ἅμα καὶ τραχεῖαν, καὶ διὰ ταῦτα ὑποδέξασθαι τὰ βρέφη δείσαντα ἥκιστός ἐστιν· ἑλκώσει γὰρ αὐτὸν καὶ κακώσει [1] ἐσπίπτοντα τὴν αὖθις. ἀλλὰ καὶ τικτόμενα ἂν καὶ ἐξιόντα εἰργάζετο παραπλήσια. οὔτε οὖν εὐώδινες ἐς ζῴων γένεσίν εἰσιν οὔτε μὴν κρησφύγετα τοῖς ἐκγόνοις ἀγαθά. ᾠοῦ δὲ τὴν φύσιν ἢ ἰδιότητα οὐχ ὁμολογεῖ τὸ τῶν βατράχων, τραχὺ δέ ἐστι καὶ ἐκεῖνο, καὶ ἔχει φολίδας, καὶ προσαψαμένῳ φανεῖταί σοι ἀντίτυπον.

6. Οἱ πολύποδες καὶ αὐτοὶ χρόνῳ γίνονται μέγιστοι, καὶ ἐς κήτη προχωροῦσι, καὶ ἐναριθμοῦνται ἐν αὐτοῖς καὶ οὗτοι. ἀκούω γοῦν ἐν Δικαιαρχίᾳ τῇ Ἰταλικῇ πολύπουν ἐς ὄγκον σώματος ὑπερήφανον προελθόντα τὴν μὲν ἐν τῇ θαλάττῃ τροφὴν καὶ τὰς ἐκεῖθεν νομὰς ἀτιμάσαι καὶ ὑπερφρονῆσαι αὐτῶν. προῄει δὲ ἄρα οὗτος καὶ ἐς τὴν γῆν, καὶ ἐλῄζετο καὶ τῶν χερσαίων ἔστιν ἅ. οὐκοῦν διά [2] τινος ὑπονόμου κρυπτοῦ ἐκβάλλοντος ἐς τὴν θάλατταν τὰ ἐκ τῆς πόλεως τῆς προειρημένης ῥυπαρὰ ἐσνέων καὶ ἀνιὼν ἐς οἶκόν τινα πάραλον, ἔνθα ἦν ἐμπόρων Ἰβηρικῶν φόρτος καὶ ταρίχη τὰ ἐκεῖθεν ἐν σκεύεσιν ἁδροῖς, εἶτα τὰς πλεκτάνας περιχέων καὶ σφίγγων τὸν κέραμον ἐρρήγνυ τὰ ἀγγεῖα καὶ κατεδαίνυτο τὰ ταρίχη. οἱ δὲ ἐσιόντες ὡς ἑώρων τὰ ὄστρακα, πολὺν δὲ τοῦ φόρτου ἀριθμὸν ἀφανῆ κατελάμβανον, ἐξεπλήττοντο καὶ τίς ἦν ὁ κεραΐζων αὐτοὺς συμβαλεῖν οὐκ εἶχον, τῶν μὲν θυρῶν ἀνεπιβουλεύτων βλεπομένων, τοῦ δὲ ὀρόφου ὄντος

[1] ἑλκοῦσι . . . κακῶς. [2] καὶ διά.

have a large, rough head, and for that reason it is incapable of taking them back when they are frightened. For their re-entry will lacerate and injure the parent, and were they to be born alive and to emerge so, they would produce the same effect. And so they are not well adapted to producing their young alive nor are they a secure place of refuge for them. The egg of the Fishing-frog does not conform to the nature and character of an egg, for even that is rough and has scales, and you will find it hard if you touch it.

A monstrous Octopus

6. Octopuses naturally, with the lapse of time, attain to enormous proportions and approach cetaceans and are actually reckoned as such. At any rate I learn of an octopus at Dicaearchia in Italy which attained to a monstrous bulk and scorned and despised food from the sea and such pasturage as it provided. And so this creature actually came out on to the land and seized things there. Now it swam up through a subterranean sewer that discharged the refuse of the aforesaid city into the sea and emerged in a house on the shore where some Iberian merchants had their cargo, that is, pickled fish from that country in immense jars: it threw its tentacles round the earthenware vessels and with its grip broke them and feasted on the pickled fish. And when the merchants entered and saw the broken pieces, they realised that a large quantity of their cargo had disappeared; and they were amazed and could not guess who had robbed them: they

ἀσινοῦς καὶ τῶν τοίχων μὴ διεσκαμμένων· ἑωρᾶτο δὲ καὶ τῶν ἰχθύων τῶν ταρίχων λείψανα ὑπολειφθέντα [1] ὑπὸ τοῦ ἀκλήτου δαιτυμόνος. ἔκριναν δή τινα τῶν οἰκείων τὸν μάλιστα εὐτολμότατον ἔνδον ὡπλισμένον καταλιπεῖν ἐλλοχῶντα. νύκτωρ οὖν ἐπὶ τὴν συνήθη δαῖτα ὁ πολύπους ἀνέρπει, καὶ περιχυθεὶς τοῖς σκεύεσιν ὥσπερ ἐς πνῖγμα ἀθλητὴς συλλαβὼν τὸν ἀντίπαλον ἐγκρατῶς τε καὶ μάλα εὐλαβῶς, εἶτα συνέτριβε τὸν κέραμον ληστὴς ὡς εἰπεῖν ὁ πολύπους ῥᾷστα. ἦν δὲ διχόμηνος, καὶ κατελάμπετο ὁ οἶκος, καὶ πάντα ἦν εὐσύνοπτα. ὁ δὲ οὐκ ἐπεχείρει μόνος, δείσας τὸν θῆρα (καὶ γὰρ μόνου μείζων ὁ ἐχθρὸς ἦν) περιηγεῖται δὲ ἕωθεν τοῖς ἐμπόροις τὰ πεπραγμένα· ἀκούοντες δὲ ἠπίστουν. εἶτα οἱ μὲν τῆς ζημίας τῆς τοσαύτης μνήμῃ τὸν κίνδυνον [2] ἀνερρίπτουν, καὶ συνελθεῖν [3] τῷ ἐχθρῷ ἔσπευδον,[4] οἱ δὲ τῆς καινῆς καὶ ἀπίστου θέας διψῶντες συναπεκλείοντο αὐθαίρετοι σύμμαχοι. εἶτα ἑσπέρας ὁ φὼρ ἐπιφοιτᾷ, καὶ ὁρμᾷ ἐπὶ τὴν συνήθη τράπεζαν. ἐνταῦθα οἱ μὲν ἀπέφραττον τὸν ὀχετόν, οἱ δὲ ὡπλίζοντο ἐπὶ τὸν πολέμιον,[5] καὶ κοπίσι καὶ ξυροῖς τεθηγμένοις αὐτοῦ διέκοπτον τὰς πλεκτάνας, ὡς δρυὸς κλάδους ἀκροτάτους [6] ἀμπελουργοί τε καὶ δρυοτόμοι. καὶ τὴν ἀλκὴν αὐτοῦ περικόψαντες καθεῖλον ὀψὲ καὶ μόγις οὐκ ὀλίγα πονήσαντες, καὶ τὸ καινότατον, ἐν τῇ γῇ τὸν ἰχθὺν ἐθηράσαντο ἔμποροι. τὸ [7]

[1] ἀπολειφθέντα.
[2] *Ges*: τὸν κίνδυνον μνήμῃ.
[3] *Schn*: συνεισελθεῖν.
[4] συνέσπευδον.
[5] *Ges*: πόλεμον.
[6] ἁβροτάτους *or* ἁδρο-.

saw that no attempt had been made upon the doors; the roof was undamaged; the walls had not been broken through. They saw also the remains of the pickled fish that had been left behind by the uninvited guest. So they decided to have their most courageous servant armed and waiting in ambush in the house. Well, during the night the Octopus crept up to its accustomed meal and clasping the vessels, as an athlete puts a strangle-hold upon his adversary with all his might gripping firmly, the robber—if I may so call the Octopus—crushed the earthenware with the greatest ease. It was full moon, and the house was full of light, and everything was quite visible. But the servant was not for attacking the brute single-handed as he was afraid, moreover his adversary was too big for one man, but in the morning he informed the merchants what had happened. They could not believe their ears. Then some of them remembering how heavily they had been mulcted, were for risking the danger and were eager to encounter their enemy, while others in their thirst for this singular and incredible spectacle voluntarily shut themselves up with their companions in order to help them. Later, in the evening the marauder paid his visit and made for his usual feast. Thereupon some of them closed off the conduit; others took arms against the enemy and with choppers and razors well sharpened cut the tentacles, just as vine-dressers and woodmen lop the tips of the branches of an oak. And having cut away its strength, at long last they overcame it not without considerable labour. And what was so strange was that merchants captured the fish on dry land. Mis-

---

[7] καὶ τό κ.

κακοῦργον δὴ τοῦδε τοῦ ζῴου καὶ τὸ δολερὸν ἀναπέφηνεν ἡμῖν ἴδιον ὄν.

7. Τῶν τεθηραμένων ἐλεφάντων ἰῶνται τὰ τραύματα οἱ Ἰνδοὶ τὸν τρόπον τοῦτον. καταιονοῦσι μὲν αὐτὰ ὕδατι χλιαρῷ, ὥσπερ οὖν τὸ τοῦ Εὐρυπύλου παρὰ τῷ καλῷ Ὁμήρῳ ὁ Πάτροκλος· εἶτα μέντοι διαχρίουσι βουτύρῳ [1] αὐτά· ἐὰν δὲ ᾖ βαθέα, τὴν φλεγμονὴν πραΰνουσιν ὕεια κρέα θερμὰ μὲν ἔναιμα δὲ ἔτι προσφέροντες καὶ ἐντιθέντες. τὰς δὲ ὀφθαλμίας θεραπεύουσιν αὐτῶν βόειον γάλα ἀλεαίνοντες εἶτα αὐτοῖς ἐγχέοντες, οἱ δὲ ἀνοίγουσι τὰ βλέφαρα, καὶ ὠφελούμενοι ἥδονταί τε καὶ αἰσθάνονται, ὥσπερ ἄνθρωποι. καὶ ἐς τοσοῦτον ἐπικλύζουσιν, ἐς ὅσον ἂν ἀποπαύσωνται λημῶντες. μαρτύριον δὲ τοῦ παύσασθαι τὴν ὀφθαλμίαν τοῦτό ἐστι. τὰ δὲ νοσήματα ὅσα αὐτοῖς προσπίπτει ἄλλως, ὁ μέλας οἶνός ἐστιν αὐτοῖς ἄκος. εἰ δὲ μὴ γένοιτο ἐξάντης τοῦ κακοῦ τῷ φαρμάκῳ τῷδε, ἄσωστά οἵ ἐστιν.

8. Ἐλέφαντι ἀγελαίῳ μὲν τετιθασευμένῳ [2] γε μὴν ὕδωρ πῶμά ἐστι, τῷ δὲ ⟨τὰ⟩ [3] ἐς πόλεμον ἀθλοῦντι οἶνος μέν, οὐ μὴν ὁ τῶν ἀμπέλων, ἐπεὶ τὸν μὲν ἐξ ὀρύζης χειρουργοῦσι, τὸν δὲ ἐκ καλάμου. προΐασι δὲ καὶ ἄνθη σφίσιν ἀθροίσοντες· εἰσὶ γὰρ ἐρασταὶ εὐωδίας, καὶ ἄγονταί γε ἐπὶ τοὺς λειμῶνας, ὀσμῇ πωλευθησόμενοι τῇ ἡδίστῃ. καὶ ὁ μὲν ἐκλέγει κρίνας τῇ ὀσφρήσει τὸ ἄνθος, τάλαρον δὲ ἔχων ὁ πωλευτὴς τρυγῶντος καὶ ἐμβάλλοντος

[1] τῷ βουτύρῳ. [2] *Reiske* : εἰθισμένῳ.

chief and craft are plainly seen to be characteristics of this creature.

Remedies for sick Elephants

7. The people of India heal the wounds of Elephants which they have captured in the following manner. They foment them with warm water, just as Patroclus fomented the wound of Eurypylus in our noble Homer [*Il.* 11. 829], and then anoint them with butter. But if they are deep, they reduce the inflammation by applying and laying on them pigs' flesh hot and with the blood still in it. Their ophthalmia they treat by warming some cow's milk and pouring it into their eyes, and the Elephants open their eyelids and are gratified just as men are, to perceive what benefit they derive. And the Indians continue the bathing until the inflammation ceases; this is evidence that the ophthalmia has been arrested. As for other diseases that afflict them, black [a] wine is the cure for them. But if this medicine does not rid them of their complaint, then nothing will save them.

The Elephant and its love of flowers

8. An Elephant belonging to a herd but which has been tamed drinks water; but an Elephant that fights in war drinks wine, not however that made from grapes, for men prepare a wine from rice or from cane. And these tame Elephants go out to gather flowers for themselves, for they love a sweet smell and are led to the meadows to be trained by the most fragrant scent. And an Elephant using its sense of smell will pick out a flower, while the trainer, basket in hand, holds it out beneath the

[a] *I.e.* dark red.

[3] ⟨τά⟩ *add. H.*

ὑπέχει. εἶτα ὅταν ἐμπλήσῃ τοῦτον, ὥσπερ οὖν ὀπώραν δρεπόμενος λοῦται, καὶ ἥδεται τῷ λουτρῷ κατὰ τοὺς τῶν ἀνθρώπων ἁβροτέρους. εἶτα ἐπανελθὼν τὰ ἄνθη ποθεῖ, καὶ βοᾷ βραδύνοντος, καὶ οὐχ αἱρεῖται τροφὴν πρὶν ἢ κομίσῃ τίς οἱ ὅσα ἐτρύγησεν. εἶτα μέντοι τῇ προβοσκίδι ἀναιρούμενος ἐκ τοῦ ταλάρου τῆς φάτνης καταπάττει τὰ χείλη, ἥδυσμα τοῦτό γε τῇ τροφῇ διὰ τῆς εὐοσμίας ἐπινοῶν, ὡς εἰπεῖν. κατασπείρει δὲ καὶ τοῦ χώρου ἔνθα αὐλίζεται τῶν ἀνθέων πολλά, ἡδυσμένον αἱρεῖσθαι γλιχόμενος ὕπνον. Ἰνδοὶ δὲ ἐλέφαντες ἦσαν ἄρα πήχεων ἐννέα τὸ ὕψος, πέντε δὲ τὸ εὖρος. μέγιστοι δὲ ἄρα τῶν ἐκεῖθι ἐλεφάντων οἱ καλούμενοι Πράσιοι,[1] δεύτεροι δ' ἂν τῶνδε τάττοιντο οἱ Ταξίλαι.[2]

9. Ἵππον δὲ ἄρα Ἰνδὸν κατασχεῖν καὶ ἀνακροῦσαι προπηδῶντα καὶ ἐκθέοντα οὐ παντὸς ἦν, ἀλλὰ τῶν ἐκ παιδὸς ἱππείαν πεπαιδευμένων. οὐ[3] γὰρ αὐτοῖς ἐστιν ἐν ἔθει χαλινῷ ἄρχειν αὐτῶν καὶ ῥυθμίζειν αὐτοὺς καὶ ἰθύνειν, κημοῖς δὲ ἄρα κεντρωτοῖς· ἀκόλαστόν τε[4] ἔχουσι τὴν γλῶτταν καὶ τὴν ὑπερῴαν ἀβασάνιστον· ἀναγκάζουσι δὲ αὐτοὺς ὅμως οἵδε οἱ τὴν ἱππείαν σοφισταὶ [περικυκλεῖν καὶ][5] περιδινεῖσθαι ἐς ταὐτὸν στρεφομένους.[6] δεῖ δὲ ἄρα τῷ τοῦτο δράσοντι καὶ ῥώμης χειρῶν καὶ ἐπιστήμης εὖ μάλα ἱππικῆς. πειρῶνται δὲ οἱ προήκοντες ἐς ἄκρον τῆσδε τῆς σοφίας καὶ ἅρμα οὕτως περικυκλεῖν καὶ περιάγειν· εἴη δ' ἂν

[1] Πραίσ- MSS *always*.
[2] Ταξιλαῖοι? *Warmington*.
[3] τοῦτο.
[4] γάρ *Jac*, *H*.
[5] [περικυκλεῖν καί] *del*. *H*.

picker as he throws it in. Later when it has filled the basket, like a fruit-gatherer it has a bath and takes as much pleasure in the bath as the more luxurious of mankind do. Then on its return it wants the flowers, and if the keeper delays, it trumpets and refuses food until somebody brings it the flowers it has gathered. Then it picks them out of the basket with its trunk and sprinkles them along the rim of its manger, for it regards them as imparting a flavour, as it were, to its food by means of their scent. And it scatters a quantity of flowers over its stall, as it desires a fragrant sleep. It seems that Indian Elephants are nine cubits high and five wide, and the largest are those they call Prasian; next to these one may reckon those from Taxila.[a]

The Indian Horse

9. To control an Indian Horse, to check him when he leaps forward and would gallop away, has not, it seems, been given to every man, but only to those who have been brought up from childhood to manage horses. For it is not the Indian custom to rule them, to bring them to order, and to direct them by means of the rein but by spiked muzzles; thus their tongue goes unpunished and the roof of their mouth untormented. Still, those who are skilled in horsemanship compel them to go round and round, returning to the same point. Now if a man would do this he requires strength of hand and a thorough understanding of horses. Those who have attained the summit of this science even try by these means to drive a chariot in circles. And it would be no con-

[6] στρεφομένους, καὶ ᾗπερ εἶδον ἀστόμους.

[a] City in the extreme NW of India.

ἆθλος οὐκ εὐκαταφρόνητος ἀδηφάγων ἵππων τέτρωρον περιστρέφειν ῥᾳδίως· φέρει δὲ τὸ ἅρμα παραβάτας δύο. ὁ δὲ στρατιώτης ἐλέφας ἐπὶ τοῦ καλουμένου θωρακίου ἢ καὶ νὴ Δία τοῦ νώτου γυμνοῦ καὶ ἐλευθέρου φέρει πολεμιστὰς μὲν τρεῖς . . . [1] παρ' ἑκάτερα βάλλοντας καὶ τὸν τρίτον κατόπιν, τέταρτον δὲ τὸν τὴν ἅρπην ἔχοντα [2] διὰ χειρῶν καὶ ἐκείνῃ τὸν θῆρα ἰθύνοντα, ὡς οἴακι ναῦν κυβερνητικὸν ἄνδρα καὶ ἐπιστάτην τῆς νεώς.

10. Θήρα δὲ παρδάλεων Μαυρουσία εἴη ἄν.[3] καὶ ἔστιν αὐτοῖς οἰκοδομία λίθων πεποιημένη, καὶ ἔοικε ζωγρείῳ [4] τινί, καὶ ἔστι μὲν ὁ λόγος ὅδε ὁ πρῶτος· ὅ γε μὴν δεύτερος, ἐνδοτέρω σαπροῦ κρέως καὶ ὀδωδότος μοῖραν μηρίνθου τινὸς μακροτέρας ἐξαρτῶσι, θύραν δὲ ἐκ ῥιπίδων καί τινων καλάμων ἀραιὰν ἐπέστησαν, καὶ μέντοι καὶ δι' αὐτῶν ἐκπνεῖται ἡ τοῦ κρέως τοῦ προειρημένου ὀσμὴ διαρρέουσα. αἰσθάνονται [5] δὲ αἱ θῆρες, καὶ γάρ πως τοῖς κακόσμοις φιληδοῦσι· προσβάλλει γὰρ αὐτὰς [6] ὁ τῶν [7] προειρημένων ἀήρ, ἐάν τε ἐν ἄκροις [8] τοῖς ὄρεσιν ἐάν τε ἐν φάραγγι, καὶ μέντοι καὶ ἐν αὐλῶνι. εἶτα ἀνεφλέχθη τῇ ὀσμῇ ἐντυχοῦσα, καὶ ὑπὸ τῆς ἄγαν ὁρμῆς ἐς τὴν θοίνην τὴν φίλην ᾄττει φερομένη· ἕλκεται δὲ ὑπ' αὐτῆς ὡς ὑπό τινος ἴυγγος. εἶτα ἐμπίπτει τῇ θύρᾳ καὶ ἀνατρέπει αὐτὴν καὶ ἔχεται τοῦ δυστυχοῦς δείπνου. τῇ γάρ τοι μηρίνθῳ τῇ προειρημένῃ συνυφάνθη

[1] *Lacuna.* [2] κατέχοντα.
[3] εἴη ἂν <τοιάδε> *add. Grasberger, cp.* 13. 14 *ad fin.*, 15. 1.
[4] *Schn* : ζωαγρίᾳ. [5] *Schn* : αἴσθονται.
[6] αὐταῖς.

temptible achievement to make a team of four ravenous horses circle about with ease. And the chariot holds two beside the driver. But a War-elephant in what is called the tower, or even, I assure you, on its bare back, free of harness, carries as many as three armed men. . . .[a] who hurl their weapons to left and right, and a third behind them, while a fourth holds the goad with which he controls the beast, as a helmsman or pilot of a vessel controls a ship with the rudder.

The War-elephant

10. The hunting of Leopards seems to be a Moorish practice. The people build a stone structure, and it resembles a kind of cage: this is the first part of the ambush; and the second part is this: inside they fasten a piece of meat that has gone bad and smells, by a longish cord and set up a flimsy door made of plaited reeds of some kind, and through them the smell of the aforesaid meat is exhaled and spreads abroad. The animals notice it, being for some reason fond of ill-smelling objects, because the scent from them assails them whether they are on mountain tops or in a ravine or even in a glen. Then when the Leopard encounters the smell it gets excited and in its excessive desire comes rushing to the feast it loves: it is drawn to it as though by some spell. Then it dashes at the door, knocks it down, and fastens upon the fatal meal—fatal, because on to the aforesaid cord there has been woven a noose most dexterously contrived, and as the meat is being eaten

Leopard-hunting in Mauretania

[a] Lacuna. The context demands: 'two in front who . . .'

[7] ὁ ⟨ἐκ⟩ τῶν? *H*

[8] *Reiske*: ἀγρίοις.

πάγη [1] καὶ μάλα σοφή, ἥπερ οὖν ἐσθιομένου τοῦ κρέως κινεῖται, καὶ περιλαμβάνει τὴν λίχνον πάρδαλιν. καὶ ἑάλω, γαστρὸς ἀδηφάγου καὶ μυσαρᾶς ἑστιάσεως δίκας ἐκτίνουσα ἡ δυστυχής.

11. Αἱροῦνται δὲ οἱ λαγῳ ὑπὸ ἀλωπέκων οὐχ ἧττον [2] ἀλλὰ καὶ μᾶλλον τέχνῃ· σοφὸν γὰρ ἀπατᾶν ἀλώπηξ, καὶ δόλους οἶδεν. ὅταν γοῦν νύκτωρ ἐς ἴχνος ἐμπέσῃ τοῦ λαγὼ καὶ αἴσθηται τοῦ θηρίου, σιγῇ τε ἐπιβαίνει καὶ ποδὶ ἀψόφῳ, καὶ ἀναστέλλει τὸ ἆσθμα, καὶ καταλαβοῦσα ἐν τῇ κοίτῃ πειρᾶται αἱρεῖν ὡς ἀδεᾶ καὶ ἄφροντιν. ὁ δὲ οὐ τρυφῶν οὐδὲ ῥᾳθύμως καθεύδει, ἀλλ' ἅμα τε ᾔσθετο τοῦ ζῴου τοῦ προσιόντος καὶ τῆς εὐνῆς ἐξεπήδησε καὶ θεῖ· καὶ ὁ μὲν ἀνύτει [3] τὸν δρόμον καὶ μάλα ὠκέως, ἡ δὲ ἀλώπηξ καὶ αὐτὴ κατ' ἴχνος ἵεται [4] καὶ τοῦ δρόμου ἔχεται. καὶ ὁ μὲν πολλὴν ὁδὸν διανύσας, ὡς ἤδη κρείττων καὶ οὐκ ἂν ἁλούς, ἐμπεσὼν ἐς λόχμην ἀσμένως ἀναπαύεται· ἡ δὲ ἀλώπηξ ἐφίσταται, καὶ ἀτρεμεῖν οὐκ ἐπιτρέπει, πάλιν τε αὐτὸν ἐγείρει, καὶ ἐς δρόμον ἐξηνέμωσεν ἕτερον. εἶτα οὐχ ἥττων τῆς προτέρας ὁδὸς καὶ δὴ διηνύσθη, καὶ ὁ μὲν ἀναπαύσασθαι διψᾷ πάλιν, ἡ δὲ ἐφίσταται, καὶ σείουσα τὸν θάμνον ἀγρυπνίαν ἐνεργάζεται αὐτῷ. ὁ δὲ πάλιν ἐκθεῖ, καὶ ἡ ἀλώπηξ οὐχ ὑστερεῖ. συνεχέστερον δὲ ὅταν αὐτὸν δρόμος ἐκ δρόμου διαλάβῃ καὶ ἀγρυπνία διαδέξηται, ὁ μὲν ἀπεῖπε, [5] ἡ δὲ ἐπελθοῦσα κατέσχεν αὐτόν, οὐ μὰ Δία δρόμῳ ἀλλὰ τῷ χρόνῳ καὶ τῷ

[1] ἡ πάγη.
[2] ἐνίοτε οὐχ ἧττον δρόμῳ.
[3] ἀνύει.

this is dislodged and encircles the gluttonous Leopard. So it is caught and pays the penalty for its ravenous belly and its foul feasting, the poor wretch.

11. Hares are caught by Foxes more often than not through an artifice, for the Fox is a master of trickery and knows many a ruse. For instance, when by night it comes upon the track of a Hare and has scented the animal, it steals upon it softly and with noiseless tread, and holds its breath, and finding it in its form, attempts to seize it, supposing it to be free of fear and anxiety. But the Hare is not a luxurious creature and does not sleep carefree, but directly it is aware of the Fox's approach it leaps from its bed and is off. And it speeds on its way with all haste: but the Fox follows in its track and continues its pursuit. And the Hare after covering a great distance, under the impression that it has won and is not likely to be caught, plunges into a thicket and is glad to rest. But the Fox is after it and will not allow it to remain still, but once again rouses it and stimulates it to run again. Then a second course no shorter than the first is gone through, and the Hare again longs to rest, but the Fox is upon it and by shaking the thicket contrives to keep it from sleeping. And again it darts out, but the Fox is hard after it. But when it is driven into running course after course without intermission, and want of sleep ensues, the Hare gives up and the Fox overtakes it and seizes it, having caught it not indeed by speed but by length of time and by craft.

Fox and Hare

---

[4] ἐστι MSS, εἶσι *Schn.*

[5] ἀπεῖπε καὶ μένει.

δόλῳ καθελοῦσα. ταῦτα μὲν οὖν ἄλλως προεκθέων ὁ λόγος ὑπὲρ τοῦ δρόμου τοῦ λαγὼ ἀναβέβληται, τὰ δὲ λοιπὰ ἐν τοῖς ἑπομένοις λέγειν ἐγκαιρότερον· ὅθεν δὲ ἐξετραπόμην καὶ δὴ ἐπάνειμι αὖθις. ἦν δὲ ἄρα τοῦ διασπείρειν τὰ ἔκγονα καὶ ἄλλο ἄλλῃ τρέφειν αἰτία ἥδε. ἔστι μὲν ὁ λαγὼς φιλότεκνον δεινῶς, δέδοικε δὲ καὶ τὰς ἐκ τῶν θηρώντων ἐπιβουλὰς καὶ τὰς ἐκ τῶν ἀλωπέκων ἐπιδρομάς, πέφρικέ γε μὴν καὶ τὰς ἐκ τῶν ὀρνίθων οὐχ ἧττον, φωνὴν δὲ κοράκων καὶ ἀετῶν μᾶλλον· πρὸς γὰρ δὴ ταῦτα τῶν πτηνῶν οὐκ ἔστιν αὐτῷ ἔνσπονδα. ὑποκρύπτει δὲ ἑαυτὸν ἢ θάμνῳ κομῶντι ἢ ληίῳ βαθεῖ,[1] ἤ τινα ἄλλην ἑαυτοῦ προβάλλεται ἀναγκαίαν καὶ ἄμαχον [2] σκέπην.

12. Θηρατοῦ δὲ ἀνδρὸς καὶ τὰ ἕτερα ἀγαθοῦ, οἵου μὴ ἂν ψεύσασθαι, λόγον ἤκουσα, καὶ αὐτῷ πεπίστευκα, καὶ [3] διὰ ταῦτα εἰρήσεται. τίκτειν γὰρ δὴ καὶ ἄρρενα λαγὼν [4] ἔλεγε καὶ παιδοποιεῖσθαί τε ἅμα καὶ ὠδίνειν καὶ τῆς φύσεως μὴ ἀμοιρεῖν ἑκατέρας. καὶ ὡς ἐκτρέφει τεκὼν ἔλεγε, καὶ ὡς ἀποτίκτει καὶ δύο που καὶ τρία, καὶ τοῦτο ἐμαρτύρει, καὶ δὴ καὶ τὸν κολοφῶνα ἐπῆγε τῷδε τῷ λόγῳ παντὶ ἐκεῖνον. θηραθῆναι γὰρ λαγὼν ἄρρενα ἡμιθνῆτα, ἐξωγκῶσθαι δὲ αὐτοῦ τὴν γαστέρα ἅτε ἔγκαρπον. ἀνατμηθῆναί τε οὖν αὐτὸν ὡμολόγει καὶ μήτραν πεφωρᾶσθαι καὶ τρεῖς λαγιδεῖς.[5] τούτους οὖν ἀκινήτους τέως εἶναι ἐξαιρεθέντας καὶ κεῖσθαι οἱονεὶ κρέα ἄλλως· ἐπεὶ δὲ ὑπὸ τοῦ

[1] γηδίῳ δασεῖ. [2] ἀμήχανον τήν.
[3] καὶ δὴ καί. [4] ἄρρενας λαγώς.
[5] λαγώς.

Anyhow the account, by starting with the running of the Hare, has got too far ahead; the remainder it will be more appropriate to relate in the sequel. But I will return to the point at which I was diverted.[a] It seems that the reason why it distributes its young and rears them in different spots is as follows. The Hare is deeply devoted to its offspring and dreads both the designs of huntsmen and the attacks of foxes; and it has no less a horror of the attacks of birds, and even more so of the cry of ravens and of eagles. For there is no treaty of peace between these birds and it. And it conceals itself in some leafy bush or deep corn-field or protects itself behind some other enforced and unassailable shelter.

The Hare and its young

12. I have heard from one who is a hunter and a good man besides, the kind that would not tell a lie, a story which I believe to be true and shall therefore relate. For he used to maintain that even the male Hare does in fact give birth and produce offspring and endure the birthpangs and partake of both sexes. And he told me how it bears and rears its young ones, and how it brings perhaps two or three to birth; and he bore witness to this too, and then as the finishing touch to the whole story added the following. A male Hare had been caught in a half-dead state, and its belly was enlarged, being pregnant. Now he admitted that it had been cut open and that its womb, containing three leverets, had been discovered. These, he said, which so far were undisturbed, were taken out and lay there like lifeless flesh. When however they were warmed

The male Hare

[a] Perhaps something has been lost at the beginning of the chapter.

ἡλίου ἀλεαινόμενοι καὶ δὴ κατὰ μικρὰ ὑποθαλπόμενοι διέτριψαν,[1] ἀναφέροντες ἑαυτοὺς ἀνεβιώσκοντο, καί πού τις αὐτῶν καὶ ἐκινήθη καὶ μετὰ ταῦτα ἀνέβλεψε, τάχα δὲ καὶ γλῶτταν ἐπὶ τούτοις προὔβαλε, καὶ στόμα ἀνέῳξε τροφῆς πόθῳ. προσενεχθῆναι οὖν οἷα [2] εἰκὸς τοῖς τηλικούτοις γάλα καὶ κατ' ὀλίγον ἐκτραφῆναι αὐτούς, δεῖγμα ἐμοὶ δοκεῖν ἐς θαῦμα τοῦ τεκόντος τούτους. μὴ πιστεύειν οὖν τῷ λόγῳ πεῖσαι ἐμαυτὸν οὐ δύναμαι· τὸ δὲ αἴτιον, ἡ τοῦ ἀνδρὸς γλῶττα οὔτε ψεῦδος οὔτε κόμπον ἠπίστατο.

13. Ἦν δὲ ἄρα ὁ λαγὼς καὶ ἀνέμων τε καὶ ὡρῶν ἐπιστήμων· σοφὸν γάρ τι χρῆμα αὐτοῦ, † ἀλλ' οὐκ εὔχαρι ὄν, † [3] χειμῶνος οὖν ⟨τὸν⟩ [4] κοῖτον ἐν τοῖς προσηλίοις τίθεται· δῆλα γὰρ δὴ ὅτι θάλπεται μὲν ἀσμένως, κρύει δὲ ἐχθρῶς ἔχει· θέρους δὲ πρὸς ἄρκτον ἀποκλίνει πόθῳ ψύχους. τῆς δὲ τῶν ὡρῶν διαφορᾶς αἱ ῥῖνες αὐτῷ γνώμων. οὐ μὴν ἐπιμύει καθεύδων ὁ λαγώς, καὶ τοῦτο αὐτῷ ζῴων μόνῳ περίεστιν, οὐδὲ νικᾶται τῷ ὕπνῳ τὰ βλέφαρα· φασὶ δὲ αὐτὸν καθεύδειν μὲν τῷ σώματι,[5] τοῖς δὲ ὀφθαλμοῖς τηνικάδε ὁρᾶν. γράφω δὲ ἅπερ οὖν οἱ σοφοὶ τῶν θηρατῶν λέγουσιν. εἰσὶ δὲ αὐτῶν νύκτωρ αἱ νομαί, τοῦτο μὲν καὶ τροφῆς ξένης ἐπιθυμίᾳ ἴσως, ἐγὼ δ' ἂν φαίην ὅτι γυμνασίας ἕνεκα, ἵνα καὶ τηνικάδε ἐπὰν ἀπὸ τοῦ ὕπνου καρτερῇ ἐνεργὸς ὢν κρατύνηται τὸ τάχος. τῆς δὲ ὁδοῦ τῆς ὀπίσω ἐρᾷ δεινῶς, καὶ συντρόφου παντὸς χωρίου ἡττᾶται· ἔνθεν τοι καὶ ἁλίσκεται

[1] *Gron*: ἐξέτριψαν. [2] αὐτοῖς ἦν.
[3] ἀλλ' . . . ὄν *corrupt*.

by the sun and had spent some time slowly acquiring a little heat, they came to themselves and revived, and one of them, I suppose, stirred and looked up and presently put out its tongue as well and opened its mouth in its craving for nourishment. Accordingly some milk was brought, as was proper for such young creatures, and little by little they were reared up, to furnish (in my opinion) an astonishing proof of their birth by a male. I cannot prevail upon myself to doubt the story, the reason being that the narrator's tongue was a stranger to falsehoods and exaggeration.

13. It seems that the Hare knows about winds and seasons, for it is a sagacious creature. . . . During the winter it makes its bed in sunny spots, for it obviously likes to be warm and hates the cold. But in summertime it prefers a northern aspect, wishing to be cool. Its nostrils, like a sundial, mark the variation of the seasons. The Hare does not close its eyes when sleeping: this advantage over other animals it alone enjoys and its eyelids are never overcome by slumber. They say that it sleeps with its body alone while it continues to see with its eyes. (I am only writing what experienced hunters say.) Its time for feeding is at night, which may be because it desires unfamiliar food, though I should say that it was for the sake of exercise, in order that, while refraining from sleep all this time and full of activity, it may improve its speed. But it greatly likes to return to its home and loves every spot with which it is familiar. That, you see,

The Hare

---

[4] ⟨τόν⟩ *add. H.*
[5] τοῦ σώματος V, τὸ σῶμα *other* MSS.

τὰ πολλά, ⟨τὰ⟩[1] ἤθη τὰ οἰκεῖα ἐκλιπεῖν οὐχ ὑπομένων.

14. Θεῖ δὲ ὁ λαγὼς ὑπό τε κυνῶν καὶ ἱππέων διωκόμενος, εἰ μὲν ἐκ πεδιάδος γῆς εἴη, ὠκύτερον τῶν ὀρείων λαγών, ἅτε μικρὸς τὸ σῶμα καὶ λεπτός· ἔνθεν τοι καὶ κοῦφον αὐτὸν εἶναι οὐκ ἀπεικός. σκιρτᾷ γοῦν τὰ πρῶτα ἀπὸ τῆς γῆς καὶ πηδᾷ, διαδύεται δὲ καὶ διὰ θάμνων ὀλισθηρῶς καὶ εὐκόλως καὶ διὰ παντὸς ἑλώδους τόπου· καὶ εἴ που πόαι βαθεῖαι, καὶ διὰ τούτων διεκπίπτει ῥᾳδίως. καὶ ὅπερ τοῖς λέουσί φασι τὴν ἀλκαίαν δύνασθαι πρὸς τὸ ἐγείρειν αὐτοὺς καὶ ἐποτρύνειν, τοῦτό τοι καὶ ἐκείνῳ τὰ ὦτά ἐστι, ῥύμης συνθήματα καὶ ἐγερτήρια δρόμου. ἀνακλίνει γοῦν κατὰ τῶν νώτων αὐτά, κέχρηται δὲ αὐτοῖς πρὸς τὸ μὴ ἐλινύειν μηδὲ ὀκνεῖν οἷον μύωψι. δρόμον δὲ ἕνα καὶ εὐθὺν[2] οὐ θεῖ, δεῦρο δὲ καὶ ἐκεῖσε παρακλίνει, καὶ ἐξελίττει τῇ καὶ τῇ, ἐκπλήττων τοὺς κύνας καὶ ἀπατῶν. ὅποι ποτὲ δ' ἂν ὁρμήσῃ καὶ ἀπονεῦσαι θελήσῃ, κατ' ἐκείνην τὴν ἐκτροπὴν κλίνει τῶν ὤτων τὸ ἕτερον, οἷον ἰθύνων ἑαυτῷ διὰ τούτου τὸν δρόμον. οὐ μὴν ἀναλίσκει τὴν ἑαυτοῦ δύναμιν ἀταμιεύτως, τηρεῖ δὲ τοῦ διώκοντος τὴν ὁρμήν, καὶ ἐὰν μὲν ᾖ νωθής, οὐ πᾶν ἀνῆκε τὸ ἑαυτοῦ τάχος, ἀλλά τι καὶ[3] ἀνέστειλεν, ὡς προεκθεῖν μὲν ⟨τοῦ⟩[4] κυνός, οὐ μὴν ἀπαγορεῦσαι ὑπὸ τοῦ συντόνου τοῦ δρόμου αὐτός. οἶδε γὰρ ἀμείνων ὤν, καὶ ὁρᾷ ἐς τὸ μὴ ὑπερπονεῖσθαί οἱ τὸν καιρὸν ὄντα. ἐὰν δὲ καὶ ὁ κύων ᾖ ὤκιστος, τηνικαῦτα ὁ λαγὼς φέρεται θέων ᾗ ποδῶν ἔχει. ἤδη γοῦν καὶ πολὺ τῆς ὁδοῦ προλαβών, καὶ ἀπολιπὼν ἐκ πολλοῦ θηρατὰς καὶ

is why it is generally caught, because it cannot endure to abandon its native haunts.

The Hare of the plains

14. The Hare when pursued by hounds and horsemen runs, if it is a denizen of the plains, swifter than the Mountain Hare, as its body is small and slim. Hence it is not unnatural for it to be nimble. At any rate to begin with it leaps and bounds from the earth and slips through thickets and across marshy ground with ease, and wherever the grass is deep it escapes without difficulty. And just as they say that the tail of the lion can rouse and stimulate it, so it is with the ears of the Hare: they are signals for speed and excite it to run. At any rate it lays them back and uses them as goads to prevent it from lagging and hesitating. But its course is not uniform and straight, but it turns aside now right now left and doubles this way and that, bewildering and deluding the hounds. And in whatever direction it wants to swerve in its course, it droops one ear to that avenue of escape, as though it were steering its course therewith. It does not however squander its powers, but observes the pace of its pursuer; and if he is tardy, it does not put forth its whole strength but keeps itself in check somewhat, enough to outrun the hound but not enough to exhaust itself by intense speed. For it knows that it can run faster and realises that this is not the moment for it to over-exert itself. If however the hound is very swift, then the Hare runs as fast as its feet can carry it. And when at length it has got far ahead and has left hunters, hounds, and horsemen a long way behind,

---

[1] ⟨τά⟩ *add. Jac.* [2] ἰθύν.
[3] ἀλλὰ καί τι. [4] ⟨τοῦ⟩ *add. H.*

κύνας καὶ ἵππους, ἐπί τινα λόφον ὑψηλὸν ἀναθορὼν καὶ ἑαυτὸν ἀναστήσας ἐπὶ τῶν κατόπιν ποδῶν, οἷον ἀπὸ σκοπιᾶς ὁρᾷ τὴν τῶν διωκόντων ἅμιλλαν, καί μοι δοκεῖ ὡς ἀσθενεστέρων καταγελᾶν αὐτῶν. εἶτα ἐκ τούτου θαρρήσας [1] ὡς πλέον ἔχων, οἷον εἰρήνης καὶ γαλήνης λαβόμενος ἀσμένως ἡσυχάζει καὶ κεῖται καθεύδων. λαγὼς δὲ ὄρειος οὐχ οὕτω ταχύς, ὥσπερ οὖν οἱ τοῖς πεδίοις ἐνοικοῦντες, εἰ μή ποτε ἄρα κἀκεῖνοι πεδίον ἔχοιεν ὑποκείμενον, ἐν ᾧ κατιόντες διαθέουσι· καὶ τὸ μὲν ὄρος κατοικοῦσι, γυμνάζονται δὲ ἐνταῦθα, συνθέοντες [2] τοῖς ἐκ τῶν πεδίων πολλάκις.[3] φιλεῖ γοῦν ἐν μὲν τοῖς πεδίοις αὐτοὺς διώκεσθαι, καὶ τὰ μὲν ὑποκινεῖν, τὰ δὲ ὑπολανθάνειν, εἶτα ἐκ τῆς συνήθους διώξεως ἀνισταμένους ὑπεκφυγεῖν οὐδὲ εἷς.[4] ἐπὰν δὲ ὦσιν ὁμοῦ τῷ ἁλίσκεσθαι, τῆς πεδιάδος ὁδοῦ βραχὺ ἀποκλίναντες ἐς τὰ ἀνάντη καὶ ὄρεια ἀνέθορον, ἅτε ἐς οἰκεῖα ἤθη καὶ ἔννομά [5] σφισι σπεύδοντες, καὶ τοῦτον τὸν τρόπον ἀπιόντες οἴχονται, ἀδοκήτου [6] σωτηρίας τυχόντες· ὀρειβασίαι γὰρ καὶ ἵπποις καὶ κυσὶν ἐχθραὶ πεφύκασιν, ἀπαγορευόντων αὐτοῖς τῶν ποδῶν καὶ ἐκτριβομένων ῥᾷστα. κυνῶν δὲ ἔτι [7] καὶ μᾶλλον ἅπτεται τὸ πάθος· σαρκώδεις γὰρ αὐτῶν εἰσιν οἱ πόδες, καὶ ἔχουσιν οὐδὲν πρὸς τὴν πέτραν ἀντίτυπον, ὡς ἵπποι τὴν ὁπλήν. ὁ δὲ λαγὼς τοὐναντίον, πέφυκε γὰρ δασὺς τοὺς πόδας, καὶ δὴ καὶ τῶν τραχέων ἀνέχεται.

[1] θαρσήσας.
[2] *Jac*: ἔνθεν τοι.
[3] *H marks a lacuna here.*
[4] φιλεῖ γοῦν . . . οὐδείς? *interpolation, Ed.*
[5] τὰ ἔννομα.

it races up some high hill and sitting up on its hind legs surveys as from a watch-tower the efforts of its pursuers and, as I think, laughs at them for being feebler than itself. Then emboldened by the advantage it has gained, like one who has achieved peace and calm, it is glad to rest and lies down to sleep.

The Hare of the mountains

The Mountain Hares, however, are not so swift as those that live in the plains, unless indeed the former also have plain-land lying below into which they can descend and run about. Though their home is on a mountain they exercise themselves in the plain, often running about with the Hares there. The usual thing when they are pursued in the plain is for them to start up and to lie hid by turns, but since they are constantly forced out, not one escapes.[a] But when they are on the point of being caught they change suddenly their direction over the plain and dart uphill into the mountains, speeding of course to their native haunts, their proper domain; and in this way they escape and are gone, reaching unexpected safety, for horses and hounds dislike going up mountains, since their feet give out and are very quickly worn down, while hounds suffer even worse, their paws being fleshy and having nothing to resist the rocks, as horses have their hooves. The Hare on the contrary has naturally hairy paws and is quite content with rough ground.

[a] The strange syntax of this sentence and the fact that the words 'not one escapes' are contradicted in the sequel suggest that the sentence is an interpolation.

---

[6] *καὶ ἀδοκήτου.* [7] *Ges*: *ὅτι.*

ὅτῳ δέ εἰσι λαγῷ [1] ἐν τοῖς δάσεσι καὶ ἐν τοῖς θάμνοις διατριβαί, νωθεῖς μὲν οὗτοι ἐς τὸν δρόμον, βραδεῖς δὲ ἐς τὴν φυγήν· πεπιασμένοι [2] γὰρ οἱ τοιοίδε εἰσὶ καὶ ὑπὸ τῆς ἀργίας οὐχὶ ἠθάδες τοῦ δρόμου, ἥκιστοί τε ὡς ὅτι πορρωτάτω τῶν θάμνων ἀποφοιτᾶν. θῆραι δὲ [3] τούτων τοιαίδε. τὰ μὲν πρῶτα διαδύονται διὰ τῶν θάμνων τῶν μικρῶν, ὅσοις μὴ συνεχὴς ἡ λόχμη, τούς γε μὴν δασυτέρους αὐτῶν, ἅτε μὴ οἷοί τε ὄντες ὑπελθεῖν, εἰκότως ὑπερπηδῶσι. πεφύκασι δὲ ἄλλοι [4] συνεχεῖς καὶ δι᾿ ἀλλήλων ⟨συνυφασμένοι⟩.[5] ὅπου οὖν τοιοῦτοι, ἅτε [6] πολλάκις ἀναγκαζόμενος τοῦτο δρᾶν ὁ λαγώς, καὶ διὰ τὴν βαρύτητα τὴν τοῦ σώματος οὐκ ὢν ἁλτικός, κάμνει ῥᾷστα καὶ ἀπαγορεύει. αἵ γε μὴν κύνες τὰ πρῶτα σφάλλονταί τε αὐτοῦ καὶ ἁμαρτάνουσιν· οὐ γὰρ ὁρῶσιν αὐτὸν διὰ τὴν τῆς ὕλης πυκνότητα, πηδῶσι δὲ καὶ αὗται κατὰ τῶν θάμνων ὑπὸ τῆς ὀσμῆς ἀγόμεναι· τελευτῶσαί γε μὴν εἶδον καὶ διώκουσι καὶ ἐνδιδόασιν οὐδὲ ἕν, ὁ δὲ ἐκ τῆς τοῦ πηδᾶν συνεχείας κάμνει τε καὶ ἀπαγορεύει καὶ ἐντεῦθεν ἑάλωκε. τὰ δὲ ἀνάντη μὲν καὶ ὑψηλὰ οἱ λαγῳ ἀναθέουσι ῥᾷστα· τὰ γάρ τοι κατόπιν κῶλα μακρότερα ἔχουσι τῶν ἔμπροσθεν· καταθέουσι δ᾿ οὐχ ὁμοίως· λυπεῖ γὰρ αὐτοὺς τῶν ποδῶν τὸ ἐναντίον.

15. Πέφυκε δὲ καὶ λαγὼς ἕτερος μικρὸς τὴν φύσιν, οὐδὲ αὔξεταί ποτε· κόνικλος ὄνομα αὐτῷ. οὐκ εἰμὶ δὲ ποιητὴς ὀνομάτων, ὅθεν καὶ ἐν ⟨τῇδε⟩ [7] τῇ συγγραφῇ φυλάττω τὴν ἐπωνυμίαν

[1] λαγῷ αἵ τε.
[2] *Ges*: πεπιεσμένοι.
[3] δὲ καί.
[4] οὗτοι.

All Hares that live among thickets and bushes are sluggish runners and slow to flee, for such animals have grown plump and from sloth are not habituated to running and are quite incapable of going a long distance from their thickets. The method of hunting them is as follows. To begin with these Hares slip through the little bushes of which the foliage is not a solid mass, but where it is denser they naturally leap over them as they cannot get beneath them. But other bushes grow in a solid mass with their branches interlaced. So where the bushes are of this nature the Hare is constantly obliged to do this, and since the weight of its body does not dispose it to be good at jumping, it very soon tires and gives up. At first the hounds are baffled and lose the track, for owing to the thickness of the wood they fail to see the quarry; but they too leap over the bushes and are led by the scent. Finally however they catch sight of it and are after it, never pausing for a moment, whereas the Hare exhausted by the continual leaping gives up and so is caught.

Hare and Hounds

Hares run up steep, high ground with the utmost ease, for their hind legs are longer than the front ones. They run down less easily, for the shortness of their front legs is a handicap to them.

15. There is also another kind of Hare, small by nature, and it never grows larger. It is called a Rabbit. I am no inventor of names, which is the reason why in this account I preserve the original

The Rabbit

[5] ⟨συνυφασμένοι⟩ *add. H, cp.* 13. 8 κλάδοι δι' ἀ. σ.
[6] ὅπου τοιοῦτοι ἅτε οὖν.
[7] ⟨τῇδε⟩ *add. H.*

τὴν ἐξ ἀρχῆς, ἥνπερ οὖν Ἴβηρες ⟨οἱ⟩[1] Ἑσπέριοι ἔθεντό οἱ, παρ' οἷς[2] καὶ γίνεταί τε καὶ ἔστι πάμπολυς. τούτῳ τοίνυν ἡ μὲν χρόα παρὰ τοὺς ἑτέρους μέλαινα, καὶ ὀλίγην ἔχει τὴν οὐράν, τά γε μὴν λοιπὰ τοῖς προειρημένοις ἰδεῖν ἐμφερής ἐστι. διαλλάττει δὲ ἔτι καὶ τὸ τῆς κεφαλῆς μέγεθος· λεπτοτέρα γὰρ ἡ τούτου καὶ δεινῶς ἄσαρκος καὶ βραχυτέρα.[3] λαγνότερος[4] δὲ τῶν λοιπῶν· †λασαρὰ διετησίους φύσει,†[5] ὑφ' ὧν οἰστρεῖταί τε καὶ ἐκμαίνεται, ὅταν ἐπὶ τὰς θηλείας ᾄττῃ. [ἔστι δὲ καὶ ἐλάφῳ[6] ὀστοῦν ἐν τῇ καρδίᾳ αὐτοῦ,[7] ὅπερ οὖν τίνος ἀγαθὸν εἰδέναι μελήσει ἄλλῳ.]

16. Τὴν τῶν θύννων θήραν Ἰταλοί τε καὶ Σικελοὶ κητείαν[8] φιλοῦσιν ὀνομάζειν· τά τε χωρία, ἔνθα αὐτοῖς εἴωθε θησαυρίζεσθαι τά τε δίκτυα τὰ μεγάλα καὶ ἡ λοιπὴ παρασκευὴ ἡ θηρατική, καλεῖται μέντοι κητοθηρεῖα,[9] τοῦ θύννου τὸ μέγεθος ἐς τὰ κήτη βουλομένων τὸ λοιπὸν ἀποκρίνειν. ἀκούω δὲ Κελτοὺς καὶ Μασσαλιώτας καὶ τὸ Λιγυστικὸν πᾶν ἀγκίστροις τοὺς θύννους θηρᾶν· εἴη δ' ἂν ταῦτα ἐκ σιδήρου μὲν πεποιημένα, μέγιστα δὲ καὶ παχέα ἰδεῖν. καὶ τά γε ὑπὲρ τῶν θύννων νῦν πρὸς τοῖς ἤδη προειρημένοις τοσαῦτα ἔστω μοι.

17. Περὶ τὰς καλουμένας νήσους Τυρρηνικὰς θηρῶσιν οἱ κατὰ τὴν ἁλιείαν ἔχοντες τὸν ἐκεῖθι

[1] ⟨οἱ⟩ *add. Ges.* [2] *Schn* : πάρος.
[3] βραχυτέρα δηλονότι κατὰ τὸ πᾶν σῶμα.
[4] *Jac* : λευκότερος MSS, *H.*

name given to it by the Iberians of the west in whose country the Rabbit is produced in great numbers. Its colour compared with that of hares is dark; it has a small tail, but in other respects it is like them. A further difference is in the size of its head, for it is smaller and curiously scant of flesh and shorter. But it is more lustful than the hare . . .[a] which cause it to go raving mad when it goes after the female. [The stag also has a bone in its heart, and someone else shall make it his business to discover what purpose it serves.][b]

Fishing for Tunny

16. The pursuit of the Tunny is commonly designated as 'big fishing' by the people of Italy and Sicily, and the places in which they are in the habit of storing their huge nets and other fishing gear are called 'big-fishing tackle stores,' for they wish henceforward to segregate the huge Tunny into the class of 'big fishes.' And I learn that the Celts and the people of Massalia and all those in Liguria catch Tunny with hooks; but these must be made of iron and of great size and stout. So much then for Tunnies in addition to what I have already said earlier on.

The 'Aulopias' fish

17. Those who are in the habit of fishing round the Tyrrhenian islands,[c] as they are called, hunt a

[a] The Greek is corrupt. Accepting Post's conjecture, render: 'It is by nature incontinent throughout the year.'
[b] The sentence is out of place here.
[c] The 'Aeoliae Insulae' (modern Lipari isl.) off the N coast of Sicily.

[5] λασαρὰ . . . φύσει *corrupt*: λαίσθα διετήσιος *Post*.
[6] καὶ ἐλάφῳ *del. H.*
[7] αὐτῷ.
[8] κητίαν.
[9] *Jac*: κητοθηρία.

κητώδη ἰχθύν, καὶ καλοῦσιν αὐτὸν αὐλωπίαν, καὶ περιηγήσασθαί γε τούτου τὰ ἴδια οὐ χεῖρόν ἐστι. μέγεθος μὲν ἡττᾶται τῶν μεγίστων θύννων ὁ μέγιστος αὐλωπίας, ῥώμην δὲ καὶ ἀλκὴν τὰ πρῶτα φέροιτο ἂν πρὸς ἐκείνους ἀντικρινόμενος. ἄλκιμον μὲν γὰρ ἰχθύων φῦλόν ἐστι καὶ οἱ θύννοι, ἀλλὰ τῷ παραταξαμένῳ καὶ προθύμως ἀνταγωνισαμένῳ μετὰ τὴν πρώτην ὁρμὴν ἀφίσταται τοῦ κράτους τοῦ αἵματος αὐτῷ πηγνυμένου, ⟨καὶ⟩ [1] παρειμένος ὤκιστα εἶτα ἑάλω. διακαρτερεῖ γε μὴν ὁ αὐλωπίας ἐπὶ μακρόν, ὅταν ἐπίθηταί οἱ κατὰ τὸ καρτερόν, καὶ ὡς πρὸς ἀντίπαλον ἀνθίσταται [2] τὸν ἁλιέα, καὶ κρατεῖ τὰ πλεῖστα, [3] ἐπὶ [4] μᾶλλον ἑαυτὸν πιέσας καὶ κάτω νεύσας τὴν κεφαλὴν καὶ ὠθήσας κατὰ τοῦ βυθοῦ· πέφυκέ τε τὴν γένυν ἰσχυρὸς καὶ τὸν αὐχένα καρτερός, καὶ ῥώμης ἔχει κάλλιστα. ὅταν δὲ αἱρεθῇ, ἰδεῖν ὡραιότατός ἐστι, τοὺς μὲν ὀφθαλμοὺς ἔχων ἀνεῳγότας καὶ περιφερεῖς καὶ μεγάλους, οἵους Ὅμηρος τοὺς τῶν βοῶν ᾄδει· ἡ δὲ γένυς, ὥσπερ οὖν εἶπον, καρτερὰ οὖσα, ὅμως καὶ ἐς ὥραν οἱ συμμάχεται. καὶ τὰ μὲν νῶτα αὐτῷ [5] κυάνου μεμίμηται χρόαν τοῦ βαθυτάτου, [6] ὑπέζωσταί γε μὴν [7] λευκὴν τὴν νηδύν· ἄρχεται δὲ ἀπὸ τῆς κεφαλῆς αὐτῷ γραμμὴ χρυσῖτις τὴν χρόαν, κατιοῦσα δὲ ἐς τὸ οὐραῖον μέρος ἀπολήγει ἐς κύκλον. εἰπεῖν δὲ καὶ τὴν δολερὰν ἐπ' αὐτοῖς θήραν, ἥνπερ οὖν ἀκούσας οἶδα, ἐθέλω. προελόμενοι χώρους ἐκ πολλοῦ, ἐς οὓς ἀθροίζεσθαι τοὺς αὐλωπίας ὑπολαμβάνουσιν, εἶτα μέντοι κορα-

---

[1] ⟨καί⟩ *add. H.*
[2] ἵσταται.
[3] πλεῖστα καί.

gigantic fish which they call the *Aulopias*, and it is worth while to describe its characteristics. In the matter of size the largest Aulopias yields to the largest Tunnies, but if matched against them it would take the prize for strength and courage. True, the Tunny also is a powerful species of fish, but after its first onset against its adversary and vigorous opponent [a] it forgoes its strength, and as its blood congeals, it very soon surrenders and is then caught. The Aulopias on the contrary carries on the struggle for a long time when it is attacked with vigour, and withstands the fisherman as it would an adversary, and on most occasions gets the better of him by gathering itself together, bowing its head, and thrusting down into the depths; it has a forceful jaw and a powerful neck and is exceedingly strong. But when it is captured it is a most beautiful sight: it has wide open eyes, round and large, such eyes as Homer sings of in oxen.[b] And the jaw, though powerful, as I remarked, contributes to its beauty. Its back is like the colour of the deepest lapis lazuli, its belly underneath is white. A stripe of a golden hue starts at the head and descending to the region of the tail ends in a circle.

I wish to speak also of the artifices employed in hunting it which I remember to have heard. The fishermen previously select spots from a large area where they suppose the Aulopiae to be congregating

how caught

[a] *I.e.* the fisherman.

[b] βοῶπις is a frequent epithet of Hera in Homer's *Iliad*.

---

[4] *Reiske* : ἔτι.

[5] αὐτοῦ.

[6] τὴν βαθυτάτην.

[7] γε μήν] μέν.

κίνους ταῖς ὑποχαῖς πολλοὺς συλλαβόντες, τὴν ἑαυτῶν ἄκατον ἐπ' ἀγκυρῶν ὁρμίσαντες καὶ συνεχῶς κτύπον τινὰ ὑποδρῶντες διατείνουσι τοὺς κορακίνους ἅμμασι [1] σφηκοῦντες. οἱ δὲ ἀκούοντες τοῦ κτύπου καὶ τὸ δέλεαρ ὁρῶντες ἄλλος ἀλλαχόθεν ἀνανέουσι [2] καὶ ἀθροίζονται καὶ περιέρχονται τὴν ἁλιάδα, πραΰνονταί τε ἐς τοσοῦτον τῷ κρότῳ καὶ ⟨τῷ⟩ [3] πλήθει τῆς τροφῆς, ὡς καὶ προτεινόντων τὰς χεῖρας παραμένειν. ἀνέχονται δὲ ἀνθρωπίνης ἐπιψαύσεως, ὡς μὲν κρίνειν ἐμέ, τῇ βορᾷ δεδουλωμένοι, ἤδη δέ, ὡς οἱ θηρατικοί φασι, καὶ τῇ ἀλκῇ [4] ⟨θαρροῦντες⟩.[5] εἰσὶ δὲ ἐν αὐτοῖς καὶ χειροήθεις, οὕσπερ οὖν οἱ ἁλιεῖς ὡς εὐεργέτας καὶ ἑταίρους γνωρίζουσιν, εἶτα μέντοι τὰ πρὸς αὐτοὺς ἔχουσιν ἔνσπονδα. ἕπονται δὲ τούτοις οἷον ἡγεμόσι καὶ ἄλλοι ξένοι, καὶ τούτους μὲν ὡς ἂν εἴποι τις ἐπήλυδας καὶ θηρῶσι καὶ ἀποκτείνουσι, πρός γε μὴν τοὺς τιθασούς, οἵπερ οὖν [6] ἐοίκασι ταῖς παλευτρίαις πελειάσιν, ἀθηρία τε αὐτοῖς ἐστι καὶ ἐκεχειρία. οὐδ' ἂν ἁλιέα σοφὸν τοσαύτη ποτὲ καταλάβοι ἀπορία, ὡς ἐξ ἐπιβουλῆς ἑλεῖν αὐλωπίαν ἥμερον· ἐκ γάρ τινων αἰτιῶν αἱρεθεὶς κατὰ τύχην καὶ λυπεῖ. ἁλίσκεται δὲ ἢ ἀγκίστρῳ [7] περιπαρεὶς ἢ τρωθεὶς ἐς θάνατον. ὁρῶμεν δὲ καὶ τοὺς ὀρνιθοθήρας μὴ ἂν τῶν ἐλλοχώντων ὀρνίθων ἀποκτείναντάς τινας ἢ ἐπὶ πράσει ἢ ἐπὶ δείπνῳ. καὶ ἄλλαι δὲ θῆραι τῶνδε τῶν ἰχθύων εἰσίν.

[1] ἅμα.
[2] *Schn* : ἀνανεύουσι.
[3] ⟨τῷ⟩ *add. H.*
[4] τῷ πλήθει τῆς ἀλκῆς.
[5] ⟨θαρροῦντες⟩ *add. Schn.*

and after catching a number of Crow-fish [a] in their bag-nets [b] they anchor their boat and maintain a continuous din; the Crow-fish they make fast in a noose and let out on a line. Meanwhile the Aulopiae hearing the din and observing the bait, come swimming up from all sides and congregate and circle about the boat. And the din and the quantity of food have such a soothing effect upon them that, even though men reach out their hands, they remain and submit to the human touch because, as I judge, they are slaves to food, and in fact, as their pursuers maintain, because their strength gives them confidence. There are also tame ones among them which the fishermen recognize as their benefactors and comrades, so with them they maintain a truce. And other strange fishes follow them like leaders, and these aliens, as one might call them, the men hunt and kill, but the tame fish, which may be likened to decoy-doves, they do not hunt but spare, nor would any prudent fisherman ever be reduced to such straits as to catch a tame Aulopias deliberately, for if by some mischance one happens to be caught it brings trouble. The fish is captured either by being pierced with a hook or by being mortally wounded.

We see bird-catchers also abstaining from killing birds that decoy others, whether for sale or for the table. There are other methods besides of catching these fish.

[a] Not certainly identified, but may be *Chromis castanea*; not identical with the Danubian fish of 14. 23 and 26.
[b] See A. W. Mair, *Oppian &c.* (Loeb Cl. Lib.), pp. xl ff.

---

[6] *οἵπερ οὖν*] *οἵ γε μήν*.
[7] *τῷ ἀγκίστρῳ*.

18. Ἐν δὲ τοῖς βασιλείοις τοῖς Ἰνδικοῖς, ἔνθα ὁ μέγιστος τῶν βασιλέων διαιτᾶται τῶν ἐκεῖθι, πολλὰ μὲν καὶ ἄλλα ἐστὶ θαυμάσαι ἄξια, ὡς μὴ αὐτοῖς ἀντικρίνειν μήτε τὰ Μεμνόνεια [1] Σοῦσα καὶ τὴν ἐν αὐτοῖς πολυτέλειαν μήτε τὴν ἐν τοῖς Ἐκβατάνοις μεγαλουργίαν· ἔοικε [2] γὰρ κόμπος εἶναι Περσικὸς ἐκεῖνα, εἰ πρὸς ταῦτα ἐξετάζοιτο. καὶ τὰ λοιπὰ μὲν περιελθεῖν τῷ λόγῳ οὐ τῆσδε τῆς συγγραφῆς ἐστιν, ἐν δὲ τοῖς παραδείσοις τρέφονται μὲν καὶ ταῶς ἥμεροι καὶ χειροήθεις φασιανοί, ἔχουσι δὲ . . . [3] ἐν τοῖς φυτοῖς τοῖς ἠσκημένοις, ἅπερ οὖν οἱ μελεδωνοὶ οἱ βασίλειοι τῆς δεούσης ἀξιοῦσι κομιδῆς. καὶ γάρ εἰσιν ἄλση σκιερὰ καὶ νομὴ σύμφυτος καὶ κλάδοι δι᾽ ἀλλήλων συνυφασμένοι σοφίᾳ τινὶ δενδροκομικῇ. καὶ τὸ σεμνότερον τῆς ὥρας τῆς ἐκεῖθι, τὰ δένδρα αὐτὰ τῶν ἀειθαλῶν ἐστι, καὶ οὔποτε γηρᾷ καὶ ἀπορρεῖ τὰ φύλλα· καὶ τὰ μὲν ἐπιχώριά ἐστι, τὰ δὲ ἀλλαχόθεν σὺν πολλῇ κομισθέντα τῇ φροντίδι, ἅπερ οὖν κοσμεῖ τὸν χῶρον καὶ ἀγλαΐαν δίδωσι, πλὴν ἐλαίας· οὐ γὰρ αὐτὴν ἡ Ἰνδῶν φέρει, οὔτε αὐτή, οὔτε ἥκουσαν ἀλλαχόθεν τρέφει. ὄρνιθες οὖν καὶ ἕτεροι ἐλεύθεροι καὶ ἀδούλωτοι, καὶ ἐλθόντες αὐτομάτως ἔχουσι κατ᾽ αὐτῶν κοίτας καὶ εὐνάς· ἐνταῦθά τοι καὶ οἱ ψιττακοὶ τρέφονται καὶ εἰλοῦνται περὶ τῷ βασιλεῖ. σιτεῖται δὲ Ἰνδῶν οὐδὲ εἷς ψιττακόν, καίτοι παμπόλλων ὄντων τὸ πλῆθος· τὸ δὲ αἴτιον,[4] ἱεροὺς αὐτοὺς εἶναι πεπιστεύκασιν οἱ Βραχμᾶνες, καὶ μέντοι καὶ τῶν ὀρνίθων ἁπάντων προτιμῶσι. καὶ ἐπιλέγουσι δρᾶν τοῦτο εἰκότως· μόνον γὰρ

[1] Μεμνόνια. [2] ἐοίκασι.

The royal parks of India and their birds

18. In the royal residences in India where the greatest of the kings of that country lives, there are so many objects for admiration that neither Memnon's city of Susa with all its extravagance, nor the magnificence of Ecbatana is to be compared with them. (These places appear to be the pride of Persia, if there is to be any comparison between the two countries.) The remaining splendours it is not the purpose of this narrative to detail; but in the parks tame peacocks and pheasants are kept, and they ⟨live⟩ in the cultivated shrubs to which the royal gardeners pay due attention. Moreover there are shady groves and herbage growing among them, and the boughs are interwoven by the woodman's art. And what is more remarkable about the climate of the country, the actual trees are of the evergreen type, and their leaves never grow old and fall: some of them are indigenous, others have been imported from abroad after careful consideration. And these, the olive alone excepted, are an ornament to the place and enhance its beauty. India does not bear the olive of its own accord, nor if it comes from elsewhere, does it foster its growth.

Well, there are other birds besides, free and unenslaved, which come of their own accord and make their beds and resting-places in these trees. There too Parrots are kept and crowd around the king.

The Parrot

But no Indian eats a Parrot in spite of their great numbers, the reason being that the Brahmins regard them as sacred and even place them above all other birds. And they add that they are justified in so

[3] *Lacuna*: ⟨τὰ ἤθη⟩ *or* ⟨τὰς διατριβάς⟩ *H*, ⟨δίαιταν⟩ *Schn*.
[4] αἴτιον δέ.

τὸν ψιττακὸν ἀνθρώπου στόμα εὐστομώτατα ὑποκρίνεσθαι. εἰσὶ δὲ ἄρα ἐν τοῖσδε τοῖς βασιλείοις καὶ λίμναι χειροποίητοι ὡραῖαι, καὶ ἰχθύας ἔχουσι μεγέθει μεγίστους καὶ πραεῖς· καὶ θηρᾷ αὐτοὺς [1] οὐδεὶς ὅτι μὴ οἱ τοῦ βασιλέως υἱεῖς παῖδες ἔτι ὄντες, ἐν ἀκλύστῳ καὶ ἥκιστα ἐπικινδύνῳ τῷ ὕδατι ἁλιεύοντές τε καὶ παίζοντες καὶ ἅμα καὶ πλεῖν [2] μανθάνοντες.

19. Ἐν τῷ Ἰονίῳ πελάγει κατὰ τὸν Λευκάτην καὶ τὴν πρὸς τῷ Ἀκτίῳ θάλατταν, ἔνθα τοι καὶ τὸν χῶρον καλοῦσιν Ἤπειρον, κεφάλων εἰσὶ κατὰ ἴλας ὡς ἂν εἴποι τις ἄφθονοι νήξεις καὶ πλήθη πάμπολλα. οὐκοῦν θηρῶνται καὶ μάλα ἐκπληκτικῶς· ὁ δὲ τρόπος τῆς θήρας οὗτός ἐστι. νύκτα ἀσέληνον οἱ ἐκεῖθι ἁλιεῖς παραφυλάξαντες, ἀπὸ δείπνου γενόμενοι κατὰ δύο ἀπῆραν [3] σκάφος, οὐκ ὄντος κύματος ἀλλὰ ἀκλύστου καὶ γαληναίας τῆς θαλάττης, εἶτα ἡσυχῇ καὶ κατὰ μικρὰ προερέττουσι· [4] καὶ ὁ μὲν αὐτὴν ὑποκινεῖ τῷ κωπίῳ, προάγων τὴν πορθμίδα βάδην ὡς ἂν εἴποις· ἅτερος δὲ κατακλινεὶς ἐπ' ἀγκῶνος τὸ καθ' ἑαυτὸν μέρος ἐπιβρίθει τῆς πορθμίδος, καὶ ἐς τοσοῦτον ἐπικλίνει, ἐς ὅσον [5] τὸ χεῖλος αὐτῆς προσπελάζει τῷ ὕδατι. οἱ κέφαλοι δὲ καὶ οἱ τούτοις ὁμοειδεῖς κεστρεῖς,[6] ἤτοι τῇ νυκτὶ τερπόμενοι ἢ χαίροντες τῇ γαλήνῃ, τοὺς μὲν χηραμοὺς τοὺς ἑαυτῶν καὶ τοὺς φωλεοὺς ἀπολείπουσιν, ἀνανέουσι δέ, καὶ τὰ ἄκρα γε τοῦ προσώπου ὑπὲρ τὸ ὕδωρ φαίνουσι, καὶ τοσοῦτον τῆς ἐς τὸ ἄνω [7] νήξεως ἐπιλαμβάνουσι,

[1] αὐτῶν. [2] νεῖν Cobet.

doing, for the Parrot is the only bird that gives the most convincing imitation of human speech. There are also in these royal domains beautiful lakes, the work of man's hands, which contain fish of immense size and tame. And nobody hunts them, only the king's sons during their childhood; and in calm waters, quite free from danger, they fish and sport and even learn the art of sailing as well.

Fishing for Mullet

19. In the Ionian sea off Leucatas[a] and in the waters round Actium (the country there they call Epirus) Mullet abound, swimming, so to say, in companies and vast multitudes. These fish are hunted, and in a most astounding manner. The method is as follows. The local fishermen watch for a moonless night and after supper pair off and launch a skiff while there is neither wave nor swell but the sea is calm, and then row forward quietly by slow degrees. One of the men gently agitates the water with his oar, propelling the boat step by step, so to speak, while the other propped on his elbow weighs down his end of the boat, depressing it until the gunwale is nearly at the water-level. And the Mullet and others of their kind,[b] either because they enjoy the night or because they delight in the calm, quit their holes and lairs, swim up, and show the tip of their head above the water and are so occupied in swimming to the surface that they draw near to the

[a] Promontory at the S end of the island of Leucas.
[b] κέφαλος and κεστρεύς both signify the Grey Mullet; see Thompson, *Gk. fishes*, s.vv.

[3] *Ges*: ἐπῆραν. [4] *Reiske*: προσερέττουσι.
[5] ἐς ἄκρον ὅσον. [6] *Ges*: κεστρέες.
[7] ἐς τὸ ἄνω τῆς.

καὶ γίνονται τῆς ἠόνος πλησιον. θεασάμενοι δὲ οἱ θηραταὶ πλέουσι, καὶ τὸ ῥόθιόν γε τῆς πορθμίδος ἡσυχῇ πως [1] ὑποκυμαίνειν ἄρχεται. φεύγοντες οὖν τὴν γῆν καὶ ὑποστρέφοντες ἐς [2] τὸ ἐπικλινὲς τῆς ἀκάτου σφᾶς αὐτοὺς ὑπὸ πλήθους ὠθοῦσι,[3] καὶ ἔσω παρελθόντες ἑαλώκασιν.

20. Τῶν δὲ κητῶν τὰ ὑπέρογκα ἄγαν καὶ τὸ μέγεθος ὑπερήφανα νήχεται μὲν ἐν τοῖς πελάγεσι μέσοις, ἤδη γε μὴν καὶ σκηπτοῖς βάλλεται. πρὸς τούτοις μὲν οὖν ἔστι καὶ ἕτερα ἐπάκτια [4] τοιαῦτα, καὶ ὄνομα τροχὸς αὐτοῖς. καὶ νεῖ [5] κατ' ἀγέλας ταῦτα, μάλιστα μὲν ἐν δεξιᾷ τοῦ Ἄθω τοῦ Θρᾳκίου, ἔν τε [6] τοῖς κόλποις τῷ ἀπὸ Σιγείου πλέοντι,[7] ἐντυχεῖν δέ ἐστιν αὐτοῖς καὶ κατὰ τὴν ἀντιπέρας [8] ἤπειρον παρά τε τὸν Ἀρταχαίου [9] καλούμενον τάφον καὶ τὸν Ἀκάνθιον [10] ἰσθμόν, ἔνθα τοι καὶ ἡ τοῦ Πέρσου φαίνεται διατομή, ᾗ διέτεμε τὸν Ἄθω. τὰ κήτη δὲ ταῦτα, ἃ καλοῦσι τροχούς, ἄλκιμα μὲν οὔ φασιν εἶναι, λοφιὰν δὲ ὑποφαίνει καὶ ἀκάνθας ὑπερμήκεις, ὡς καὶ πολλάκις ὁρᾶσθαι ἐξάλους αὐτάς. ἀκούσαντα δὲ εἰρεσίας κτύπου περιστρέφεταί τε καὶ κατειλεῖται ὡς ὅτι κατωτάτω ἑαυτὰ ὠθοῦντα· ἔνθεν τοι καὶ τοῦδε τοῦ ὀνόματος μετείληχεν. ἀναπλεῖ δὲ ἀνελιχθέντα καὶ κυλιόμενα ἔμπαλιν.

[1] *Ges*: ὅπως.
[2] ἄνευ δικτύων εἰς.
[3] *Reiske*: ὑποπλήθουσι.
[4] ἔστι . . . ἐπάκτια] *Jac*: καὶ ὅσα ἕτερα σπάνια MSS, *H*.
[5] *Ges*: ἔνι.
[6] *Gow*: γε MSS, *H*.
[7] τῷ . . . πλέοντι] *Jac*: τοῦ . . . πλέοντα MSS, *H*.
[8] ἀντιπέραν.
[9] *Voss*: Ἀρτακαίου.
[10] *Voss*: Ἀκανθαῖον.

shore. So the fishermen observing this, begin to sail, and the rush of the boat starts a gentle ripple. Therefore the fish in fleeing from the shore turn and owing to their numbers jostle one another into the portion of the boat sloping toward them, and once inside are caught.

The 'Trochus'

20. Sea-monsters of excessive bulk and of prodigious size swim in mid-ocean, and are at times struck by lightning. Besides these there are others of the same kind that come close to the shore, and their name is *Trochus* (wheel).[a] These swim in droves, especially on the right side of Thracian Athos and in the bays as one sails from Sigeum, and one may encounter them along the mainland opposite, close to what is called the Tomb of Artachaees [b] and the isthmus of Acanthus where the canal which the Persian King cut through Athos is to be seen. And they say that these monsters which they call Trochus are timid, though they expose their crest and spines of enormous length so that they are often seen above the water. But at the sound of oars they revolve and contract and plunge as deep as they can go. It is from this, you see, that they derive their name. And again they uncoil and with a rolling motion swim up to the surface.

[a] E. de Saint-Denis, *Vocabulaire des animaux marins en latin* s.v. *Rota:* 'monstre indéterminé . . . le fabuleux et le réel s'embrouillent . . . dans les descriptions de Pline [9. 8] et d'Elien.'

[b] Persian general who superintended the construction of Xerxes's canal through the promontory of Athos; see Hdt. 7. 117. His 'Tomb' has not been certainly identified.

21. Τριτώνων πέρι σαφῆ μὲν λόγον καὶ ἀπόδειξιν ἰσχυρὰν οὐ μάλα τί φασιν εἰπεῖν ἔχειν τοὺς ἁλιέας· λέγει[1] δ' οὖν φήμη διαρρέουσα ναὶ μὰ Δία πολλὴ[2] γίνεσθαί τινα ἐν τῇ θαλάττῃ κήτη ἀνθρωπόμορφα τὰ ἀπὸ κεφαλῆς ὅσα ἐς ἰξὺν λήγει. λέγει δὲ Δημόστρατος ἐν λόγοις ἁλιευτικοῖς ἐν Τανάγρᾳ θεάσασθαι τάριχον Τρίτωνα. καὶ τὰ μὲν ἄλλα ἦν φησι καὶ τοῖς πλαττομένοις ὅμοιος καὶ τοῖς γραφομένοις, τὴν δέ οἱ κεφαλὴν ὑπὸ χρόνου διεφθαρμένην οὐ πάνυ ⟨τι⟩[3] σαφῆ ἔφατο εἶναι οὐδὲ οἵαν συνιδεῖν[4] τε καὶ γνωρίσαι ῥᾷστα· 'προσαψαμένου δέ μου φολίδες ἀπέπιπτον τραχεῖαι καὶ μέντοι καὶ ἀντίτυποι εὖ μάλα. τῶν δέ τις ἐκ τῆς βουλῆς ἁρμοζόντων κλήρῳ τὴν Ἑλλάδα καὶ πεπιστευμένων τὴν ἀρχὴν ἑνὸς ἔτους, οἷα δὴ βασανιῶν καὶ ἐλέγξων[5] τοῦ βλεπομένου τὴν φύσιν,[6] τοῦ δέρματος παρελὼν ὀλίγον καθήγισεν ἐπὶ πυρός, καὶ ὀσμὴ μὲν βαρεῖα καομένου τοῦ ἐμβληθέντος προσέβαλε τῶν παρόντων τὰς ῥῖνας. οὐ μὴν συμβαλεῖν φησιν εἴτε χερσαῖον τὸ ζῷον εἴτε θαλάττιον εἴη τὴν φύσιν εἴχομεν. ἀλλ' ἥ γε πεῖρα οὐ χρηστόν οἱ τὸν μισθὸν ἀπέδωκεν. οὐ γὰρ μετὰ μακρὸν[7] τὸν βίον κατέστρεψε, περαιούμενος ὀλίγον καὶ στενὸν πορθμὸν ἐξήρει πορθμείῳ καὶ βραχεῖ. καὶ ἔλεγόν γε, ὡς ἐκεῖνος λέγει, Ταναγραῖοι παθεῖν αὐτὸν ταῦτα ἀνθ' ὧν ἐς τὸν Τρίτωνα ἠσέβησε, τεκμηριοῦντες ὅτι ἀποψύχων μὲν ἐξῃρέθη τῆς θαλάττης, ἰχῶρα δὲ ἠφίει παραπλήσιον τὴν ὀσμὴν τῇ τοῦ Τρίτωνος δορᾷ, ὅτε αὐτὴν ἐκεῖνος

[1] ἔχει.
[2] πολλῶν.
[3] ⟨τι⟩ *add.* *H.*
[4] συντυχεῖν.
[5] βασανίζων . . . ἐλέγχων.
[6] φύσιν, εἶτα.

The Triton

21. Concerning Tritons, while fishermen assert that they have no clear account or positive proof of their existence, yet there is a report very widely circulated of certain monsters in the sea, of human shape from the head down to the waist. And Demostratus in his treatise on fishing says that at Tanagra he has seen a Triton in pickle. It was, he says, in most respects as portrayed in statues and pictures, but its head had been so marred by time and was so far from distinct that it was not easy to make it out or recognize it. 'And when I touched it[a] there fell from it rough scales, quite hard and resistant. And a member of the Council, one of those chosen by lot to regulate the affairs of Greece and entrusted with the government for a single year, intending to test and prove the nature of what he saw, removed a small piece of the skin and burnt it in the fire; whereupon a noisome smell from the burning object thrown into the flames assailed the nostrils of the bystanders. But' he says, 'we were unable to guess whether the creature was born on land or in the sea. The experiment however cost him dear, for shortly afterwards he lost his life while crossing a small, narrow strait in a short, six-oared ferry-boat. And the inhabitants of Tanagra maintained,' so he says, 'that this befell him because he profaned the Triton, and they declared that when he was taken lifeless from the sea he disgorged a fluid which smelt like the hide of the Triton at the time when the man cast it into the fire and burnt it.'

[a] Ael. was never out of Italy (see vol. I, p. xii): he is quoting the words of Demostratus.

---

[7] οὗτος γὰρ . . . μικρόν.

ἔκαε καὶ ἐνεπίμπρα.' ὁπόθεν δὲ ἄρα ὁ Τρίτων οὗτος ἐπλανήθη, καὶ ὅπως δεῦρο ἐξεβράσθη, Ταναγραῖοί τε λεγέτωσαν καὶ Δημόστρατος. ἐπὶ τούτοις δὲ αἰδοῦμαι τὸν θεόν, καὶ ἄξιον πείθεσθαι τῷ μάρτυρι τῷ τοσῷδε· εἴη δ' ἂν ὁ ἐν Διδύμοις Ἀπόλλων τεκμηριῶσαι ἱκανὸς παντί, ὅτῳ νοῦς τε ὑγιαίνει καὶ ἔρρωται ἡ φρήν. Τρίτωνα γοῦν θρέμμα θαλάττιόν φησιν εἶναι, καὶ ἃ λέγει ταῦτά ἐστι

θρέμμα Ποσειδάωνος, ὑγρὸν τέρας, ἠπύτα Τρίτων,
νηχόμενος γλαφυρῆς [1] ὁρμήμασι σύντυχε νηός.

εἰ τοίνυν ὁ πάντα εἰδὼς καὶ Τρίτωνας εἶναί φησιν, ἡμᾶς ὑπὲρ τούτου διαπορεῖν οὐ χρή.

22. Τὸν Ἰνδῶν βασιλέα προϊόντα ἐπὶ δίκαις προσκυνεῖ ὁ ἐλέφας πρῶτος, δεδιδαγμένος τοῦτο, καὶ μάλα γε δρῶν μνημόνως τε καὶ εὐπειθῶς αὐτό (παρέστηκε δὲ καὶ ἐκεῖνος, ὅσπερ οὖν ἐνδίδωσίν οἱ τοῦ παιδεύματος τὴν ὑπόμνησιν τῇ ἐκ τῆς ἅρπης κρούσει καὶ φωνῇ τινι ἐπιχωρίῳ, ἧσπερ οὖν ἐλέφαντες ἐπαΐειν εἰλήχασι φύσει τινὶ ἀπορρήτῳ καὶ μάλα γε ἰδίᾳ τοῦ ζῴου τοῦδε)· καὶ μέντοι καὶ κίνησίν τινα ὑποκινεῖται πολεμικήν, οἷον ἐνδεικνύμενος ὅτι καὶ τοῦτο τὸ μάθημα ἀποσῴζει. τέτταρες δὲ καὶ εἴκοσι τῷ βασιλεῖ φρουροὶ παραμένουσιν ἐλέφαντες ἐκ διαδοχῆς, ὥσπερ οὖν οἱ φύλακες οἱ λοιποί, καὶ αὐτοῖς παίδευμα τὴν φρουρὰν ⟨ἔχειν⟩ [2] οὐ κατανυστάζουσι· διδάσκονται γὰρ

[1] *Voss*: γλαφυροῖς.

As to the quarter from which the Triton strayed and how he came to be cast ashore here, the inhabitants of Tanagra and Demostratus must explain. In view of these facts I bow to the god, and a witness of such authority claims our belief; and Apollo of Didyma [a] must be a sufficient guarantee to every man of sound mind and strong intelligence. At any rate he says that the Triton is a creature of the sea, and his words are

'A child of Poseidon, portent of the waters, a clear-voiced Triton, encountered as he swam the rush of a hollow vessel.'

If then the omniscient god says that Tritons do exist, we should entertain no doubts on the subject.

22. When the Indian King sets forth to administer justice an Elephant first bows down before him: it has been taught to do so and remembers perfectly and obeys. (At its side stands the man who teaches it to remember its instruction by a stroke from his goad and by some words in his native speech which thanks to a mysterious gift of nature peculiar to this animal the Elephant can understand.) Moreover it executes some warlike motion, as though it would show that it recollects this part of its teaching also. Four and twenty Elephants take it in turn to stand sentry over the King, just like the other guards, and are taught to keep watch and not to fall asleep: for this lesson also they are taught by Indian skill. And

The Elephant as bodyguard

[a] In the territory of Miletus; it was also known as Branchidae.

[2] ⟨ἔχειν⟩ *add.* H.

τοι σοφίᾳ τινὶ Ἰνδικῇ καὶ τοῦτο. καὶ λέγει μὲν Ἑκαταῖος ὁ Μιλήσιος Ἀμφιάρεων τὸν Οἰκλέους κατακοιμίσαι τὴν φυλακὴν καὶ [1] παθεῖν ὅσα λέγει. οὗτοι δὲ ἄρα ἄγρυπνοι καὶ ὕπνου [2] μὴ ἡττώμενοι, πιστότατοι τῶν ἐκεῖθι φυλάκων μετά γε τοὺς ἀνθρώπους εἰσίν.

23. Ἐγὼ δὲ ἄρα ὡς εἶχον ὁρμῆς ἐπὶ μακρότατον ταῦτά τε καὶ τὰ ὑπὲρ τούτων ἀνασκοπούμενός τε καὶ ἀνιχνεύων πέπυσμαι καὶ σκολόπενδραν εἶναί τι θαλάττιον κῆτος, μέγιστον κητῶν καὶ τοῦτο, καὶ ἐκβρασθεῖσαν μὲν θεάσασθαι οὐκ ἄν τις θρασύνοιτο. λέγουσι δὲ οἱ ἀκριβοῦντες ἄνθρωποι τὰ θαλάττια ὁρᾶσθαι αὐτὰς πλωτάς, καὶ πᾶν μὲν ὅσον ἐστὶ κεφαλή, τοῦτο ὑπερτείνειν ἔξαλον, καὶ μέντοι καὶ μυκτήρων τρίχας ἐξεχούσας καὶ μάλα γε ὑψηλὰς ἐπιδεικνύναι,[3] πλατεῖαν δὲ τὴν οὐρὰν καὶ οἵαν δοκεῖν καράβου. ἤδη δὲ ἄρα αὐτῆς καὶ τὸ λοιπὸν σῶμα ἐπιπολάζον τοῖς κύμασιν ὁρᾶται, ὅσον ἀντικρῖναι τριήρους τελείας αὐτὸ μεγέθει. νήχονται δὲ ἄρα πολλοῖς τοῖς ποσὶ καὶ κατὰ στοῖχον ἐντεῦθεν [4] καὶ ἐκεῖθεν οἱονεὶ σκαλμοῖς παρηρτημένοις (εἰ καὶ τραχύτερον ἀκοῦσαι) ἑαυτὰς [5] ἐρέττουσαι. λέγουσιν οὖν οἱ δεινοὶ ταῦτα καὶ ὑπηχεῖν τὸ ῥόθιον ἡσυχῇ, καὶ πείθουσι λέγοντες.

24. Ξενοφῶν δὲ ὑπὲρ κυνῶν λέγει καὶ ταῦτα. δεῖν ἐς τὰ ὄρη πολλάκις ἄγειν αὐτάς, τὰ δὲ ἔργα ἧττον· τοὺς γάρ τοι τριμμοὺς ⟨τοὺς⟩ [6] ἐκ τῶν ἐνεργῶν χωρίων λυπεῖν τε αὐτὰς καὶ σφάλλειν.

[1] καὶ ὀλίγου.
[2] *Reiske* : ὕπνῳ.
[3] ἀποδεικνύναι.
[4] αὐτοῖς καὶ ἐντεῦθεν.

Hecataeus of Miletus says that Amphiaraus, the son of Oicles, went to sleep during his watch and suffered the fate which he describes.[a] These animals however are wakeful and are not overcome by sleep; they are the most trustworthy of the guards there, at any rate next to human beings.

The Sea Scolopendra

23. Now in the course of examining and investigating these subjects and what bears upon them, to the utmost limit, with all the zeal that I could command, I have ascertained that the Scolopendra is a sea-monster, and of sea-monsters it is the biggest, and if cast up on the shore no one would have the courage to look at it. And those who are expert in marine matters say that they have seen them floating and that they extend the whole of their head above the sea, exposing hairs of immense length protruding from their nostrils, and that the tail is flat and resembles that of a crayfish. And at times the rest of their body is to be seen floating on the surface, and its bulk is comparable to a full-sized trireme. And they swim with numerous feet in line on either side as though they were rowing themselves (though the expression is somewhat harsh) with tholepins hung alongside. So those who have experience in these matters say that the surge responds with a gentle murmur, and their statement convinces me.

Xenophon on Hounds

24. Xenophon has also the following remarks touching Hounds [*Cyn.* 4. 9]. You should take them to the mountains frequently, but less frequently on to fields. For the beaten tracks on cultivated

[a] The allusion has not been explained.

---

[5] *Reiske*: ἑαυτοῖς. [6] ⟨τοὺς⟩ *add. H.*

λῷον δὲ εἶναι ὁ αὐτός φησιν ἐς τὰ τραχέα ἄγειν, καὶ κέρδος γε ἐκεῖνο πρὸς τούτῳ [1] διδάσκει, εὔποδάς τε αὐτὰς γίνεσθαι καὶ ἀλτικωτέρας ἐκπονούσας τὸ σῶμα. ἴχνη δὲ ἄρα λαγὼ τοῦ μὲν χειμῶνος μακρὰ ὁρᾶσθαι λέγει διὰ τὸ μῆκος τῶν νυκτῶν, τοῦ δὲ θέρους οὐκέτι διὰ τοὐναντίον.[2] καὶ σαφὲς ἐκ τοῦ προειρημένου τί βούλεται τὸ ἐναντίον.

25. Ἵππους καὶ ἐλέφαντας ἅτε ζῷα καὶ ἐν ὅπλοις καὶ ἐν πολέμοις λυσιτελῆ τιμῶσιν Ἰνδοί, καὶ μάλα γε ἰσχυρῶς. τῷ γοῦν βασιλεῖ κομίζουσι καὶ κώμυθας, ἃς ἐμβάλλουσι ταῖς φάτναις, καὶ χιλόν, καὶ ἐπιδεικνύουσι νεαρόν τε καὶ ἀσινῆ. καὶ ἐὰν μὲν ᾖ [3] τοιοῦτος, ἐπαινεῖ ὁ βασιλεύς· εἰ δὲ μή, κολάζει τούς τε τῶν ἐλεφάντων μελεδωνοὺς καὶ τοὺς ἱπποκόμους πικρότατα. οὐκ ἀτιμάζει δὲ οὐδὲ τὰ ἄλλα τὰ μικρότερα [4] ζῷα, ἀλλὰ καὶ ἐκεῖνα προσίεται δῶρά οἱ κομιζόμενα. Ἰνδοὶ γὰρ οὐκ ἐκφαυλίζουσι ζῷον οὔτε ἥμερον οὔτε μὴν [5] ἄγριον οὐδέν. αὐτίκα γοῦν δωροφοροῦσι τῶν ὑπηκόων οἱ διὰ τιμῆς ἰόντες γεράνους τε καὶ χῆνας ἀλεκτορίδας τε καὶ νήττας καὶ τρυγόνας τε καὶ ἀτταγᾶς προσέτι, πέρδικάς τε καὶ σπινδάλους (ἔστι δὲ ἐμφερὲς τῷ ἀτταγᾷ τοῦτό γε) καὶ [6] ἐπὶ τούτοις τῶν προειρημένων βραχύτερα, βωκκαλίδας τε καὶ συκαλίδας καὶ τὰς καλουμένας κεγχρῆδας. ἐπιδεικνύουσι δὲ αὐτὰ ἀναπτύξαντες, τὸν ἐς βάθος αὐτῶν ἐλέγχοντες πιασμόν. καὶ πλοῦτον [7] πεπια-

[1] *Reiske* : τοῦτο.
[2] τοὐναντίον δε τούτου τοῦ θ. οὐκέτι.
[3] *Jac* : εἴη.
[4] μικρότερα *H* (1858), μικρότατα MSS, τὰ μ. *del. H* (1864).

lands injure and mislead them. And the same writer says that it is better to take them on to rough ground, and points out the additional advantage of so doing, viz that by exercising their bodies their legs gain in strength and ability to jump. He also says [*ib*. 5. 1] that in winter the Hare's scent is perceptible for a long time because of the length of the nights, but in summer this is so no more, for the opposite reason. The meaning of 'the opposite' is clear from what has been said above.

Animals presented to the Indian King

25. The Indians value Horses and Elephants as animals serviceable under arms and in warfare; and they value them very highly. At any rate they bring to the King trusses of hay which they throw into the mangers, and fodder which they show to be fresh and undamaged. And if it is so, the King thanks them; if it is not, he punishes the keepers of the Elephants and the grooms most severely. But he does not reject even other and smaller animals but accepts the following also when brought to him as presents. For the Indians do not disparage any animal whether tame or wild. For example, those of his subjects who hold high office bring him presents of cranes, geese, hens, ducks, turtle-doves, francolins also, partridges, spindaluses [a] (this bird resembles the francolin), and even smaller birds than the aforenamed, the boccalis,[a] beccaficos, and what are called ortolans. And they uncover their gifts and display them, to prove how thoroughly plump they are.

[a] Unidentified.

[5] οὔτε μὴν ἥμερον οὐδέ. [6] καὶ τά.
[7] τούτων.

σμένων ἐλάφων [1] τε καὶ βουβαλίδων καὶ δορκάδων καὶ ὀρύγων καὶ τῶν ὄνων τῶν ἐχόντων ἓν κέρας, ὧν καὶ ἀνωτέρω που μνήμην [2] ἐποιησάμην, καὶ ἰχθύων δὲ γένη διάφορα κομίζουσι καὶ ταῦτα.

26. Ἔστι δὲ ἄρα καὶ τέττιξ ἐνάλιος. καὶ ὁ μὲν μέγιστος αὐτῶν ἔοικε καράβῳ σμικρῷ, κέρατα δὲ οὐκ ἔχει μεγάλα κατ᾽ ἐκείνους οὐδὲ κέντρα. ἰδεῖν δέ ἐστι τοῦ καράβου ὁ τέττιξ ζοφωδέστερος, καὶ ἐπὰν αἱρεθῇ, προσέοικε τετριγότι. πτέρυγες δὲ ὀλίγαι τὸ μέγεθος ὑπὸ τοῖς ὀφθαλμοῖς αὐτοῦ ἐκπεφύκασι, καὶ εἶεν ἂν κατὰ τὰς τῶν χερσαίων καὶ αὗται. οὐ σιτοῦνται δὲ αὐτὸν οἱ πολλοί, νομίζοντες ἱερόν. Σεριφίους δὲ ἀκούω καὶ θάπτειν νεκρὸν ἑαλωκότα· ζῶντα δὲ ἐς δίκτυον ἐμπεσόντα οὐ κατέχουσιν, ἀλλὰ ἀποδιδόασι τῇ θαλάττῃ αὖθις. θρηνοῦσι δὲ ἄρα αὐτοὺς [3] ἀποθανόντας, καὶ λέγουσι Περσέως τοῦ Διὸς ἄθυρμα αὐτοὺς εἶναι.

27. Ὕαινα ἰχθὺς ὁμώνυμος τῇ χερσαίᾳ ὑαίνῃ ἐστί. ταύτης οὖν τὴν δεξιὰν πτέρυγα εἰ ὑποθείης ἀνθρώπῳ καθεύδοντι, εὖ μάλα ἐκταράξεις αὐτόν· δέα γάρ τινα καὶ ἰνδάλματα καὶ φάσματα ὄψεται, καὶ ἐνύπνια ἕτερα οὐδαμῶς εὐμενῆ καὶ φίλα. τραχούρου γε μὴν ζῶντος ἐὰν ἀποκόψῃς τὴν οὐράν, καὶ τὸν τράχουρον αὖθις ἐλεύθερον ἀπολύσῃς ἐς τὴν θάλατταν, τήν γε μὴν προειρημένην οὐρὰν ἐξαρτήσῃς ἵππου κυούσης, οὐ μετὰ μακρὸν [4] ἐκπεσεῖται τὸ ἔμβρυον, καὶ ἐξαμβλώσει ἡ ἵππος.

[1] *Ges*: ἐλεφάντων.
[2] τὴν μνήμην.
[3] τούς.

They bring also a wealth of fattened stags, of antelopes,[a] of gazelles, and one-horned asses,[b] which I have mentioned somewhere earlier on, and different kinds of fish also.

26. There is also a Cicada that lives in the sea, and the largest one is like a small crayfish, though neither its horns nor its stings are as long as those of the crayfish. The Sea-cicada is of a darker hue than the crayfish, and when caught appears to squeak. From beneath its eyes there grow small wings, and these also resemble those of the land-cicada. But few people eat it, since they regard it as sacred. And I have heard that the inhabitants of Seriphus even bury any that is dead when caught; if however a live one falls into their nets, they do not keep it but return it to the sea. And they even mourn for these creatures when dead and assert that they are the darlings of Perseus the son of Zeus.

The Sea-Cicada

27. The Hyena fish[c] has the same name as the land-hyena. Now if you put its right-hand fin under a man asleep, you will give him a considerable shock. For he will see fearful sights, forms and apparitions, dreams too, sinister and unwelcome. Further, if you cut off the tail of a live Horse-mackerel and let the fish go again in the sea, and then attach the aforesaid tail to a mare in foal, she will presently drop her foetus and will miscarry.

The Hyena fish

[a] βούβαλις and ὄρυξ both signify *antelope*; but ὄ. may stand for the four-horned species mentioned in *NA* 15. 14.

[b] See 10. 40

[c] Unidentified.

[4] *Reiske*: μικρόν.

μειρακίου γε μὴν δεομένου ἐπὶ μήκιστον τριχῶν ἀπορίας τῶν ἐπὶ τοῦ γενείου, αἷμα ἐπιχρισθὲν θύννου ἀωρόλειον τὸ μειράκιον ἀπεργάζεται.[1] δρᾷ δὲ ἄρα καὶ νάρκη καὶ πνεύμων τὸ αὐτό· ἐν ὄξει γὰρ διασαπεῖσαι αἱ τούτων σάρκες καὶ ἐπιχρισθεῖσαι τοῖς γενείοις φυγὴν τριχῶν ἐνεργάζονται[2] φασι.[3] τί πρὸς ταῦτα Ταραντῖνοί τε καὶ Τυρρηνοὶ σοφισταὶ κακῶν, δαίδαλον[4] ἐκεῖνό γε ἀνιχνεύσαντές τε καὶ πειράσαντες τὴν πίτταν, ὡς ἐξ ἀνδρῶν ἐς γυναῖκας ἀποκρίνειν;

28. Ὁ ⟨δὲ⟩[5] χρύσοφρυς ἄρα ἰχθύων ἁπάντων δειλότατος ἦν. ἐν δὴ ταῖς παλιρροίαις τῆς θαλάττης, ὅταν ᾖ ὥρα Ἀρκτούρῳ[6] σύνδρομος, ὑπονοστεῖ μὲν[7] ἡ θάλαττα περὶ τὸ ἄκτιον, ψιλὴ δὲ ἡ ψάμμος ὑπολείπεται, καὶ αἱ ναῦς πολλάκις ἐπὶ τῆς γῆς ἑστήκασιν ὕδατος χῆραι. οὐκοῦν οἱ ἐπιχώριοι ὅρπηκας αἰγείρων χλωροὺς καὶ κομῶντας ὀξύναντες δίκην σκολόπων καὶ ἐμπήξαντες τῇ ψάμμῳ ὑπαναχωροῦσιν, εἶτα ὑποστρέψαν τὸ κῦμα ἐπισύρει ἰχθύων τῶν προειρημένων πλῆθος ἄμαχον, ὑπονοστεῖ ⟨δὲ⟩[8] αὖθις, καὶ ὑπολείπονται πολλοὶ χρυσόφρυες ἐν ὀλίγῳ ὕδατι, ἔνθ' ἂν[9] καθήμενα εὑρεθῇ καὶ κοῖλα,[10] εἶτα ὑπὸ τοῖς κλάδοις πτήξαντες ἡσυχάζουσι· διασειομένους γὰρ αὐτοὺς καὶ διακινουμένους ὑπὸ τοῦ προσπίπτοντος πνεύματος ὀρρωδοῦσι, καὶ οὔτε σπαίρουσιν οὔτε ἀναπάλλονται.

[1] *Reiske*: ἐνεργάζεται.
[2] *Ges*: ἐνεργάζεται.
[3] *Schn*: φησί.
[4] δαιδάλων.
[5] ⟨δέ⟩ *add. H.*
[6] *Abresch*: Ἀρκτούρου.
[7] μάλιστα.
[8] ⟨δέ⟩ *add. Reiske.*
[9] *Jac*: ἔνθα.
[10] καθειμένον . . . κοῖλον.

Again, if a youth wants to keep his chin hairless for as long as possible, the blood of a Tunny rubbed on renders him beardless. And the Torpedo and the Jelly-fish have the same effect, for if their flesh is dissolved in vinegar and rubbed on the cheeks, they say that it banishes hair. What have those contrivers of evil from Tarentum and Etruria to say to this, men who after experimenting with pitch have discovered that artifice whereby they differentiate men and turn them into women?

Depilatorie

28. Of all fishes the Gilthead is the most timid. When the season of neap-tides coincides with Arcturus,[a] the sea recedes from the beach and the sand is left bare and vessels frequently stand high and dry for want of water. Accordingly the inhabitants take branches of poplar-trees, green and in leaf, and after sharpening them like stakes, fix them in the sand and withdraw. Later the returning tide draws in a countless multitude of the aforesaid fishes; again it ebbs, leaving a great number of Giltheads in shallow water wherever low-lying or hollow spots may be found, and the fish cower beneath the branches and remain still. For they are terrified by the branches when the oncoming wind stirs and shakes them, and neither quiver nor dart about. It is quite easy, you might say, for anyone who sets

The Gilthead

[a] The phrase *ὥρα Ἀρκτούρῳ σύνδρομος* is borrowed from Plato, *Legg.* 8. 844 D [figs and grapes are not to be gathered] *πρὶν ἐλθεῖν τὴν ὥραν τὴν τοῦ τρυγᾶν Ἀρκτούρῳ σύνδρομον.* The morning rising of Arcturus in the region of Rome was on Sept. 20, the evening rising on Feb. 27. Ael. appears to think that Arcturus has some effect upon the tides, but does not tell us which date we are to understand.

πάρεστι δὴ συλλαβεῖν ὡς αἰχμαλώτους καὶ παίειν παντὶ τῷ προσπεσόντι δειλῶν ἰχθύων δῆμον εἴποι τις ἄν. αἱροῦσι γοῦν αὐτοὺς οὐ τεχνῖται μόνοι, ἀλλὰ κἂν ἰδιώτης παρατυχὼν ᾖ, καὶ παῖδες καὶ γυναῖκες.

upon the mob of timorous fish to capture and strike them. At any rate it is not only skilled fishermen that can catch them, but any inexperienced person who chances to be at hand, even children and women.

BOOK XIV

ΙΔ

1. Ἐν δὲ τῷ Ἰονίῳ πελάγει πλησίον Ἐπιδάμνου, ὅπου καὶ Ταυλάντιοι παροικοῦσι, νῆσός ἐστι καὶ Ἀθηνᾶς κέκληται, καὶ οἰκοῦσιν ἐνταῦθα ἁλιεῖς. ἔστι δὲ καὶ λίμνη αὐτόθι, καὶ σκόμβρων ἠθάδων καὶ ἡμέρων ἀγέλαι τρέφονται. καὶ τούτοις μὲν τροφὰς ἐμβάλλουσιν οἱ ἁλιεῖς, καὶ ἔστιν αὐτοῖς πρὸς αὐτοὺς ἔνσπονδα, καί εἰσιν ἐλεύθεροι, καὶ ἀθηρίαν εἰλήχασι, καὶ προΐασιν ἐς χρόνου πλῆθος, καὶ ζῶσιν αὐτόθι σκόμβροι καὶ γέροντες. οὐ μὴν ἀργοὶ σιτοῦνται, οὐδὲ ὑπὲρ ὧν τρέφονταί εἰσιν ἀχάριστοι, λαβόντες δὲ ἐκ τῶν ἁλιέων τροφὰς τὰς ἑωθινὰς εἶτα μέντοι καὶ αὐτοὶ ἐπὶ τὴν θήραν ἴασιν, ὥσπερ οὖν τροφεῖα ἐκτίνοντες. καὶ τοῦ λιμένος προελθόντες ἐπὶ τοὺς ξένους στέλλονται σκόμβρους, καὶ ἐντυχόντες ὡς ἴλῃ τινὶ ἢ φάλαγγι, ἅτε ὁμοφύλοις καὶ τῆς αὐτῆς φύσεως οὖσι προσνέουσι, καὶ οὔτε τούτους ἐκεῖνοι φεύγουσιν οὔτε οὗτοι ἐκκλίνουσιν ἐκείνους,[1] ἀλλὰ συνίασιν. εἶτα οἱ τιθασοὶ τοὺς ἐπήλυδας περιελθόντες καὶ κυκλόσε γενόμενοι καὶ ἑαυτοὺς συμφράξαντες ἀπειλήφασι μέσους πολύ τι πλῆθος, καὶ οὐκ ἐῶσι διαδιδράσκειν, ἀναμένουσι δὲ τοὺς τροφέας, καὶ ἀνθ' ὧν ἐκορέσθη-

[1] αὐτούς.

---

[a] Seemingly unknown to geographers. There are, however,

BOOK XIV

1. In the Ionian Sea close to Epidamnus where the Taulantii live, there is an island [a] and it is called 'Athena's Isle,' and fisher folk live there. There is also a lagoon in the island where shoals of tame Mackerel are fed. And the fishermen throw in food to them and observe a treaty of peace with them; so the fish are free and immune from pursuit and attain to a great age; there are even ancient Mackerel living there. Yet they do not feed without making any return, nor do they fail in gratitude for their food, but after they have been fed by the fishermen in the morning they too of their own accord go to join the pursuit, as though they were paying for their maintenance. And advancing beyond the harbour they set out to meet the strange Mackerel. When they have encountered them as it were in a company or in line of battle, they swim up to them as being of the same family and the same kind, nor do the strangers flee from them, nor do the tame fish attempt to divert them but bear them company. Presently the tame fish surround the newcomers, and having encircled them, close their ranks and cut off the fish in their midst, amounting to a great number, and prevent them from escaping; they wait for their keepers and provide the fishermen with a

Tame Mackerel

two lagoons, one 30 mi., the other about 55 mi. S of Epidamnus.

σαν ἀνθεστιῶσι τοὺς ἁλιέας· ἐπελθόντες γὰρ αἱροῦσιν αὐτοὺς καὶ πολὺν ἐργάζονται φόνον. οἱ δὲ τιθασοὶ ἐπανίασι σπεύδοντες ἐς τὸν λιμένα, καὶ τοὺς ἑαυτῶν χηραμοὺς ὑπελθόντες ἀναμένουσι τὸ δειλινὸν δεῖπνον. οἱ δὲ ἥκουσι κομίζοντες, εἰ βούλονται συνθήρους ἔχειν καὶ φίλους πιστούς. ὁσημέραι μέντοι πράττεται ταῦτα.

2. Σκάρου τὴν χολὴν [1] ἐὰν δῷς ἐμφαγεῖν ἀνθρώπῳ νοσοῦντι τὸ ἧπαρ καὶ ἴκτερον ἔχοντι, σωθήσεται, ὡς οἱ σοφοὶ τῶν ἁλιέων διδάσκουσιν.

3. Ἁλίσκεται δὲ ἄρα ἰχθὺς καὶ ἄνευ κύρτων καὶ ἀγκίστρων καὶ δικτύων τὸν τρόπον τοῦτον. κόλποι θαλάττιοι πολλοὶ τελευτῶσιν ἐς τενάγη τινά, καὶ ἔστι ταῦτα ἐπιβατά. ὅταν οὖν ᾖ γαλήνη καὶ εἰρήνη πνευμάτων, οἱ τεχνῖται τῶν ἁλιέων ἄγουσι πολλοὺς ἐνταῦθα, εἶτα αὐτοὺς προστάττουσι βαδίζειν καὶ πατεῖν τὴν ψάμμον, ὡς ὅτι μάλιστα ἀπερείδοντας [2] τὸ πέλμα ἰσχυρῶς. εἶτα ἴχνη καταλείπεται βαθέα, ἅπερ οὖν ἐὰν φυλαχθῇ, καὶ μή ποτε συμπεσοῦσα ἡ ψάμμος συγχέῃ αὐτά, μηδὲ ἐκταραχθῇ [3] ὑπὸ πνεύματος τὸ ὕδωρ, ὀλίγον διαλείπουσιν οἱ ἁλιεῖς, καὶ ἐμβάντες καταλαμβάνουσιν ἐν τοῖς κοιλώμασι τῶν βημάτων καὶ τοῖς ἴχνεσι τοὺς ἰχθῦς τοὺς πλατεῖς εὐναζομένους, ψήττας τε καὶ ῥόμβους καὶ στρουθοὺς καὶ νάρκας καὶ τὰ τοιαῦτα.

[1] *Jac*: σκάρου, διαχυθείσης τῆς χολῆς περὶ πᾶν τὸ σῶμα, ἐάν MSS; *if the words* διαχυθείσης . . . σῶμα *are to be regarded as genuine, Jac would place them after* ἔχοντι.

[2] ἐπερείδοντες.

feast in return for the satisfaction of their own appetites. For the fishermen arrive, catch the strangers, and perpetrate a massacre. But the tame fish return with all haste to the lagoon, dive into their lairs, and wait for their afternoon meal, which the fishermen bring, if they want allies and loyal friends as fellow-hunters. And this happens every day.

A cure for jaundice

2. Experienced fishermen teach us that if you give a man whose liver is out of order and who is afflicted with jaundice, the gall of a Parrot Wrasse, he will be cured.

Fishing in shallow waters

3. Fish are caught without weels or hooks or nets in the following manner. There are many bays in the sea which end in shallows, and one can walk in them. When, therefore, it is calm and the winds are at rest, skilled fishermen bring a number of people to the spot and then direct them to walk about and trample the sand, throwing all their weight on to the soles of their feet. As a result deep footprints are left, and if they are preserved and the sand does not collapse and obliterate them, and if the water is not agitated by the wind, after a short interval the fishermen enter and in the trodden hollows and footprints capture flat fish asleep, viz flounders, turbot, plaice,[a] torpedo-fish, and the like.

[a] Thompson has omitted *στρουθός* from his *Glossary*; L-S[9] give 'flounder'; E. de Saint-Denis gives 'plaice.'

---

[3] ἐνταραχθῇ.

4. Ἐχίνου θαλαττίου πέρι εἶπον ⟨καὶ⟩ [1] ἀνωτέρω καὶ νῦν δὲ εἰρήσεται ὅσα προσακήκοα. ἔστι γὰρ καὶ στομάχῳ ἀγαθόν· τὸν τέως γὰρ κακόσιτον ὄντα καὶ πᾶν ὅ τι οὖν βδελυττόμενον ὁ δὲ ἀναρρώννυσιν. ἔστι δὲ καὶ κύστεως κενωτικός, ὡς οἱ τούτων λέγουσι σοφοί. εἰ δὲ αὐτοῦ ἐπιχρίσειας σώματι ψωριῶντι, ὁ δὲ σῶν ἐργάζεται τὸν τέως νοσοῦντα τὴν νόσον τὴν προειρημένην. καυθεὶς δὲ ἄρα ὀστράκοις αὐτοῖς ἐκκαθαίρει τὰ ῥυπῶντα τῶν τραυμάτων. χερσαίου δὲ ἐχίνου καυθέντος ἡ σποδιὰ πίττῃ προσανακραθεῖσα εἶτα μέντοι καταχρίεται τῶν λειψοτρίχων μερῶν, καὶ αἱ τέως φυγάδες (ἵνα τι καὶ παίσω [2]) ὑπαναφύονται. πινομένη δὲ οἴνῳ νεφροῖς ἀγαθόν ἐστι, σώζει δὲ ἄρα καὶ ὑδεριῶντας ποθεῖσα, ὥσπερ δὴ καὶ προεῖπον. τὸ δὲ ἧπαρ ἄρα τοῦ ἐχίνου τούτου ἰᾶται ὑφ' ἡλίου γενόμενον αὖον τοὺς τῇ νόσῳ τῇ τοῦ καλουμένου ἐλέφαντος κατειλημμένους.

5. Ἐλέφας ὁ θῆλυς, τιμιώτερα εἶναι τὰ τούτου κέρατα οἱ σοφοὶ ταῦτα ὑμνοῦσι, καὶ ἐκεῖνά γε ἡμᾶς διδάσκουσιν. ἐν τῇ Μαυρουσίᾳ γῇ οἱ ἐλέφαντες, δεκάτῳ ἔτει πάντως αὐτοῖς τὰ κέρατα ἐκπεσεῖν φιλεῖ,[3] ὥσπερ οὖν καὶ ⟨τὰ⟩ [4] τῶν ἐλάφων, ἀλλὰ τούτων ἀνὰ πᾶν ἔτος. οἱ τοίνυν ἐλέφαντες οἵδε γῆν πεδιάδα καὶ ἔνδροσον προαιροῦνται τῆς ἄλλης, καὶ ἀπερείδουσιν [5] ἐς αὐτὴν ⟨τὰ κέρατα⟩,[6] ὀκλὰξ ἐπικύψαντες, ἐκδῦναι αὐτὰ δεινῶς σφριγῶντες. τοσοῦτον δὲ ἄρα ἐπωθοῦσιν, ὡς καὶ τελέως [7] αὐτὰ

[1] ⟨καί⟩ *add. H.*
[2] παίξω.
[3] δεῖ V, σπεύδει *other* MSS.
[4] ⟨τά⟩ *add. H.*
[5] ἐπερείδουσιν.
[6] ⟨τὰ κέρατα⟩ *add. H.*

Medicinal properties of Sea-urchin and Hedgehog

4. I have spoken earlier on[a] about the Sea-urchin and I will now mention what more I have heard. It is also good for the stomach: it helps a man who has been suffering from loss of appetite and loathing every kind of food to regain his strength; it is also a diuretic, according to those who know about these things. And if you rub it on one who is suffering from the itch, it cures a man hitherto afflicted with the aforesaid disease. And if you burn a Sea-urchin, shell and all, it cleanses suppurating wounds. If you burn a Hedgehog and mingle the ashes with pitch and then rub them on those parts where the hair has fallen off, the fugitives (if I may be allowed the joke) will sprout again. If drunk with wine, it is good for the kidneys; it is also a cure for dropsy when drunk, as in fact I remarked before. Further, the liver of a Hedgehog, if desiccated by the sun, is a cure for those who suffer from the disease known as elephantiasis.

Hunting for Elephants' tusks

5. Those who are learned in these matters constantly assert that the tusks of the female Elephant are more valuable than those of the male, and this is what they teach us. In Mauretania Elephants are in the habit of dropping their tusks every tenth year, just as stags drop their horns, though with stags it is every year. Now these Elephants prefer a level, well-watered country to any other, and they go down upon their knees and rest their tusks upon the ground in their passionate desire to shed their tusks. And they thrust with such force as finally

[a] See 7. 33; 9. 47.

[7] *Reiske*: τέως.

ἀποκρύψαι· εἶτα μέντοι ὑποψήσαντες τοῖς ποσὶ λεῖον τὸν χῶρον ἀπέφηναν τὸν φρουροῦντα τὸ θησαύρισμα αὐτοῖς. γονιμωτάτη δὲ ἄρα ἡ γῆ οὖσα εἶτα ὤκιστα πόαν ἀναφύει, καὶ ἀφανίζει τὴν ὄψιν τοῦ γεγενημένου τοῖς ὁδῷ χρωμένοις. οἱ δὲ [1] ταῦτα ἀνιχνεύοντες τὰ φώρια καί τινα σοφίαν τῆς ⟨ἐξ⟩ [2] ἐκείνων ἐπιβουλῆς ἔχοντες ἐν ἀσκοῖς αἰγείοις ὕδωρ κομίζουσιν, εἶτα αὐτοὺς διασπείρουσι πεπληρωμένους ἄλλους ἀλλαχόσε, καὶ αὐτοὶ καταμένουσι. καὶ καθεύδει τις, καὶ ἄλλος ὑποπίνει, καί πού τις καὶ μεταξὺ ἐπιρροφῶν τῆς κύλικος ὑπαναμέλπει καὶ μέμνηται διὰ τοῦ μέλους ἧς ἐρᾷ· εἰ δὲ καὶ νέον ὡρικὸν ὑποπειρᾷ παρόντα τις καὶ αὐτὸν τῆς ἰχνεύσεως κοινωνόν, οὐκ ἂν θαυμάσαιμι· εἰσὶ γὰρ Μαυρούσιοι καὶ καλοὶ καὶ μεγάλοι, καὶ ἀνδρικὸν ὁρῶσι, καὶ ἔργων ἔχονται θηρατικῶν, καὶ μέντοι καὶ πολλοὺς [3] ἀναφλέγουσι, μειράκια ἔτι καὶ τηλικοίδε ὄντες. οὐκοῦν εἰ τὰ κέρατα εἴη κατορωρυγμένα [4] πλησίον ἐκεῖνα, τὰ δὲ ἴυγγί τινι ἀπορρήτῳ καὶ θαυμαστῇ τὸ ὕδωρ ἐκεῖνο ἐκ τῶν ἀσκῶν ἕλκει, καὶ ἀποδείκνυσι κενοὺς αὐτούς. ἐνταῦθά τοι [5] σμινύαις τε καὶ μακέλλαις διασκάπτουσι τὸν χῶρον, καὶ ἔχουσι τὸ θήραμα ῥινηλατήσαντες ἄνευ κυνῶν· ἐὰν δὲ ἔμπλεῳ μείνωσιν οἱ ἀσκοὶ οὗπερ οὖν καὶ κατέθεσαν αὐτοὺς οἱ τῶν κεράτων τῶνδε θηραταί, οἱ δὲ ἀπίασιν ἐπ' ἄλλην θήραν, καὶ μέντοι καὶ ἀσκοὺς καὶ ὕδωρ ἐπάγονται πάλιν, τὰ θήρατρα τῆς ἄγρας τῆς προειρημένης.

6. Λέγεται δὲ καὶ ἐλέφας διπλῆν ἔχειν καρδίαν καὶ διπλᾶ νοεῖν, καὶ τῇ μὲν θυμοῦσθαι, πραΰνεσθαί

to bury them in the ground. Next, with their feet they gently scrape and make smooth the spot that guards their treasure. Now the soil is extremely fertile and in a very short while sends up a crop of grass and effaces the evidence of what occurred for those who pass by. But those who track down these secreted objects and who have some knowledge of the Elephants' designs, bring water in goatskins and disperse them, well filled, in different places, and themselves remain where they are. And one sleeps while another drinks a little, and I dare say that in the intervals of quaffing from his cup he sings to himself and remembers his sweetheart in his song. (Nor should I be surprised if a man tries to seduce some well-grown boy who is with him and is his companion in the quest, for the Moors are handsome, stalwart, and of manly aspect, and are devotees of the chase: and many a heart do they inflame too, while still boys, though they are so big). So then if those tusks have been buried near by, by some mysterious and amazing spell they draw the aforesaid water out of the skins and leave them empty. Thereupon the men dig up the ground with mattocks and picks, and the spoil which they have tracked down without the aid of dogs is theirs. If however the skins remain filled in the place where the tusk-hunters laid them, they go off on a fresh quest and again bring the skins and the water, the instruments of the hunt which I have described.

6. The Elephant is even said to possess two hearts and to think double: one heart is the source of anger,

The Elephant

[1] δὲ *καί*. [2] ⟨ἐξ⟩ *add. H.* [3] *Jac* : *πολλοί*. [4] *κατωρυγμένα*. [5] *τοίνυν*.

γε μὴν τῇ ἑτέρᾳ· Μαυρουσίοις δὲ ἄρα ἕπομαι λόγοις λέγων ταῦτα. ἐπεί τοι καὶ ἐκεῖνα οἱ αὐτοὶ ὑμνοῦσι, λύγκας [1] εἶναι. φασὶ δὲ αὐτὰς παρδάλεως μὲν ἔτι καὶ πλέον σιμάς, ἄκρα ⟨γε⟩ [2] μὴν τὰ ὦτα λασίους. θηρίον ⟨δὲ⟩ [3] τοῦτο ἀλτικὸν δεινῶς, καὶ κατασχεῖν βιαιότατά τε καὶ ἐγκρατέστατα καρτερόν. ἔοικε δὲ ἄρα τῷ θηρίῳ τούτῳ μαρτυρεῖν καὶ Εὐριπίδης τὸ ἀπρόσωπον, ὅταν που λέγῃ

ἥκει δ' ἐπ' ὤμοις ἢ συὸς φέρων βάρος
ἢ τὴν ἄμορφον λύγκα,[4] δύστοκον δάκος.

ὑπὲρ ὅτου δὲ λέγει δύστοκον τοὺς κριτικοὺς ἐρέσθαι λῷον.

7. Περὶ στρουθοῦ δὲ τῆς μεγάλης εἴποι τις ἂν καὶ ἐκεῖνα. ἡ γαστὴρ αὐτῆς ἀνῃρημένης εὑρίσκεται καθαιρομένη λίθους ἔχουσα, οὕσπερ οὖν καταπιοῦσα ἡ στρουθὸς ἐν τῷ ἐχίνῳ φυλάττει καὶ πέττει τῷ χρόνῳ. εἶεν δ' ἂν οὗτοι καὶ ἀνθρώπων πέψεως [5] ἀγαθόν,[6] νεῦρα δὲ τὰ ταύτης καὶ λίπος ἀνθρωπείων νεύρων ἀγαθά ἐστιν.[7] ἁλίσκεται μὲν οὖν αὕτη ὑπὸ ἵππων· [8] θεῖ μὲν γὰρ ἐς κύκλον, ἀλλ' ἐξωτέρω περιθέουσα· οἱ δὲ ἱππεῖς τῷ ἐνδοτέρω [9] ὑποτέμνονται κύκλῳ, καὶ ἔλαττον περιιόντες ἀπειποῦσαν τῷ δρόμῳ ἀγρεύουσιν αὐτὴν χρόνῳ. λαμβάνεται δὲ καὶ τοῦτον τὸν τρόπον. καλιὰν ἐργάζεται ταπεινὴν ἐν τῷ δαπέδῳ, τὴν ψάμμον διαγλύψασα τοῖς ποσί. καὶ τὸ μὲν μεσαί-

[1] λύγγας.
[2] ⟨γε⟩ *add. Ges.*
[3] ⟨δέ⟩ *add. H.*
[4] λύγγα.
[5] *Ges* : ὄψεως.
[6] ἀγαθά.
[7] εἰσιν.

the other of gentleness. In saying this I am following accounts given by the Moors. Moreover the same people constantly affirm the following, namely that there are lynxes, and that they are even more snub-nosed than the leopard, and that the tips of their ears are hairy. The Lynx has a wonderful spring and can maintain the most vigorous and overpowering grip on its catch. So it seems that Euripides bears witness to the unloveliness of this beast when he says somewhere [*fr.* 863 N]

The Lynx

> 'And he comes bearing upon his shoulders either the burden of a boar, or the mis-shapen lynx, a ravening brute ill-conceived.'

But why he says 'ill-conceived' is rather a question for the grammarians.

7. Concerning the Ostrich one may also mention the following facts. If you kill an Ostrich and wash out its stomach it will be found to contain pebbles which the bird has swallowed and keeps in its gizzard and in time digests. And these pebbles are an aid to the human digestion; its sinews also and its fat are good for the human sinews.

The Ostrich

Now the capture of this bird is effected by means of horses, for it runs in a circle keeping to the outer edge, but the horsemen intercept it by keeping on the inner side of the circle, and by wheeling in a narrower compass at length overtake it when it is exhausted with running. And here is another way to catch it. It builds itself a nest low down on the

method of capture

[8] ἵππων ἀπειποῦσα τῷ δρόμῳ.
[9] ἐνδοτέρῳ.

τατον αὐτῆς κοῖλόν ἐστι, τὰ χείλη δὲ τὰ κύκλῳ ὑψηλὰ ἐργάζεται, ἀποτειχίζουσα τρόπον τινά, ἵνα τὸ ἐκ Διὸς ὕδωρ ἀποστέγῃ τὰ χείλη, καὶ μὴ ἐσρέῃ τῇ καλιᾷ, καὶ ἐπικλύζῃ τῆς στρουθοῦ τοὺς νεοττοὺς ὄντας ἁπαλούς. τίκτει δὲ καὶ ὑπὲρ τὰ ὀγδοήκοντα, οὐ μὴν ἀθρόα ἐκγλύφει, οὐδὲ ἐν ταὐτῷ χρόνῳ πάρεισιν[1] ἐς τὸ φῶς πάντα, ἀλλὰ τὰ μὲν ἤδη τέτεκται, ἄλλα δὲ ἔτι ἐν τοῖς ᾠοῖς ὑποπήγνυται,[2] τὰ δὲ ὑποθάλπεται. ὅταν οὖν ἐν τούτοις ᾖ ἐκείνη, θεασάμενος ἀνήρ τις οὐκ ἄφρων ἀλλὰ τῆς τοιαύτης θήρας πεπειραμένος, αἰχμὰς περὶ τὴν καλιὰν πήγνυσι τεθηγμένας, ὀρθὰς δὲ ἄρα κατὰ τοῦ σαυρωτῆρος πήγνυσι, καὶ ὁ σίδηρος ἐκλαμπει, καὶ ἀναχωρήσας ἐλλοχᾷ τὸ πραττόμενον. ἐπάνεισιν οὖν ἐκ τῆς νομῆς ἡ στρουθὸς ἐρῶσα τῶν νεοττῶν ἰσχυρῶς καὶ διψῶσα αὐτῶν τῆς συνουσίας. καὶ τὰ μὲν πρῶτα περιβλέπει δεῦρο καὶ ἐκεῖσε καὶ ἑλίττει τὸ ὄμμα, δεδοικυῖα μή τις αὐτὴν θεάσηται· εἶτα μέντοι νικωμένη ὑπὸ τοῦ ἱμέρου καὶ οἰστρουμένη, τὰς πτέρυγας ἁπλώσασα ὡς ἱστίον, δρόμῳ φερομένη συντόνῳ καὶ ῥοίζῳ ἐσήλατο ἐς τὴν ἑαυτῆς καλιὰν καὶ οἴκτιστα ταῖς[3] αἰχμαῖς ἐμπαλαχθεῖσα καὶ περιπαρεῖσα ἀποθνήσκει. ἐφίσταται οὖν ὁ θηρατὴς καὶ ᾕρηκε σὺν τῇ μητρὶ τοὺς ἐκγόνους.

8. Πόλις ἐστὶν ἐν τοῖς ὑπὸ τὴν ἑσπέραν χωρίοις Ἰταλική. ὄνομα αὐτῇ Πατάβιον. Ἀντήνορος ἔργον εἶναι λέγουσι τοῦ Τρωὸς τὴν πόλιν. ταύτην δὲ ᾤκισεν[4] ἄρα οἴκοθεν σωθείς, ὅτε ἀπηλλάγη τῆς

[1] παρίασιν.
[2] ὑποπήγνυται μέν.

ground after scooping out the sand with its feet. The centre of the nest is hollow, but it builds up the lips all round and walls off the nest so that the lips may keep out the rain and prevent it from streaming into the nest and deluging the young at a tender age. It lays over eighty eggs, but does not hatch them simultaneously, nor do they all emerge to daylight at the same time, but while some have already been born, others are still acquiring consistency within the shell. Others again are being kept warm. When therefore the Ostrich is so engaged, a man—not a witless person but one who has experience of this kind of hunting—who has seen her, fixes some sharp spears round the nest, planting them upright by the ferrule; and the iron shines. Then he withdraws and lies in wait to see the result. So the Ostrich returns from her feeding-ground full of love for her chicks and yearning to be with them. And first of all she casts her eyes around, looking this way and that for fear someone should catch sight of her. And then overcome and stimulated by her longing, she spreads her wings like a sail and rushing at full speed leaps into her nest to die a most pitiful death entangled and impaled upon the spears. Then the hunter is at hand and seizes the young birds with their mother.

Eels in the Eretaenus

8. There is an Italian city in the regions towards the west, and its name is Patavium.[a] They say that the city was the work of Antenor the Trojan. He founded it, having escaped with his life from his

[a] Mod. Padua, about 20 mi. inland from Venice.

---

[3] *Jac*: *ταῖς γάρ.* [4] *ᾤκησεν.*

πατρίδος ἁλούσης τῆς Ἰλίου, αἰδεσθέντων αὐτὸν [1] τῶν Ἑλλήνων, ἐπεὶ πρεσβεύοντα τὸν Μενέλεων σὺν τῷ Ὀδυσσεῖ ὑπὲρ τῆς Ἑλένης ἔσωσεν, Ἀντιμάχου συμβουλεύσαντος ἀποκτεῖναι αὐτούς. ἔλεγε δὲ ἄρα οὗτος ταῦτα

χρυσὸν Ἀλεξάνδροιο δεδεγμένος, ἀγλαὰ δῶρα,

ὡς Ὅμηρός φησιν. οὐκοῦν τῷδε Παταβίῳ πόλις γειτνιᾷ ἑτέρα, καὶ Βικετίαν [2] καλοῦσιν αὐτήν, καὶ παραρρεῖ ποταμὸς αὐτῇ Ἠρέταινος [3] ὄνομα, καὶ παραμείβεται οὗτος γῆν οὐκ ὀλίγην εἶτα ἐς τὸν Ἠριδανὸν ἐμβάλλει, καὶ ἀνακοινοῦταί ⟨οἱ⟩ [4] τὸ ὕδωρ. ἐν δὴ τῷ Ἠρεταίνῳ [5] ἐγχέλεις γίνονται μέγισταί τε καὶ τῶν ἀλλαχόθεν πιότεραι [6] μακρῷ, ἁλίσκονται δὲ ἄρα τὸν τρόπον τοῦτον. ἐπὶ πέτρας προβλῆτος κάθηται ὁ θηρατὴς ἔν τινι κολποειδεῖ χωρίῳ, ὅπου καὶ πλατύνεται τὸ ῥεῦμα ἐπὶ [7] μᾶλλον, ἢ καὶ ἐπί τινος δένδρου κάθηται προρρίζου πλησίον τῆς ὄχθης ῥιφέντος ὑπὸ πνεύματος σκληροῦ, ὅπερ οὖν ὑποσήπεται μέν, ἀχρεῖον δέ ἐστι κατακοπῆναί [8] τε καὶ ἐκκαῦσαι αὐτό. οὐκοῦν ἑαυτὸν ἐγκαθίσας ὁ τῶν ἐγχέλεων [9] ἁλιεὺς τῶνδε, καὶ λαβὼν ἔντερον νεοσφαγοῦς ἀρνὸς τριῶν μὲν ἢ τεττάρων πήχεων, πεπιασμένον δὲ ἰσχυρῶς, τὴν μὲν ἀρχὴν αὐτοῦ καθίησιν ἐς τὸ ὕδωρ, καὶ εἰλεῖται

[1] αὐτὸν αἰδεσθέντων.
[2] Βιγητίαν, Βικεντίαν etc.
[3] Ἠρέτενος.
[4] ⟨οἱ⟩ add. *Reiske*.
[5] Ἠρετένῳ.
[6] πιόταται.
[7] *Reiske* : ἔτι.
[8] *Lobeck* : κατακῆναι.
[9] ἐγχελύων.

home when he left his native land after the capture of Troy, because the Greeks had compassion on him, since he saved Menelaus who came with Odysseus as ambassador to treat about Helen,[a] when Antimachus advised that they should be put to death. These were Antimachus's words:

'He had accepted the gold of Paris, splendid gifts,'

as Homer says [*Il.* 11. 124]. Well, there is another city not far away which they call Vicetia,[b] and past it there flows a river of the name of Eretaenus:[c] it traverses a considerable area and then falls into the Eridanus, to which it imparts its waters. Now in the Eretaenus there are Eels of very great size and far fatter than those from any other place, and this is how they are caught. The fisherman sits upon a rock jutting out in some bay-like spot on the river where the stream widens out, or else upon a tree which a fierce wind has uprooted and thrown down close to the bank—the tree is beginning to rot and is no use for cutting up and burning. So the eel-fisher seats himself and taking the intestine of a freshly slaughtered lamb which measures some three or four cubits and has been thoroughly fattened, he lowers one end into the water, and keeps it turning

[a] He tried to persuade the Trojans to give back Helen to Menelaus.

[b] Mod. Vicenza, 22 mi. to the NW of Padua.

[c] Mod. Retrone; below Vicenza it joins the Bacchiglione and together they flow into the sea at Venice. Ael. seems unaware that the Eridanus (Lat. Padus, mod. Po) is some 30 mi. farther south and that the river Athesis (mod. Adige) flows between the Bacchiglione and the Po.

ἐν ταῖς δίναις στρεφόμενον, τό γε μὴν τέλος διὰ χειρῶν ἔχει, ἐμβέβληται δὲ ἐς αὐτὸ καλάμου τρύφος, ὅσον κώπην εἶναι τὸ μῆκος ξίφους. οὐ μὴν λανθάνει τὰς ἐγχέλεις ἡ τροφή· χαίρουσι γὰρ τῷδε τῷ ἐντέρῳ. καὶ ἥ γε πρώτη προσελθοῦσα, οἰστρουμένη ὑπὸ τοῦ λιμοῦ καὶ περιχανοῦσα, ἐμφύει τοὺς ὀδόντας γυρούς τε καὶ ἀγκιστρώδεις καὶ δυσεξελίκτους ὄντας, καὶ συνεχῶς ἐπισκαίρει τε καὶ πειρᾶται καθέλκειν τὸ δέλεαρ. ὁ δὲ κραδαινομένου τοῦ ἐντέρου συνεὶς ἔχεσθαι τὴν ἔγχελυν, τὸν κάλαμον ᾧ τὸ ἔντερον προσήρτηται ἐνθεὶς τῷ ἑαυτοῦ στόματι καὶ ὅσον[1] σθένει καταπνέων, φυσᾷ τὸ ἔντερον καὶ μάλα γε ἰσχυρῶς, τὸ δὲ ἐκ τοῦ καταρρέοντος πνεύματος πίμπραται καὶ οἰδαίνει.[2] ὁ τοίνυν ἄνεμος κατολισθάνει[3] ἐς τὴν ἔγχελυν, καὶ πληροῖ μὲν τοῦ πνεύματος αὐτῆς τὴν κεφαλήν, πληροῖ δὲ τὴν φάρυγγα, καὶ ἐμφράττει τῷ θηρίῳ τὸ ἆσθμα. καὶ ἀναπνεῦσαι μὴ δυναμένη μηδὲ μὴν ἐξελεῖν τοῦ σπλάγχνου τοὺς ἐμπεφυκότας ὀδόντας ἀποπνίγεται, καὶ ἀνασπᾶται ἁλοῦσα ὑπὸ τοῦ ἐντέρου καὶ τοῦ πνεύματος καὶ τοῦ καλάμου τρίτου. καθ' ἑκάστην μὲν οὖν δρᾶται τοῦτο, ἁλίσκονται δὲ ὑπὸ πολλῶν πολλαί. ἔστω δή[4] μοι καὶ ταῦτα τῶνδε τῶν ἰχθύων λεχθέντα ἴδια.

9. Λέοντα θαλάττιον ἐοικέναι καράβῳ ἁμηγέπη καὶ ἡμεῖς ἴσμεν, λεπτότερον δὲ τὴν ἕξιν τοῦ σώματος ὁρῶμεν αὐτὸν καὶ ὑπό τι καὶ κυάνου[5] προσβάλλοντα, νωθῆ δὲ καὶ ἔχοντα χηλὰς μεγίστας καὶ ταῖς τῶν καρκίνων προσεοικυίας κατὰ σχῆμα.

[1] *Schn*: οἷον. [2] οἰδάνει *H*.

in the eddies; the other end he holds in his hands, and a piece of reed, the length of a sword-handle, has been inserted into it. The food does not escape the notice of the Eels, for they delight in this intestine. And the first Eel approaches, stimulated by hunger and with open jaws, and fastening its curved, hook-like teeth, which are hard to disentangle, in the bait, continues to leap up in its efforts to drag it down. But when the fisherman realises from the agitation of the intestine that the Eel is held fast, he puts the reed to which the intestine has been attached to his mouth and blows down it with all his might, inflating the intestine very considerably. And the downflow of breath distends and swells it. And so the air descends into the Eel, fills its head, fills its windpipe, and stops the creature's breathing. And as the Eel can neither breathe nor detach its teeth which are fixed in the intestine, it is suffocated, and is drawn up, a victim of the intestine, the blown air, and thirdly of the reed. Now this is a daily occurrence, and many are the Eels caught by many a fisherman. This then is what I have to say of the habits peculiar to these fishes.

The Sea-Lion

9. We also know that the Sea-lion [a] is in some respects like the crayfish, though we see that the shape of its body is slimmer, with an added dash of dark blue colour; but it is sluggish though possessed of enormous claws resembling those of crabs. And it

[a] A kind of large lobster.

---

[3] κατολισθαίνει L. [4] δέ.
[5] κυανοῦ μέρη τῶν ὀστράκιον.

λέγεται δὲ ὑπὸ τῶν σοφωτέρων ἁλιέων ἔχειν τινὰς ὑμένας προσηρτημένους τοῖς ὀστράκοις, ὑφ' οἷς ὑμέσιν εἶναι σαρκία ἁπαλά, καὶ καλεῖσθαι ἐκείνου τοῦ λέοντος στέαρ ταῦτα. ὀνίνασθαι δὲ τοὺς ἀνθρώπους ἐξ αὐτῶν [1] ἐκεῖνα. προσώπου καθαίρει [2] χρῶτα θολερόν, καὶ ἐλαίῳ ῥόδοις ἀνακραθέντι ἐμβληθέντα καὶ γενόμενα χρῖμα [3] ἐς ὥραν καὶ ἀγλαΐαν συμμάχεται.[4] προσακήκοα δὲ καὶ ἐκεῖνο, τὸν ἐπὶ τῆς γῆς λέοντα δεδιέναι ἰσχυρῶς τοῦ θαλαττίου τῆς ὄψεως τὸ ἐκτράπελον, καὶ μὴ φέρειν αὐτοῦ τὴν ὀσμήν· ὡς δέδοικε δὲ καὶ ἀλεκτρυόνα ὁ αὐτός,[5] ἀνωτέρω μοι λέλεκται. λέγουσι δὲ καὶ συντριβέντων αὐτοῦ τῶν ὀστράκων καὶ ἐμβληθείσης τῆς κόνεως ἐς ὕδωρ, πιόντα τὸν χερσαῖον λέοντα ἐξάντη [6] γίνεσθαι νόσου λυπούσης αὐτοῦ τὴν κοιλίαν. εἰρήσθω δή [7] μοι καὶ ταῦτα τοῦ θαλαττίου λέοντος ἴδια.

10. Ὄνοι δὲ Μαυρούσιοι, ὤκιστοι δραμεῖν, παρά γε τὴν πρώτην ὁρμήν εἰσιν ὀξύτατοι, ὡς αὔρας τινὰ ἐμβολὴν ⟨δοκεῖν⟩ εἶναι ἢ καὶ νὴ Δία [8] πτερὸν αὐτόχρημα ὄρνιθος· ταχέως δὲ κάμνουσι, καὶ αὐτοῖς οἱ πόδες ἀπαγορεύουσι, καὶ τὸ πνεῦμα ἐπιλείπει, καὶ τῆς [9] ὠκύτητος εἰλήφασι λήθην καὶ ἑστᾶσι [10] πεπεδημένοι, καὶ ἀφιᾶσι δάκρυα θαλερά, οὔ μοι δοκεῖν [11] ἐπὶ τῷ μέλλοντι θανάτῳ τοσοῦτον, ὅσον ἐπὶ τῇ τῶν ποδῶν ἀσθενείᾳ. τῶν μὲν οὖν ἵππων ἀποπηδήσαντες εἶτα μέντοι περιβάλλουσιν αὐτοῖς βρόχους περὶ τὴν δέρην, καὶ τῷ ἵππῳ

[1] αὐτοῦ.
[2] καθαίρουσι.
[3] χρίσμα.
[4] *Jac*: συμμάχεσθαι.
[5] ὁ λέων αὐτός.
[6] *Jac*: ἔξω ἄν.

is said by the more experienced fishermen to have certain membranes attached to its shell, and beneath them are some portions of tender flesh which are called 'lobster-lard.' And these benefit mankind: they cleanse a muddy complexion, and if added to oil-of-roses and applied as an ointment, they contribute to a person's beauty and adornment. And I have also heard the following: that the Land-lion is terrified of the monstrous appearance of the Sea-lion and cannot endure the smell of it. And how the same Lion dreads a cock I have explained earlier on.[a] They say also that if the Sea-lion's shell be ground down and the powder cast into water, and the Land-lion drinks it, he becomes immune from troubles of the stomach. This then is what I have to say of the peculiarities of the Sea-lion.

The Wild Ass of Mauretania

10. The Asses of Mauretania gallop at a very great speed, at least at the start they are extremely swift: they seem like a rushing wind or, I do declare, the very wings of a bird. But they quickly tire; their feet weary; their breath fails; they forget their speed; they stand chained to the spot and shed copious tears, not, I think, so much from any fear of impending death as on account of the weakness of their feet. And so the men leap from their horses and throw halters round the Asses' necks, and each

[a] See 3. 31; 6. 22.

---

[7] δέ.

[8] ὡς αὔρας . . . Δία] ἢ ὡς καὶ νὴ Δία αὔρας τινος ἐμβ. εἶναι ἢ MSS, ἤ (before ὡς) del. Reiske, καὶ νὴ Δία transposed by Jac, τινά H, ⟨δοκεῖν⟩ add. Jac.

[9] τῆς τε. [10] ἑστᾶσι νωθεῖς. [11] Schn: δοκεῖ.

προσαρτήσας ἕκαστος ἄγει ὡς αἰχμάλωτον ⟨τὸν⟩[1] ἑαλωκότα. ὅτι δὲ μικροὶ μὲν ἰδεῖν εἰσιν οἱ Λίβυες ἵπποι, δραμεῖν δὲ ὤκιστοι, ἀνωτέρω εἶπον.

11. Βοῶν δὲ Λιβύων πλῆθος ἦν ἄρα καὶ πλέον ἀριθμοῦ, καί εἰσιν ὤκιστοι οἵ ἄγριοί τε καὶ ἐλεύθεροι. καὶ οἵ γε θηραταὶ πολλάκις σφάλλονται ἕνα διώκοντες, καὶ ἐμπίπτουσιν ἐς ἑτέρους ἀκμῆτας· καὶ ὁ μὲν ἐσδὺς ἐς θάμνον ἢ νάπην ἠφανίσθη, ἕτεροι δὲ ἀναφαίνονται ὅμοιοι καὶ ἀπατῶσι τὴν ὄψιν. καὶ εἴ γέ τις ὑπάρξαιτο τούτων διώκειν τινά, προαπερεῖ[2] αὐτῷ ἵππῳ· τὸν μὲν γὰρ ἤδη καμόντα αἱρήσει[3] τῷ χρόνῳ, τοὺς δὲ αὐτῶν ἀρχομένους δρόμου προπονήσαντός οἱ τοῦ ἵππου οὐχ αἱρήσει. ἁλίσκονται δὲ ἀνὰ πᾶν ἔτος πολλοὶ καὶ ἀποθνήσκουσιν, ἥ γε μὴν ἐπιγονὴ αὐτῶν διαδέχεται καὶ μάλα ἀφθόνως. ἁλῶνται δὲ σὺν τοῖς μόσχοις καὶ οἱ ταῦροι κοινῇ καὶ αἱ θήλειαι, αἱ μὲν κύουσαι, αἱ δὲ ἀρτιτόκοι.[4] εἰ δὲ ἕλοι τις μόσχον ἔτι νεαρόν, καὶ μὴ παραχρῆμα ἀποκτείνειε,[5] διπλοῦν κέρδος ἕξει· συνῄρηκε γὰρ καὶ τὴν τεκοῦσαν αὐτόν, δράσας γε ἐκεῖνα ἅπερ εἰπεῖν οὐκ ἔστιν ἄτοπον. τὸν μὲν καταδήσας σχοίνῳ ἀπολέλοιπε καὶ ἀναχωρεῖ αὐτός, ἡ δὲ τῷ πόθῳ τοῦ τέκνου τείρεται καὶ φλεγομένη οἰστρᾶται, καὶ βουλομένη λύσασα ἀπάγειν ἐμβάλλει τὰ κέρατα, ἵνα διαξήνῃ[6] τε καὶ διαστήσῃ τὰ δεσμά. ὅ τι[7] δ' ἂν τῶν κεράτων ἐς τὴν τῆς σχοίνου συμπλοκὴν

[1] ⟨τόν⟩ *add. Schn.* [2] *Abresch*: προαπαίρει.
[3] αἱρήσει τις. [4] ἄτοκοι.
[5] ἀποκτείνει. [6] διαξάνῃ. [7] ὅτῳ.

[a] See 3. 2.

one securing an Ass to his horse, leads the one he has caught like a prisoner of war.

I have said earlier on that the horses of Libya are small in appearance but can gallop at very great speed.[a]

The Wild Cattle of Libya

11. It seems that of Libyan Cattle there are multitudes past numbering, and those that are wild and roam at large are exceedingly swift. And it often happens that hunters in pursuit of one animal go astray and fall in with others, fresh and untired. Meantime the hunted animal has plunged into a thicket or a glen and vanished, and others appear, exactly like it, and deceive the sight of the hunter. And if he should start to pursue one of these, he and his horse as well will be the first to give up the chase, for though in course of time he will overtake an animal already weary, he will not overtake those just starting to run: his horse will tire before they do.

Every year these Cattle are caught and slaughtered in great numbers, but their offspring take their place, and they are abundant. And they roam the land with their calves, the bulls along with the cows, some in calf, others with a calf lately born. If a man captures a calf while still young and does not slaughter it forthwith, he reaps a double advantage, because he captures the mother at the same time if he does what may fittingly be described here. He makes the calf fast with cord and then leaves it and withdraws. But the cow is wasted with yearning for her child and is goaded with ardent longing, and in her desire to release and carry it off attacks the bonds with her horns, hoping to fret them away and burst them. But whichever horn she inserts into

διείρῃ, κατέχεται καὶ πεδηθεῖσα σὺν τῷ μόσχῳ καταμένει, ἐκεῖνον μὲν οὐκ ἀπολύσασα, ἑαυτήν γε μὴν ἀφύκτῳ τῷ δεσμῷ περιβαλοῦσα. ταύτης οὖν ὁ θηρατὴς ἐξελὼν τὸ ἧπαρ αὐτῷ καὶ τὰ οὔθατα σφριγῶντα ἔτι ἐκτεμὼν καὶ τὴν δορὰν δείρας τὰ κρέα ἀφῆκεν ὄρνισι καὶ θηρίοις δαῖτα. τὸν δὲ μόσχον οἴκαδε κομίζει πάντα· ἔστι γὰρ καὶ ἐδωδὴν ἥδιστος, καὶ πῆξαι γάλα παρέξει ὀπὸν δούς.

12. Ὁ δράκων ⟨ὁ⟩ [1] θαλάττιός ἐστι μὲν παραπλήσιος τοῖς ἰχθύσι τοῖς ἄλλοις ὅσα ἐς τὸ λοιπὸν σῶμα, τήν γε μὴν κεφαλὴν ἔοικε τῷ χερσαίῳ δράκοντι καὶ τῶν ὀφθαλμῶν τὸ μέγεθος (εἰσὶ γὰρ μεγάλοι καὶ τούτῳ), καὶ μέντοι καὶ αἱ γένυς [2] τοῖς χερσαίοις προσβάλλουσι τὴν ἑαυτῶν ἁμωσγέπως μορφήν. ἔχει δὲ καὶ φολίδας, καὶ τραχεῖαί εἰσι, καὶ τῆς δορᾶς τῆς δρακοντείου οὐ πόρρω δοκοῦσιν, εἴ τις προσάψαιτο· ἐκπέφυκε δὲ καὶ κέντρα χαλεπὰ αὐτοῦ, καὶ ἰὸν φέρει τὰ κέντρα, καὶ ἔστι τῷ θιγόντι οὐ χρηστά.

13. Ζῴων δ' ἂν εἴη με εἰπεῖν καὶ τὸ ἴδιον αὐτῶν . . . [3] ὁ τῶν Ἰνδῶν βασιλεὺς ἐπιδόρπια σιτεῖται ταὐτὰ [4] οἷα δήπου Ἕλληνες ἐντραγεῖν αἰτοῦσι· φοινίκων ⟨δὲ⟩ [5] τῶν χαμαιζήλων ἐκεῖνος σκώληκά τινα ἐν τῷ φυτῷ τικτόμενον σταθευτὸν ἐπιδειπνεῖ γλύκιστον, ὡς Ἰνδῶν λέγουσι λόγοι, καί φασιν οἱ τὴν ἡδονὴν τὴν τοσαύτην ἐκ τοῦ

[1] ⟨ὁ⟩ *add. H.*
[2] μεγάλοι καὶ καλοί· καὶ τοῦτο μέντοι καὶ αἱ γ. μέν *most* MSS, τούτῳ V.
[3] *Some words are missing.*

the tangle of cord she is caught and held fast and remains by her calf, having failed on the one hand to release it, and on the other having entangled herself in bonds from which there is no escape. So then the hunter after removing the liver for his own use and cutting off the udder, which is still swollen, and flaying the hide, leaves the flesh for the birds and beasts to feed upon. But the calf he takes home entire, for it is extremely pleasant to eat, and also affords rennet which will curdle milk.

12. The Weever resembles other fishes in all other parts of its body excepting its head, and that is like the python both in the size of its eyes (those of the python also are large) and in its jaws, which to some extent are shaped like the python's. It has scales too and they are rough, and if one handles them they feel not unlike the skin of the python. Sharp spines spring from its body, which contain poison and cause harm if one touches them.

The Weever

13. . . . . .[a] The Indian King by way of dessert eats the same things as, no doubt, the Greeks would desire to eat. But according to Indian accounts he feasts with the greatest relish upon a certain worm that is begotten in the date-palm, when fried; and they say that he derives such pleasure from the eating. . . . And their accounts convince me. The

The Indian King, his food

[a] The first sentence is defective; the general sense was perhaps: 'There are countless details that I might relate touching the characteristics of animals.' (Gow.)

---

[4] *Gow* : ταῦτα MSS, *H.*

[5] ⟨δέ⟩ *add. Gow, punctuating after* αἰτοῦσι αἰτ., φ. τῶν χαμαιζήλων· *edd.*

σιτεῖσθαι . . .[1] καὶ ἐμέ γε αἱροῦσι λέγοντες. ἐπάϊκλα[2] δέ οἱ καὶ ἐκεῖνά ἐστι, κύκνων τε ᾠὰ καὶ τὰ τῶν χερσαίων στρουθῶν καὶ χηνῶν. τὰ μὲν οὖν ἄλλα οὐ μέμφομαι αὐτῷ,[3] κύκνων γε μὴν Ἀπόλλωνι μὲν λατρευόντων ᾠδικωτάτων δὲ ὡς ἡ φήμη διαρρέουσα λέγει[4] ἐπιβουλεύειν ἐκγόνοις καὶ διαφθείρειν τὰ ᾠά, ὦ[5] Ἰνδοὶ φίλοι,[6] οὐκέτι.

14. Λιβυστίνων γε μὴν περὶ δορκάδων καὶ κεμάδων τῶν ἐκεῖθι εἰπεῖν αἱρεῖ με θυμὸς τὰ νῦν ταῦτα. ὤκισται μέν εἰσιν αἱ δορκάδες, καὶ ὅμως τοὺς ἵππους τοὺς Λίβυας οὐ διαδιδράσκουσι. λαμβάνονται δὲ καὶ ἄρκυσι. φαιαὶ δ' εἰσὶ τὴν γαστέρα, καὶ αὐταῖς ἥδε ἡ χρόα ἐς τὰς λαπάρας ἄνεισι· παρ' ἑκάτερα δὲ τῆς νηδύος μέλαιναι ταινίαι καθέρπουσιν αὐταῖς. ξανθαί γε μὴν τὸ λοιπὸν σῶμά εἰσι, μακραὶ τοὺς πόδας, μέλαιναι τὸ ὄμμα, τὴν κεφαλὴν κέρασι κεκοσμημέναι, τὰ ⟨δὲ⟩[7] ὦτα αὐταῖς ἐστι[8] μήκιστα. ἥ γε μὴν καλουμένη ὑπὸ[9] τῶν ποιητῶν κεμὰς δραμεῖν μὲν ὠκίστη θυέλλης δίκην, ἰδεῖν δὲ ἄρα πυρρόθριξ καὶ λασιωτάτη· τὴν δὲ οὐρὰν λευκὴν ἔχει. εἴκασται δὲ τοὺς ὀφθαλμοὺς κυάνου βαφῇ. τὰ δὲ ὦτα τριχῶν ἀνάπλεω[10] δασυτάτων.[11] τὰ κέρατά τε αὐτῆς ἀντία καὶ ὡραῖα, ὡς ἐπιέναι μὲν τὴν θῆρα,[12] ἐν ταὐτῷ δὲ καὶ φοβεῖν ἅμα καὶ † βλάπτεσθαι καλήν.†[13] αὕτη δὲ ἄρα ἡ κεμὰς οὐκ ἐπὶ γῆς μόνης τὴν τῶν ποδῶν ὠκύτητα ἐπιδείκνυται,[14]

[1] *Lacuna.*
[2] *Schn*: ἔπεκλα.
[3] πω.
[4] διαρρεῖ λέγουσα.
[5] οἱ.
[6] Ἰνδοί, φιλῶ? *H.*
[7] ⟨δέ⟩ *add. H.*
[8] εἰσι.

following also are additions to his meals, the eggs of swans, of ostriches, and of geese. Now I find no fault with the others, but that he should plot against the offspring and destroy the eggs of swans, the servants of Apollo and, as the common report has it, the most tuneful of birds, is a thing, my dear Indians, that I cannot approve.

14. I have a mind now to relate the following facts touching the Gazelles and Prickets of Libya. The Gazelles are very swift-footed; for all that they cannot outrun the Libyan horses. They are also caught with nets. The belly is grey, and this colour extends upwards to their flanks; and on either side of the belly black stripes creep down their bodies. The rest of the body however is light-brown; the legs are long; the eyes black; the head is adorned with horns; the ears are very long. But the Pricket, as poets call it, 'runneth very swiftly, even as the hurricane'; in appearance it is red and very shaggy, but its tail is white; its eyes are the colour of dark blue dye; its ears are filled with very thick hair; its horns incline forwards and are graceful, so that the creature comes on and while inspiring fear, is a thing of beauty.[a] Now this Pricket does not display its speed only on land, but

The Gazelle of Libya

[a] With Triller's correction the sense will be 'so that it . . . is to be admired for its beauty.' Jac. compares Ael. *VH* 13.1 [Atalanta] *δύο δὲ εἶχεν ἐκπληκτικά, κάλλος ἄμαχον. καὶ συν τούτῳ καὶ φοβεῖν ἐδύνατο.*

---

[9] *Reiske*: *καὶ ὑπό.*
[10] *ἀνάπλεως.*
[11] *βαθυτάτων.*
[12] *Schn*: *θήραν* MSS, *H.*
[13] *Corrupt*: *βλέπεσθαι Triller.*
[14] *ἀποδείκνυται.*

ἀλλὰ ἐμπεσοῦσα καὶ ἐς ῥεῦμα ποταμοῦ ταῖς χηλαῖς τῶν ποδῶν ὡς εἰπεῖν ἐρέττουσα εἶτα μέντοι διακόπτει τὸ ῥεῦμα. χαίρει δὲ καὶ ἐν λίμνῃ νήξασθαι, καὶ ἐνταῦθά τοι καὶ τροφὴν ἴσχει, τεθηλός τε ἀεὶ θρύον καὶ κύπειρον δειπνεῖ. οὐκοῦν καὶ τὴν γαστέρα ἦρος ἀρχομένου πεπληρωμένην ὑπολαπάττει, καὶ τὰ οὔθατα [1] καθῆκε καὶ μέντοι καὶ ἐξέθρεψε τὰ ἑαυτῆς βρέφη ἡ κεμάς.

15. Μῦρος [2] δὲ ἄρα ἰχθὺς πυνθάνομαί ἐστιν. ἐξ ὅτου μὲν οὖν ἐσπάσατο τὴν ἐπωνυμίαν ἐκείνην, εἰπεῖν οὐκ οἶδα· κέκληται δ' οὖν ταύτῃ. λέγουσι δὲ αὐτὸν εἶναι θαλάττιον ὄφιν. ὀφθαλμὸς δὲ ἄρα ὁ τούτου ὁπότερος οὖν ἐξαιρεθεὶς καὶ περίαπτον γενόμενος ἀπαλλάττει ξηρᾶς ἄνθρωπον ὀφθαλμίας· τῷ δὲ ἄρα μύρῳ τῷδε ἀναφύεταί φασιν ὀφθαλμὸς ἕτερος. δεῖ δὲ αὐτὸν ἀπολῦσαι τὸν ἰχθὺν ζῶντα, ἢ μάτην τὸν ὀφθαλμὸν ἔχων φυλάττεις.

16. Αἶγες ἄγριοι ⟨οἱ⟩[3] τὰς Λιβύων ἄκρας ἐπιστείβοντές εἰσι κατὰ τοὺς βοῦς τὸ μέγεθος ἰδεῖν, τούς γε μὴν μηροὺς καὶ τὰ στέρνα καὶ τοὺς τραχήλους κομῶσι θριξὶ δασυτάταις, καὶ σὺν τούτοις καὶ τὴν γένυν. τὰ μέτωπα μὲν ἀγκύλοι καὶ περιφερεῖς, καὶ τὰ ὄμματα χαροποί, σκέλη δὲ αὐτοῖς ἐστι κολοβά. κέρατα μετὰ τὴν πρώτην συμφυὴν [4] ἀλλήλων ἀπηρτημένα καὶ πλάγια· οὐ γάρ τί που κατὰ τοὺς ὀρειβάτας αἶγας τοὺς ἄλλους ὀρθά ἐστι, κάτεισι δὲ ἐγκάρσια καὶ ἐς τοὺς ὤμους προήκοντα. οὕτως ἄρα μήκιστά ἐστιν. ἐκ δὲ τῶν λόφων τῶν

[1] *Bernhardy*: ταυθοταν. [2] *Ges*: μύρον.

will plunge into a running river and cleave the stream by rowing, so to speak, with its hooves. And it loves to swim in a lake, and there, let me tell you, it obtains food and feasts upon the ever-flowering rush and galingale. So at the beginning of spring it empties its full belly; its udder drops and it suckles its young.

The 'Myrus' and its eye

15. There is, I learn, a fish called *Myrus*,[a] but from what source it has derived its name I cannot say. At any rate that is the name by which it is called. And they say that it is a sea-snake. Now if one takes out either of its eyes and wears it as an amulet, it cures a man of dry ophthalmia; but the Myrus, they say, grows a fresh eye. But you must let the fish go alive, otherwise you will preserve its eye to no purpose.

The Ibex of Libya

16. The Wild Goats [b] which tread the mountain heights of Libya are about the size of oxen, but their thighs, breasts, and necks are covered with long and very shaggy hair, and so too are their jaws. Their foreheads are curved and rounded; their eyes are yellow, and their legs stumpy. Their horns, united at the beginning, part asunder and grow aslant: for they are not straight like those of other mountain goats but turn downwards obliquely and extend as far as the shoulders. Consequently they are of considerable length. And these Goats spring with

[a] Perhaps the *Muraena serpens*, a larger relation of the Moray.
[b] The 'Udad,' *Ovis lervia*.

---

[3] ⟨οἱ⟩ *add. Jac.* [4] *σύμφυσιν H.*

ὑπεράκρων, οὓς ἐρίπνας [1] οἵ τε νομευτικοὶ φιλοῦσιν ὀνομάζειν καὶ ποιητῶν παῖδες, ῥᾳδίως ἐς ἕτερον πάγον πηδῶσιν· ἁλτικώτατοι γὰρ αἰγῶν ἁπάντων οἵδε εἰσίν. εἴ γε μὴν καὶ πέσοι τις πορρωτέρω ὄντος τοῦ ὑποδεξομένου [2] αὐτὸν ἢ ὡς ἐκείνου ἐφικέσθαι, τῷ δὲ ἄρα μελῶν περίεστι τοσοῦτον κράτος, ὡς ἀσινῆ μένειν κατενεχθέντα αὐτόν. θραύει γοῦν οὐδὲ ἕν, εἰ καὶ πέσοι κατὰ ῥωγάδος, οὐ κέρας, οὐ βρέγμα· ἔστι δὲ καρτερὰ καὶ προσόμοια τῇ τῆς πέτρας ἀντιτυπίᾳ. οἱ πλεῖστοι μὲν οὖν τούτων ἐν ταῖς ἀκρωρείαις αὐταῖς ἄρκυσι καὶ ἀκοντίοις καὶ ποδάγραις αἱροῦνται, σοφίᾳ δὲ ἄρα τῇ τε ἄλλῃ ⟨ἐν⟩ [3] κυνηγέταις ἀνδράσι καὶ οὖν [4] καὶ αἰγοθηρικῇ· [5] θηρῶνται δὲ καὶ ἐν πεδίοις, καὶ φυγεῖν ἐνταῦθα ἀσθενεῖς εἰσιν. αἱρήσει οὖν [6] αὐτοὺς καὶ ὅστις ἐστὶ βραδὺς τοὺς πόδας. ἦν δὲ ἄρα ἀγαθὸν δορά τε καὶ κέρατα· ἡ μὲν ⟨γὰρ⟩ [7] δορὰ ἐν χειμῶσι τοῖς σφοδροτάτοις τὸν κρυμὸν πελάζειν οὐκ ἐᾷ [8] νομευτικοῖς καὶ ὑλουργοῖς ἀνδράσι· κέρατα δὲ ἐκεῖνα ἀρύσασθαι καὶ πιεῖν ἐκ ποταμοῦ παραρρέοντος ἢ πηγῆς τινος ἀνατελλούσης ἐν ὥρᾳ θερείῳ χρηστὰ καὶ δίψος ἀκέσασθαι λυσιτελῆ· παρέχει γὰρ ἀμυστὶ πιεῖν τῶν ἁδρῶν κυλίκων μεῖον οὐδὲ ἕν, ἕως ἂν ψύξῃς [9] τὸ ἆσθμα καὶ σβέσῃς [10] τὸ ὑπεκκαῖόν τε πᾶν καὶ ἀναφλέγον. [11] οὐκοῦν εἰ τὰ ἔνδον καθαρθείη ὑπό τινος ξέειν κέρατα δεινοῦ, [12] καὶ τρία μέτρα ῥᾳδίως αὐτοῖν δέξαιτο τὸ ἕτερον ἄν.

[1] *Bochart*: ἐπιπλάς.
[2] πορρωτέρω τοῦ ὑ. ὄντος.
[3] ⟨ἐν⟩ *add. Jac.*
[4] γοῦν.
[5] αἰγοθήμαις.
[6] γοῦν.
[7] ⟨γάρ⟩ *add. H.*

ease from towering pinnacles—'crags' as pastoral and poetical folk like to call them—on to another height, for they are far better at leaping than all other kinds of goat. If, however, one should happen to fall owing to the spot which should receive it being beyond its reach, it has such a reserve of strength in its limbs that it remains uninjured on landing. At any rate not a thing does it break, even though it falls down a cleft rock, neither horn nor front of the skull. But these creatures are as strong and as resistant as the stone itself. Now it is on the actual ridges that most of them are caught, by means of nets, spears, and snares, and by the general skill of a huntsman, but especially by skill in hunting the Goat. They are also caught in the plains, and there they cannot run strongly enough to escape. So even a man who is slow of foot will take them. And it seems that their hide and horns are serviceable. Thus, in the severest winters their hide keeps out the cold for herdsmen and woodcutters, while those famous horns of theirs are useful in summer time for drawing water and drinking from a flowing stream or some bubbling spring, and help to quench thirst, for they allow you to drink at one draught not a drop less than the contents of the largest cups, until you have cooled your panting heat and quenched all the fire and flame. And so if the inside is cleaned out by some skilled polisher of horns, either horn will easily contain as much as three measures.

---

[8] *οὐ πελάζει καί* MSS, *οὐ πελάζειν ἐᾷ Jac.*
[9] *ἄξῃ.*
[10] *σβέσῃ.*
[11] *καὶ τὸ ἀ.*
[12] *Jac* : *νου* V, *ξέειν εἰδότος κ. other* MSS.

17. Εἰσὶ δὲ ἄρα καὶ χελῶναι θρέμμα Λιβύης, οὐλόταται ὅσα ἰδεῖν, ὄρειοι δὲ αὗται, καὶ ἔχουσι τὸ χελώνιον ἐς βάρβιτα ἀγαθόν.

18. Ἵππος ὅταν τέκῃ, τοῦ βρέφους ἐκπεφυκυῖαν σάρκα οὐ πολλὴν ἀλλὰ ὀλίγην ἀπηρτῆσθαι οἱ μὲν κατὰ τοῦ μετώπου φασίν, οἱ δὲ κατὰ τῆς ὀσφύος, ἄλλοι γε μὴν κατὰ τοῦ αἰδοίου. ταύτην οὖν ἀποτραγοῦσα ἀφανίζει, καλεῖται δὲ τὸ σαρκίον τοῦτο ἱππομανές. οἴκτῳ δὲ ἄρα τῆς φύσεως καὶ ἐλέῳ ἐς τοὺς ἵππους δρᾶται τοῦτο. εἰ γὰρ ἀεί, φασί, καὶ διὰ τέλους προσήρτητο ἐκεῖνο, ἐς οἶστρον ἂν ἀκατασχέτου μίξεως ἐξήπτοντο οἵ τε ἄρρενες καὶ αἱ θήλειαι αὐτῶν. ἔστω δέ, εἰ δοκεῖ, Ἱππείου Ποσειδῶνος ἢ Ἀθηνᾶς Ἱππείας τοῦτο δῶρον ἵπποις δοθέν, ἵνα αὐτοῖς τὸ γένος διαμείνῃ,[1] μηδὲ ἀφροδισίων λύττῃ διαφθείρηται. ἴσασι δὲ ἄρα ἱπποφορβοὶ τοῦτο εὖ καὶ καλῶς, καὶ ἐάν ποτε δεηθῶσι τοῦ προειρημένου σαρκίου ἐς ἐπιβουλήν τινος, ὡς ἐξάψαι οἱ ἔρωτα, τὴν ἵππον κύουσαν παραφυλάττουσι, καὶ ὅταν τέκῃ παραχρῆμα ἁρπάζουσι τὸ πωλίον, καὶ ἀποκόπτουσι τὴν προειρημένην σάρκα, καὶ ἐς ὁπλὴν ἐμβάλλουσιν ἵππου θηλείας· ἐνταυθοῖ γὰρ καὶ μόνως ἂν φυλαχθείη καλῶς καὶ ἀποθησαυρισθείη. τὸν δὲ πῶλον ἀνίσχοντι τῷ ἡλίῳ καταθύουσιν· οὐ γὰρ ἔτι θηλάζει ἡ μήτηρ αὐτὸν τὸ γνώρισμα ἀφῃρημένον καὶ τῆς εὐνοίας οὐκ ἔχοντα τὴν ὑπόθεσιν· ἐκ γάρ τοι τοῦ κατατραγεῖν τὴν σάρκα φιλεῖν τὸ βρέφος ἡ μήτηρ ἰσχυρῶς ἄρχεται. ὅστις δ' ἂν κατά τινα ἐπιβουλὴν ἀνὴρ

[1] διαμένῃ.

17. Tortoises too are a product of Libya; they have a most cruel look, and they live in the mountains, and their shell is good for making lyres.

The Tortoise of Libya

18. When a Mare gives birth, some say that a small piece of flesh is attached to the foal's forehead, others say to its loin, others again to its genitals. This piece the Mare bites off and destroys; and it is called 'Mare's-frenzy.' It is because Nature has pity and compassion on horses that this occurs, for (they say) had this continued to be attached always to the foal, both horses and mares would be inflamed with a passion for uncontrolled mating. This may, if you like, be a gift bestowed by Poseidon or Athena, the god and the goddess of horses, upon these animals to insure that their race is perpetuated and does not perish through an insane indulgence. Now those who tend horses are fully aware of this and if they chance to need the aforesaid piece of flesh with the design of kindling the fires of Love in some person, they watch a pregnant Mare, and directly she bears the foal they seize it, cut off the piece of flesh, and deposit it in a Mare's hoof,[a] for there alone will it be securely kept and stored away. As to the foal, they sacrifice it to the rising sun, for its dam refuses to suckle it any more now that it has lost its birth-token and no longer possesses the premise of her affection. For it is by eating that piece of flesh that the dam begins to love her offspring passionately. But any man who as a result of some plot tastes of

'Mare's-frenzy'

[a] For *horn* as the only substance proof against poison, cp. 10. 40, and see Frazer on Paus. 8. 18. 6.

ἐκείνου γεύσηται τοῦ σαρκίου ἔρωτι καὶ μάλα γε ἀκρατεῖ συνέχεται καὶ ἐκφρύγεται καὶ βοᾷ, καὶ ἀκατασχέτως ὁρμᾷ καὶ ἐπὶ παιδικὰ αἴσχιστα καὶ ἐπὶ γυναῖκα ἀφήλικα καὶ ἀπρόσωπον, καὶ μαρτύρεται τὴν νόσον, καὶ τοῖς ἐντυχοῦσιν ὅπως ἐξοιστρᾶται λέγει. καὶ λείβεται μὲν τὸ σῶμα καὶ φθίνει, ἐλαύνεται δὲ τὴν ψυχὴν ἐρωτικῇ μανίᾳ. ἀκούω τοίνυν καὶ ἐν Ὀλυμπίᾳ τὴν ἵππον τὴν χαλκῆν, ἧς ἐρῶσιν ἵπποι καὶ ἐπιμαίνονται καὶ ἐγχρίμπτεσθαι ἐθέλουσι καὶ χρεμετίζουσι θεασάμενοι χρεμέτισμα ἐρωτικόν, ἔχειν τὴν ἐκ τοῦδε τοῦ ἱππομανοῦς ἐπιβουλὴν ἐν [1] τῷ χαλκῷ γεγοητευμένῳ λανθάνουσαν, καὶ κρυφίῳ τινὶ μηχανῇ τοῦ τεχνίτου ἐπιβουλεύειν τὸν χαλκὸν τοῖς ζῶσιν· μὴ γὰρ εἶναι τοσαύτην ἀκρίβειαν, ὡς οὕτως ἐξ αὐτῆς ἀπατᾶσθαί τε καὶ ἐξοιστρᾶσθαι τοὺς ἵππους τοὺς ὁρῶντας. καὶ ἴσως ⟨μὲν⟩ [2] λέγουσί τι οἱ λέγοντες, ἴσως δὲ οὐδὲν λέγουσιν· ἃ δ' οὖν ἤκουσα καὶ ὑπὲρ τούτων εἶπον.

19. Λέγεται δὲ ἐν τῇ Λιβύῃ λίμνη εἶναι ζέοντος ὕδατος, καὶ ἐν τῷδε τῷ ὕδατί φασιν ἰχθύας ζῆν καὶ νήχεσθαι καὶ τροφῆς ἐμβληθείσης ἀναπάλλεσθαι πρὸς τὴν τροφήν. εἰ δέ τις αὐτοὺς ἐς ὕδωρ ἐμβάλοι [3] ψυχρόν, ὅτι ἀποθνήσκουσι, καὶ τοῦτο προσακήκοα.

20. Λέγουσι δὲ ἄνδρες ἁλιείας [4] ἐπιστήμονες, τὴν τοῦ ἱπποκάμπου γαστέρα εἴ τις ἐν οἴνῳ κατατήξειεν [5] ἕψων καὶ τοῦτον [6] δοίη τινὶ πιεῖν,

[1] τὴν ἐν. [2] ⟨μέν⟩ *add.* H.
[3] *Jac* : ἐμβάλλοι. [4] ἁλιεῖς.

that piece of flesh becomes possessed and consumed by an incontinent desire and cries aloud, and cannot be controlled from going after even the ugliest boys and grown women of repellent aspect. And he proclaims his affliction and tells those whom he meets how he is being driven mad. And his body pines and wastes away and his mind is agitated by erotic frenzy.

Statue of Mare at Olympia

I have heard also this story of the bronze mare at Olympia: horses fall madly in love with it and long to mount it, and at the sight of it neigh amorously. Hidden away in the charmed bronze it contains the treacherous Mare's-frenzy, and through some secret contrivance of the artist the bronze works against living animals. For it could not possibly be so true to life that horses with their eyes open should be deceived and inflamed to that extent.

It may be that those who relate the story are speaking the truth, or it may be that they are not: I have only reported what I have heard.

A boiling lake

19. In Libya there is said to be a lake of boiling water, and in this water they say that fishes exist and swim about, and that when food is thrown into the water they leap up to get it. But I have also heard that if one casts these fish into cold water, they die.

The Sea-horse, its poisonous nature

20. Those who are expert at fishing say that if one boils and dissolves in wine the stomach of the Sea-horse and gives it to someone to drink, the wine

---

[5] κατατήξει.
[6] τοῦτο.

φάρμακον εἶναι τὸν οἶνον ἄηθες ὡς πρὸς τὰ ἄλλα φάρμακα ἀντικρινόμενον· τὸν γάρ τοι πιόντα αὐτοῦ πρῶτον μὲν καταλαμβάνεσθαι λυγγὶ σφοδροτάτῃ, εἶτα βήττειν ξηρὰν βῆχα, καὶ στρεβλοῦσθαι μέν, ἀναπλεῖν δὲ αὐτῷ οὐδὲ ἕν, διογκοῦσθαι δὲ καὶ διοιδάνειν τὴν ἄνω γαστέρα, θερμά τε τῇ κεφαλῇ ἐπιπολάζειν ῥεύματα, καὶ διὰ τῆς ῥινὸς κατιέναι φλέγμα [1] καὶ ἰχθυηρᾶς ὀσμῆς προσβάλλειν· τοὺς δὲ ὀφθαλμοὺς ὑφαίμους αὐτῷ γίνεσθαι καὶ πυρώδεις, τὰ βλέφαρα δὲ διογκοῦσθαι. ἐμέτων δὲ ἐπιθυμίαι ἐξάπτονταί φασιν, ἀναπλεῖ δὲ οὐδὲ ἕν. εἰ δὲ ἐκνικήσειεν [2] ἡ φύσις, τὸν μὲν ⟨τὸ⟩ [3] ἐς θάνατον σφαλερὸν παριέναι, ἐς λήθην δὲ ὑπολισθαίνειν [4] καὶ παράνοιαν. ἐὰν δὲ ἐς τὴν κάτω γαστέρα διολίσθῃ, μηδὲν ἔτι εἶναι, πάντως δὲ ἀποθνήσκειν τὸν ἑαλωκότα. οἱ δὲ περιγενόμενοι ἐς παράνοιαν [5] ἐξοκείλαντες ὕδατος ἱμέρῳ πολλῷ καταλαμβάνονται, καὶ ὁρᾶν διψῶσιν ὕδωρ καὶ ἀκούειν λειβομένου· καὶ τοῦτό γε αὐτοὺς καταβαυκαλᾷ καὶ κατευνάζει.[6] καὶ διατρίβειν φιλοῦσιν ἢ παρὰ τοῖς ἀενάοις ποταμοῖς ἢ αἰγιαλῶν πλησίον ἢ παρὰ κρήναις ἢ λίμναις τισί, καὶ πιεῖν μὲν οὐ πάνυ ⟨τι⟩ [7] γλίχονται, ἐρῶσι δὲ νήχεσθαι καὶ τέγγειν τὼ πόδε ἢ ἀπονίπτειν τὼ χεῖρε. οἱ δὲ οὐκ αὐτὴν τὴν τοῦ ἱπποκάμπου γαστέρα τούτων αἰτίαν εἶναί φασιν, ἀλλὰ νέμεσθαί τι φυκίον τὸ ζῷον πικρὸν δεινῶς, οὗ [8] τὴν ποιότητα [9] ἐς ἐκείνην μεταχωρεῖν. εὑρέθη δὲ ἄρα καὶ ἐς σωτηρίαν ἱππόκαμπος ἐπιτήδειος [10] ἀγχινοίᾳ παλαιοῦ μὲν ἁλιέως, σοφοῦ δὲ τὰ

[1] λεπτά.
[2] ἐκνικήσει.
[3] ⟨τό⟩ *add. Jac.*
[4] ὑπολισθάνειν *H.*
[5] παράνοιαν δέ.
[6] κατανυστάζει.

becomes a poison abnormal in comparison with others. For the man who has tasted it is first of all seized with a most violent retching; next he is racked with a dry cough but brings up nothing at all; yet his upper stomach is enlarged and swells, while hot streams mount to his head and phlegm descends from his nose, emitting a fishy odour; his eyes turn bloodshot and fiery and the lids become puffy. He is possessed, they say, by a longing to vomit, but brings up nothing whatever. If however Nature prevails, the man escapes the threat of death but sinks gradually into a state of forgetfulness and insanity. But if the wine penetrates into his lower stomach, it is all over with him, and the victim inevitably dies. Those who survive, having drifted into insanity, are seized with a strong desire for water; they yearn to see water and to listen to it falling. This at any rate quiets them and lulls them to sleep. And they like to spend their time either by ever-flowing rivers or near the sea-shore or by the side of springs or lakes, and though they do not at all desire to drink, they love to swim and to dip their feet and to wash their hands.

But there are those who maintain that it is not the actual stomach of the Sea-horse which causes these sufferings, but that the creature feeds upon a certain kind of seaweed of extraordinary bitterness and that its essence is transferred to the Sea-horse. Notwithstanding, the Sea-horse has been found to be an efficient remedy thanks to the shrewdness of an aged fisherman who was versed in matters regarding

---

[7] ⟨τι⟩ *add. H.*
[8] ἐξ οὗ.
[9] *Ges*: πιότητα.
[10] ἐπιτήδειον.

θαλάττια. ἦν Κρὴς[1] ἁλιεὺς γέρων, καὶ παῖδας νεανίας εἶχε καὶ τούτους ἁλιέας. οὐκοῦν συνηνέχθη τὸν μὲν πρεσβύτην ἱπποκάμπους θηρᾶσαι μετὰ καὶ ἄλλων ἰχθύων, τοὺς δὲ νεανίας δηχθῆναι ὑπὸ κυνὸς λυττώσης, τῷ πρώτῳ δηχθέντι τῶν ἄλλων ἀμυνόντων[2] καὶ τῷ αὐτῷ πάθει περιπεσόντων. οἱ μὲν οὖν ἔκειντο Ῥιθύμνης[3] τῆς Κρητικῆς πρὸς ταῖς ἠόσιν (ἔστι δὲ αὕτη κώμη, ὥς φασιν), οἱ δὲ θεώμενοι συνήλγουν τῷ πάθει, καὶ τὴν κύνα ἀποκτεῖναι προσέταττον καὶ τὸ ἧπαρ δοῦναι τοῖς νεανίαις ὡς φάρμακον τοῦ κακοῦ καταφαγεῖν, οἱ δὲ ἐς τῆς Ῥοκκαίας οὕτω καλουμένης Ἀρτέμιδος ἄγειν καὶ αἰτεῖν ἴασιν παρὰ τῆς θεοῦ. ὁ δὲ γέρων καὶ μάλα ἀδεῶς τε καὶ ἀτρέπτως ταῦτα μὲν ἐπαινεῖν[4] τοὺς συμβουλεύσαντας εἴα, τῶν δὲ ἱπποκάμπων ⟨τὰς⟩[5] γαστέρας ἐκκαθήρας,[6] τὰς μὲν ὤπτησε καὶ ἔδωκεν αὐτοῖς προσενέγκασθαι, τὰς δὲ συντρίψας ἐς ὄξος καὶ μέλι, καὶ τὰ ἕλκη περιπλάσας τούτοις τὰ τοῦ δήγματος, εἶτα τῆς τῶν νεανιῶν ἐκράτησε λύττης τῷ πόθῳ τοῦ ὕδατος, ὅνπερ οὖν οἱ ἱππόκαμποι αὐτοῖς ὑπεξῆπτον. καὶ τόνδε τὸν τρόπον τοὺς παῖδας ἰάσατο, ὀψὲ μέντοι.

21. Ὑπὲρ θαλαττίων μὲν κυνῶν εἴρηται ἡμῖν καὶ πολλά· κύνες δὲ οἱ ποτάμιοι ἰδεῖν μέν εἰσι κατὰ τοὺς κύνας τοὺς χερσαίους τοὺς μικρούς, λάσιοι δέ εἰσι καὶ τὴν οὐράν. λέγονται δὲ τῷ μὲν αἵματι νεῦρα ἀνθρώπων διοιδάνοντα πραΰνειν, εἰ ἐγχέοις[7] ὕδατι καὶ ὄξει ἀναμιχθέντι· ἡ δορὰ δὲ

[1] *Gill*: Κράης. [2] ἀμυνάντων *H.* [3] Μηθύμνης.
[4] ἐπαινῶν. [5] ⟨τάς⟩ *add. H.*
[6] ἐκκαθάρας καὶ ἐκβαλών. [7] ἐγχεῖς.

the sea. There was an old fisherman of Crete and he had some young sons, also fishermen. Now it so happened that the old man caught some Sea-horses along with other fish, and that the boys were bitten by a mad dog: when the first was bitten, the others who came to help him suffered the same fate. So they lay on the beach at Rhithymna [a] in Crete (this is said to be a village), while the spectators sympathised with their plight and gave orders for the dog to be killed and its liver to be given to the boys to eat as an antidote to the poison. Others urged that they should be taken to the temple of Artemis of Rhocca and that the goddess should be implored to heal them. But the old man, without a sign of fear, without swerving from his purpose, allowed these advisers to make their recommendations, washed out the stomachs of the Sea-horses, some of which he roasted and gave to the boys to apply, while others he pounded into a mixture of vinegar and honey, and then smeared on the wounds made by the bite, and so overcame the boys' madness by that longing for water which the Sea-horses engendered in them. And in this way he cured his sons, though it took time.

The Otter

21. I have already said much regarding Dog-fish in the sea. But river Dog-fish [b] have the appearance of small dogs that live on land, and they even have hairy tails. And it is said that their blood, if poured into a mixture of water and vinegar, acts as an embrocation for swollen sinews. Their skin provides

[a] On the N coast and towards the western end of Crete.

[b] Gesner (*Hist. anim.: de quadrup. vivip.* (Francof. 1603), p. 683) explains this as meaning an otter.

ὑποδήματα δίδωσιν ἀγαθά, καὶ ταῦτα νεύρων χρηστά, ὥς φασιν.

22. Θύμαλλον δὲ ἰχθὺν οὕτω καλούμενον τρέφει Τέκινος [1] [ποταμοῦ δὲ ὄνομα τοῦτο' Ἰταλοῦ],[2] καὶ μέγεθος μὲν ὅσον καὶ ἐπὶ πῆχυν προήκει, ἰδεῖν δὲ μεταξὺ λάβρακός ἐστι καὶ κεφάλου. ἄξιον δὲ αὐτοῦ ἑαλωκότος θαυμάσαι τὴν ὀσμήν· οὐ γάρ τί που προσβάλλει ἰχθυηρὸν ἀέρα κατὰ τοὺς λοιπούς, ἀλλὰ εἴποις ἂν διὰ χειρῶν κατέχειν θύμον [3] νεωστὶ [4] τρυγηθέντα, καὶ οὖν καὶ εὔοσμός ἐστι, καί τις οὐκ ἰδὼν τὸ ζῷον οἰήσεται πόαν ἔνδον εἶναι τὴν μάλιστα μελιττῶν τροφόν,[5] ἔνθεν τοι καὶ κέκληται. λίνῳ μὲν οὖν αἱρεθείη ἂν ῥᾷστα· δελέατι δὲ καὶ ἀγκίστρῳ οὐχ αἱρήσεις αὐτόν,[6] οὐχ ὑὸς πιμελῇ, οὐ σέρφῳ, οὐ χήμῃ, οὐκ ἰχθύος ἑτέρου ἐντέρῳ, οὐ στρόμβου τένοντι. κώνωπι δὲ αἱρεῖται μόνῳ, πονηρῷ μὲν ζῴῳ καὶ μεθ' ἡμέραν καὶ νύκτωρ ἀνθρώποις ἐχθρῷ καὶ δακεῖν καὶ βοῆσαι, αἱρεῖ δὲ τὸν θύμαλλον τὸν προειρημένον· φιληδεῖ γὰρ αὐτῷ μόνῳ.

23. Ὑπὸ τῷ ποδὶ δὲ τῶν Ἀλπίων [7] ὀρῶν πρὸς ἄνεμον βορρᾶν ὑπὸ τῇ ἄρκτῳ . . .[8] οὕτω κέκληνται.[9] γένος δὲ οὗτοι [10] ἱππικοὶ ἄνδρες. ἐντεῦθέν τοι [11] πρόεισιν ὁ τῶν Εὐρωπαίων ποταμῶν μέγιστος

[1] Τέκηνος. [2] [ποταμοῦ . . . Ἰταλοῦ] *gloss, Gow.*
[3] *Schn* : θύμαλλον MSS, θύμαλον *H here and below.*
[4] νεωστὶ τῆς γῆς. [5] τροφόν τὴν προειρημένην.
[6] *Jac* : αὐτὸν ῥᾷστα. [7] Ἀλπέων *or* Ἀλπείων.
[8] *Lacuna.* [9] *Reiske* : κέκληται.
[10] τοιοῦτον. [11] τοι ⟨καί⟩ *H.*

[a] Mod. Ticino, in the NW of Italy.

excellent shoes, and these too, they say, are good for the sinews.

22. The river Tecinus [a] (this is the name of a river in Italy) breeds the fish called the Grayling. It attains to as much as a cubit in length, and in appearance is between the basse and the mullet. The odour of the fish when caught is something to astonish one, for it is not the least like the fishy odour of others, but you would say that you held in your hand some freshly plucked thyme; moreover it is sweet-scented and a man who did not notice the fish would fancy that the herb which is the bees' principal food (from which incidentally the fish *thymallus*, derives its name) was in your hand.

The Grayling

The easiest way to catch it is with a net; with a lure and hook you will not catch it, neither with hog's fat nor with a gnat nor with a clam nor with the entrails of any other fish nor with the muscle of a spiral-shell. It is only to be caught with a mosquito,[b] a troublesome insect, man's enemy by day and by night with its sting and its buzzing: that will catch the aforesaid Grayling, for this is the only bait that it delights in.

how caught

23. At the foot of the Alps, facing the north wind, and beneath the Great Bear, live the people called. . .[c] They are a nation of horsemen. It is in that region, you know, that the largest of the

The Ister and its fish

[b] See W. Radcliffe, *Fishing from the Earliest Times* (Lond. 1921), pp. 185 ff.

[c] The name of the people is lost. Ptolemy (*Geog.* 2. 11. 6) mentions a people of the name of Οὐισποί, Vispi, who appear to inhabit this region, and before οὕτω the word might well have fallen out. See G. B. Grundy's map *Germania*.

Ἴστρος, ἐκ πηγῶν μὲν οὐ πολλῶν, ταῖς δὲ τοῦ ἡλίου προσβολαῖς ταῖς πρώταις ἀντίος. εἶτα οἱονεὶ δορυφοροῦντες αὐτὸν ἅτε τῶν ἐπιχωρίων ῥευμάτων βασιλέα συνανίσχουσίν οἱ πολλοί, καὶ ἀέναον τὸ ῥεῦμα αὐτῶν ἐστι, καὶ ἴσασιν [1] ἑκάστου τὸ ὄνομα οἱ περιοικοῦντες αὐτούς. ὅταν δὲ ἐς τὸν Ἴστρον ἐμβάλωσι, τοῖς μὲν ἡ ἀπὸ γενεᾶς ἐπωνυμία πέπαυται, ἀφίστανταί [2] γε μὴν ἐκείνῳ τοῦ ὀνόματος, καὶ ἐξ αὐτοῦ καλοῦνται πάντες, καὶ συνεκβάλλουσιν ἐς τὸν Εὔξεινον. γίνεται δὲ ἐνταῦθα ἰχθύων γένη διάφορα, κορακῖνοί τε καὶ μύλλοι [3] καὶ ἀντακαῖοι καὶ κυπρῖνοι, μέλανες οὗτοι, καὶ χοῖροί τε καὶ κόσσυφοι [4] ἰδεῖν λευκοί, πέρκαι τε ἐπὶ τούτοις καὶ ξιφίαι. πρέπουσι δὲ τῷ ὀνόματι οἱ ἰχθύες οἵδε, καὶ τὸ μαρτύριον, τὸ μὲν ἄλλο πᾶν σῶμα ἁπαλοί τε εἰσὶ καὶ ἄλυποι προσαπτομένῳ, καὶ ὀδόντές οἱ οὐ πάνυ τι [5] σκολιοὶ οὐδὲ ἀπηνεῖς ἰδεῖν, οὐκ ἄκανθα ἐπὶ τῶν νώτων [6] ὀρθή, τὸ τῶν δελφίνων,[7] ἐκπέφυκεν, οὐ κατὰ τὸ οὐραῖον· ὃ δέ ἐστι θαῦμα καὶ ἀκούσαντι καὶ ἰδόντι, ὑπ᾽ αὐτὴν τὴν ῥῖνα, δι᾽ [8] ἧς καὶ ἀναπνεῖ καὶ τὸ ῥεῦμα αὐτῷ διαρρεῖ ἐς τὰ βράγχια καὶ ἐκπίπτει, ἐς ὀξύ οἱ προήκει ἡ γένυς, καὶ εὐθεῖά ἐστι καὶ αὐξάνεται κατ᾽ ὀλίγον ἐς μῆκός τε καὶ πάχος, καὶ κητουμένῳ τῷ ἰχθύι συναύξεται [9] καὶ ἐκείνη, καὶ ἔοικε

[1] ἴσασι μέν.
[2] *Reiske*: ἀφίσταται.
[3] *Ges*: μυαλοί.
[4] ἥσυχοι.
[5] πάνυ or πάντῃ.
[6] τῷ νώτῳ.
[7] τὸ τῶν δελφίνων *del. H.*
[8] ἐξ.
[9] συνεπείγεται MSS, συνεπαύξεται *Reiske*.

[a] Mod. Danube.

rivers of Europe, the Ister,[a] rises from only a few springs and moves in a direction facing the first assaults of the sun. Later, many rivers rise with one accord as though they were escorting him—for he is the King of the rivers of that country—and flow perpetually, and those who live on their banks know the name of each one. But as soon as they discharge into the Ister, the name which they had at their birth ceases to be used, they surrender it in his favour, all are called after him, and together pour their waters into the Euxine. And there there are fish of different species, crow-fish,[b] myllus, sturgeon, carp (these are black), and schall and wrasse (which are white), and besides these, perch and sword-fish. These last are suited to their name, witness the fact that the rest of their body is soft and harmless to the touch, that their teeth do not appear curved and sharp, that there are no spines springing erect from their back, as in the case of dolphins,[c] or from their tail, but what surprises one to learn and to see is this: the jaw just below its nose, through which it breathes and through which the stream flows to the gills and falls out, is prolonged to a sharp point, is straight and increases gradually in length and in bulk; it grows also as the fish grows into a monster and resembles the beak of a trireme. And the Sword-fish makes straight for fishes, kills them, and then feeds on them, and with this same sword beats off the attacks of the largest sea-monsters. No smith has forged this weapon which grows upon the fish, and Nature has made it sharp.

The Sword-fish

[b] Gossen identifies this with the Danube salmon, *Salmo hucho*.

[c] See Thompson, *Gk. fishes*, s.v. Δελφίς, p. 54 med.

τριήρους ἐμβόλῳ. καὶ διὰ εὐθείας[1] ἐμπίπτων ὁ ξιφίας ἰχθύσι καὶ ἀποκτείνας εἶτα θοινᾶται, καὶ ἀμύνεται δὲ τῷ αὐτῷ τὰ μέγιστα τῶν κητῶν. καὶ ἀχάλκευτόν γε τοῦτο τὸ ὅπλον προσπέφυκέν οἱ, καὶ τέθηκται φύσει. οὐκοῦν οἵδε οἱ ξιφίαι ἐς μέγεθος προήκοντες ἔρχονται καὶ νεὼς ἀντίοι.[2] καὶ νεανιεύονταί γέ τινες λέγοντες ναῦν Βιθυνίδα ἰδεῖν ἀνασπωμένην, ἵνα αὐτῇ πονήσασα ὑπὸ γήρως ἡ τρόπις τύχῃ τῆς δεούσης κομιδῆς, οὐκοῦν προσηλωμένην θεάσασθαι ξιφίου κεφαλήν· [3] τοῦ ⟨γὰρ⟩ [4] θηρὸς ἐμπήξαντος μὲν τῷ σκάφει τὴν αἰχμὴν τὴν συμφυῆ, ἀποσπάσαι δὲ αὐτὸν πειρωμένου [5] ὑπὸ τῆς ῥύμης [6] τῆς πολλῆς σχισθῆναι [7] μὲν ἀπὸ τοῦ τένοντος τὸ πᾶν σῶμα, τὴν δὲ ἐναπομεῖναι [8] πεπηγμένην, ὡς ἐνέπεσεν ἐξ ἀρχῆς. θηρᾶται δὲ οὗτος ἄρα καὶ ἐν τῇ θαλάττῃ καὶ ἐν τῷ Ἴστρῳ, χαίρει δὲ καὶ πικρῷ ὕδατι καὶ γλυκεῖ ῥεύματι.

24. Θέρους ἐνακμάζοντος τοῦ σφοδροτάτου οἱ θαλάττιοι κύνες καὶ τὰ ἄλλα ζῷα, οἷσπερ οὖν ἔστι συμφυὴς ἡ τόλμα, ἔς τε τοὺς αἰγιαλοὺς παραβάλλει καὶ εὐθὺ τῶν κρημνῶν ἔρχεται, καὶ τὰς ῥοώδεις ἄκρας ὑποτρέχει, καὶ ἐς τοὺς στενοὺς καὶ βαθεῖς ἐσνήχεται [9] πορθμούς. φεύγουσι δὲ τὰ πελάγια ἤθη, καὶ τῆς ἐκεῖ νομῆς τηνικάδε τῆς ὥρας ὑπερορῶσι. γίνεται δὲ ἄρα τι φῦκος ἐν τοῖς ἕρμασι τοῖς βαθέσι, καὶ τὸ μέγεθος αὐτῷ κατὰ τὴν μυρίκην ἐστί, φέρει δὲ καρπὸν τῇ μήκωνι προσεμφερῆ. καὶ τῶν μὲν ἄλλων ὡρῶν τοῦ ἔτους μέμυκε, καὶ ἔστιν ἀντίτυπος καὶ στερεὸς [10] φύσει

[1] πορείας. [2] ἀντίον.

And so when these Sword-fish have attained a considerable size they even attack ships. And there are some who boast that they have seen a Bithynian vessel drawn up on shore in order that the keel which was suffering from age might receive the necessary attention, and fixed to the keel they saw the head of a sword-fish. For the creature had planted the sword given it by Nature, in the vessel, and when it attempted to withdraw, the whole of its body was rent from the neck owing to the force of the ship's onrush, while the sword remained fixed just as it entered originally. So then this fish is caught both in the sea and in the Ister, and it delights both in salt water and in fresh streams.

24. When the summer is at its hottest, Sharks and other fish which are bold by nature approach the sea-shore and make straight for cliffs and run in under headlands where the current is strong and swim into narrow, deep straits. They forsake their haunts in the open seas and at this season neglect their feeding-ground there. Now a certain seaweed [a] grows among deep reefs: it is about the size of a tamarisk and bears fruit resembling a poppy. At other seasons of the year the fruit is closed and is resistant and hard like a shell; it opens however

A deadly seaweed

[a] This has not been identified, but there is no known seaweed that is poisonous to fish, and much of Aelian's description appears to be fanciful.

[3] *τὴν τοῦ ξ. κ. αὐτοῦ.*
[4] *⟨γάρ⟩ add. H.*
[5] *Ges: τε αὐτὸν πειρώμενον.*
[6] *Schn: ῥώμης.*
[7] *Jac: ἐνσχεθῆναι.*
[8] *ἀπομεῖναι.*
[9] *εἰσνήχονται.*
[10] *Ges: στερεά.*

ὀστρέου· ἁπλοῦταί γε μὴν μετὰ τὰς τροπὰς τὰς θερινάς, ὥσπερ οὖν αἱ ἐν ῥοδωνιαῖς κάλυκες. καὶ τὸ μὲν περικείμενον ἔλυτρον φρουρεῖ τὸ ἔνδον, καὶ δίκην ἕρκους [1] περιέρχεται· ἰδεῖν γε μὴν ξανθότατόν ἐστι,[2] τὸ δὲ ὑπὸ τούτῳ τῷ χιτῶνι κυανοῦν [3] ἐστὶ χρόᾳ καὶ χαῦνον, ὥσπερ οὖν πεπρημένη κύστις, ⟨καὶ⟩ [4] διαυγὲς [5] ἄγαν, λείβεταί τε ἐξ αὐτοῦ [6] πονηρὸν φάρμακον. καὶ νύκτωρ μὲν ἐκπέμπει τοῦτο αὐγὴν πυρὶ ἐοικυῖαν, καί τινας ἀφίησι μαρμαρυγάς· ὑπανατέλλοντος δὲ τοῦ Σειρίου ἔτι καὶ μᾶλλον κατισχύει ἡ τοῦ φαρμάκου κακία. καὶ ἐντεῦθεν ὅσον ἐστὶν ὑδροθηρικὸν παγκύνιόν οἱ ὄνομα θέμενοι εἶτα οἴονται τὴν τοῦ ἄστρου ἐπιτολὴν τίκτειν αὐτό. οἱ θαλάττιοι οὖν κύνες πρὸς τὴν νύκτωρ τοῦ ἄνθους φαντασίαν τὴν φλογώδη ἐμπεσόντες [7] ὥσπερ οὖν ἐς ἕρμαιον [8] τὴν ἔναλον μυρίκην τήνδε, τοῦ φαρμάκου τοῦ μὲν καταδεύσαντος [9] αὐτούς, τοῦ δὲ καταποθέντος, καὶ ἑτέρου διὰ τῶν βραγχίων αὐτοῖς ἐσθορόντος, εἶτα μέντοι τεθνήκασι καὶ παραχρῆμα ἀναπλέουσιν. οἱ τοίνυν δεινοὶ τὰ τοιαῦτα ἀνιχνεύειν τοῦδε τοῦ φαρμάκου ἐκ τῶν κητῶν τῶν προειρημένων τὸ μὲν ἐκ τῶν μελῶν τῶν λοιπῶν, τὸ δὲ ἐκ τοῦ στόματος τοῦ θηρὸς ἀθροίζουσι.[10] δεύτερον ⟨δὲ τὸ⟩ [11] κακὸν τοῦτο τῆς καλουμένης χερσαίας ἀγλαοφώτιδος. ὄνομα δὲ αὐτῇ ἄρα ἔθεντο καὶ κυνόσπαστον· καὶ τίς ἡ αἰτία, ἐὰν ὑπομνησθῶ εἰπεῖν, εἴσεσθε αὐτήν.

[1] *ἕρκους ὀστρακῶδες ὄν.*
[2] *περιέρχεται· ἰδεῖν . . . ἐστι, so Gow punctuates.*
[3] *κυάνεον.*
[4] *⟨καί⟩ add. H.*

after the summer solstice, like buds in rose-gardens. And the surrounding sheath protects the inside, encircling it like a barrier : it is a bright yellow colour, but the part beneath this covering is dark blue and flabby like a bladder with air in it, and is quite translucent, and from it there oozes a violent poison. By night this seaweed sends out a fiery ray and sparkles. And when the Dog-star is rising the evil power of the poison is even stronger. For that reason all fishermen have given it the name of *Pancynium* in the belief that it is the rising of the star that generates the poison. Now the Sharks fall upon the flower which by night seems to be burning, rushing at this tamarisk of the sea as if it were treasure trove, and when the poison has drenched them, some being swallowed and some having penetrated through their gills, they die and at once float up to the surface.

Now those who are skilled at investigating such matters collect this poison which emanates from the aforesaid monsters, some of it from other parts of the creature's body and some from its mouth. This poison is second only to that of the land-peony, as it is called, which people have also named *Cynospastus*. The reason for this you will learn if I remember to tell it you.[a]

[a] See below, ch. 27.

---

5 διαυγής.
6 αὐτῆς.
7 εἶτα ἐμπεσόντες.
8 *Ges* : ἕρμα MSS, *H.*
9 καταλούσαντος.
10 *Jac* : φάρμακον ἀθροίζουσι.
11 ⟨δὲ τό⟩ *add. H.*

25. Μυσοὶ δέ, οὐχ οἱ τοῦ Τηλέφου τὸ Πέργαμον κατοικοῦντες,[1] ἀλλὰ ἐκείνους τοὺς πρὸς τῷ Πόντῳ μοι νόει τοὺς κάτω, οἵπερ οὖν καὶ τῇ γῇ τῇ Σκυθίδι προσοικοῦσι τὰς ἐκείνων ἐπιδρομὰς ἀνείργοντες καὶ τῇ Ῥώμῃ τὸν χῶρον τὸν προειρημένον φρουροῦντες πάντα· ⟨τοὺς⟩ [2] Ἡρακλείας πλησίον φημὶ καὶ τῶν Ἀξίου ῥευμάτων.[3] ἐνταῦθά τοι καὶ τὴν Αἰήτου Μήδειαν οἱ ἐπιχώριοι ὑμνοῦσι τὸ ἔργον ἐκεῖνο τὸ ἐς [4] τὸν Ἄψυρτον τὸν ἀδελφὸν [5] χερσὶ κακαῖς τολμῆσαι, ναὶ μὰ Δία δυστυχῆ φήμην ἐπὶ τῇ Κόλχῳ φαρμακίδι [6] πρὸς ταῖς ἄλλαις ταῖς ⟨ἐν⟩ [7] Ἕλλησι τήνδε ᾄδοντες οἱ Μυσοί. ἀλλὰ οὗτοί γε θήραν ἰχθύων ἐκείνην θηρῶσιν. ἀνὴρ Ἰστριανὸς γένος, τὴν τέχνην ἁλιεύς, τῆς τοῦ Ἴστρου ὄχθης πλησίον ἐλαύνει βοῶν ζεῦγος, οὔ τι που δεόμενος ἀροῦν οὗτος· ὥσπερ γάρ φησιν ὁ λόγος, μηδὲν εἶναι βοῒ κοινὸν καὶ [8] δελφῖνι, οὕτω τοι φιλία χερσὶν ἁλιέων καὶ ἀρότρῳ πόθεν ἂν γένοιτο; εἰ οὖν [9] οἱ καὶ ἵππων παρείη ζεῦγος, τοῖς ἵπποις χρῆται. καὶ τὸν μὲν ζυγὸν ὁ ἀνὴρ φέρει κατὰ τῶν ὤμων, ἔρχεται δὲ ἔνθα οἱ δοκεῖ καλῶς ἔχειν ἑαυτὸν καθίσαι καὶ ἐν

[1] κατοικοῦντες Μυσοί.
[2] ⟨τούς⟩ *add. H.*
[3] ῥευμάτων τοῦ καλουμένου Τομέως πλήσιον.
[4] τὸ ἐς] ταῖς MSS, πρός *Oud.*
[5] *Oud*: τῶν Ἀψύρτων τῶν ἀδελφῶν.
[6] *Ges*: φαρμακεῖ MSS, φαρμακῷ *H.*
[7] τοῖς ἄλλοις τοῖς Ἑ.
[8] ἅμα καί.
[9] γοῦν.

The Moesians and their fishing

25. The people of Mysia[a]—not those who inhabit the Pergamum of Telephus, but you are to understand those who live by the Black Sea in the lower part and are neighbours of the Scythians whose inroads they check, and who are guardians of the aforesaid country on behalf of Rome. I am referring to those that live near Heraclea and the river Axius.[b] It is there, you know, that the inhabitants tell the tale of Medea, daughter of Aeetes, whose impious hands dared to commit that outrage upon her brother Apsyrtus,[c] for the Mysians harp on this evil report against the Colchian sorceress, besides the others that are current among the Greeks.—Well, this is the way in which these people hunt fish. An Istrian whose trade is fishing drives a pair of oxen near the bank of the Ister, but not because he has the least wish to plough, for, as the saying goes, 'an ox and a dolphin have nothing in common;' so in the same way what friendship can there be between a fisherman's hands and a plough? If however he has a pair of horses he uses horses. The man carries the yoke on his shoulders and comes to a spot where he thinks it suitable to sit down and where he be-

[a] *I.e.* Moesia Inferior, a region N of Thrace; cp. 2. 53. 'Scythia Minor' was the name given to the NE portion which lay along the Black Sea.

[b] The Axius rises in Dardania, about 145 mi. SW of M. Inferior, and flows SE into the Thermaic gulf. 'Heraclea,' whether 'Lyncestis' or 'Sintica,' is in Macedonia, and the latter is on (or near) the Strymon. Aelian's geography is confused.

[c] Apsyrtus according to one story pursued Medea when she fled with Jason from Iolcos; according to another she took him with her—he was only a child; she murdered him and scattered his limbs in the path of Aeetes in order to delay his pursuit.

καλῷ τῆς ἄγρας εἶναι πεπίστευκε. τῆς οὖν μηρίνθου στερεᾶς οὔσης καὶ ἄγαν ἑλκτικῆς[1] τὴν μὲν ἀρχὴν ἐξῆψε μέσου[2] τοῦ ζυγοῦ, ἄδην δὲ τροφῆς παρατίθησιν ἢ τοῖς βουσὶν ἢ τοῖς ἵπποις, οἱ δὲ ἐμπίπλανται. καὶ ἐκεῖνος τῇ μηρίνθῳ κατὰ θάτερα προσῆψεν ἄγκιστρον ἰσχυρὸν καὶ μέντοι καὶ τεθηγμένον δεινῶς, περιπείρας[3] δὲ ἄρα αὐτῷ πνεύμονα ταύρου † τεθηραμένου †,[4] μεθῆκε τροφὴν Ἰστριανῷ σιλούρῳ καὶ μάλα γε ἡδίστην, ὑπὲρ τοῦ συνδέοντος τὸ ἄγκιστρον λίνου ἐξάψας τὸν ἀρκοῦντα μόλιβον, οἷον ἐς τὴν ἕλξιν εἶναι ἕρμα αὐτοῦ.[5] ὁ τοίνυν ἰχθὺς ὁπόταν αἴσθηται τῆς ταυρείου βορᾶς, παραχρῆμα κατὰ τὴν ἄγραν ὁρμᾷ· εἶτα ὧν ἱμείρει τούτοις ἐντυχὼν ἀθρόως καὶ περιχανὼν ἄδην καὶ ἀταμιεύτως τὴν ἐμπεσοῦσάν οἱ κακὴν δαῖτα ἐς ἑαυτὸν σπᾷ. εἶτα ὑφ᾽ ἡδονῆς ἑλκόμενος[6] ὅδε ὁ γάστρις ἑαυτὸν διαλέληθε τῷ προειρημένῳ περιπαρεὶς ἀγκίστρῳ, καὶ ἀποδρᾶναι τὸ ἐμπεσὸν κακὸν διψῶν τὴν μήρινθον ὡς ἔχει δυνάμεως ὑποταράττει τε καὶ κινεῖ. συνίησιν οὖν ὁ θηρατὴς καὶ ἡδονῆς ὑπερεμπίπλαται, εἶτα τῆς ἕδρας ἀνέθορε, καὶ μεθῆκεν ἑαυτὸν ποταμίων τε ἔργων καὶ κυνηγεσίων ἐνύδρων, ὥσπερ δὲ ἐν δράματι ὑποκριτὴς ἀμείψας προσωπεῖον ὁ δὲ τὼ βόε ἐλαύνει ἢ τὼ ἵππω, ἀλκὴ δὲ ἄρα καὶ ἡ τοῦ κήτους καὶ ἡ τῶν ὑποζυγίων ἀντίπαλός ἐστιν. ὁ μὲν γὰρ θὴρ ὁ τοῦ Ἴστρου τρόφιμος ἕλκει κάτω ὅσον ποτὲ ἄρα τῆς ἐν αὐτῷ ῥώμης ἔχει, τὸ μέντοι ζεῦγος τὸ ἀνθέλκον ἐκτείνει τὴν μήρινθον. ἀλλά οἱ πλέον οὐδὲ ἕν· τῆς γοῦν ἐπ᾽ ἀμφοῖν ἕλξεως ὁ ἰχθὺς ἡττᾶται, καὶ ἀπειπὼν

[1] *Valck*: ἑκτικῆς.
[2] μέσου τῶν ζῴων.

lieves he is well placed for fishing. One end of his rope, which is stout and thoroughly capable of standing a strain, he attaches to the middle of the yoke. He provides ample fodder for the oxen or the horses, and they eat their fill. And to the other end of the rope he attaches a strong hook which has been well sharpened, and on this he spits the lungs of a bull, and lets them down as food, and indeed its favourite food, for the Sheat-fish in the Ister, after fastening above the point where the rope secures the hook enough lead to prevent it from being dragged away. So directly the fish notices the bulls' meat he rushes to seize it. Then, finding what he wants, all at once with jaws agape he recklessly tugs at the deadly meal which has come to him. Next, this glutton, drawn on by his enjoyment, is spitted on the aforesaid hook before he knows it, and in his eagerness to escape the disaster that has befallen him he agitates and shakes the rope with all his might. So when the hunter is aware of this he is filled with joy; he leaps from his seat, abandons his labours in the river and his watery pursuits, and like an actor in a play changing his mask, sets his pair of oxen or horses in motion, and there ensues a trial of strength between the monster and the beasts of burden. For the creature bred in the Ister exerts a downward pull with all the strength at his command, while the pair of beasts pulling in the opposite direction makes the rope taut. But it avails the fish nothing: at any rate he is defeated in the tug-of-

The Sheat-fish

[3] *περιείρας.*
[4] *Corrupt*: *εὖ τεθραμμένου Jac.*
[5] *Gow*: *αὐτόν* MSS, *H.*
[6] *Cobet*: *ἑλιττόμενος* MSS, *H.*

ἕλκεται κατὰ τῆς ᾐόνος. εἴποι ἂν Ὁμηρίδης δρυῶν στελέχη ἕλκειν ἡμιόνους τινάς, ὡς ἐπὶ τῇ Πατρόκλου ταφῇ Ὅμηρος ᾄδει ταῦτα δήπου τὰ ὑμνούμενα.

26. Ἔστι δὲ ἄρα τῷ Ἴστρῳ καὶ κόλπος οἷος βαθύτατος, καὶ ἔοικε τῇ θαλάττῃ τὴν πολλὴν περίοδον. καὶ μέντοι ⟨καὶ⟩ βάθους ⟨ὅτι⟩ [1] εὖ ἥκει ὅδε ὁ κόλπος καὶ ἐκεῖνο τεκμηριῶσαι ἱκανόν. αἱ ναῦς αἱ φορτίδες αἱ τὴν θάλατταν περῶσαι καὶ ἐνταῦθα κατακολπίζουσι,[2] πεφρίκασι δὲ καὶ τοῦτον ὡς θάλατταν, ὅταν ἀγριαίνηται ὑπὸ τῶν καταπνεόντων ἀνέμων ἐς κύματα ἐξαπτόντων τε αὐτὸν καὶ ἐκμαινόντων. πεφύκασι δὲ ἄρα ἐν αὐτῷ καὶ νῆσοι καὶ μέντοι καί τινες [3] τῆς ὄχθης ὑποδρομαὶ ἐς ἃς ἔστι καταφυγεῖν. ἀλλὰ καὶ ἀκταὶ καὶ ἄκραι προήκουσι, καὶ προσρήγνυται αὐταῖς καὶ περισχίζεται κλύδων ἄγριος, ἡνίκα ἂν [4] ἑαυτοῦ μάλιστα ὑποπλησθεὶς εἶτα ἐς τὴν θάλατταν οἱονεὶ στενοχωρούμενος ὠθῆται. φιλεῖ δὲ ἄρα δρᾶν τοῦτο ἤδη τρίτης [5] ὥρας φθινοπωρινῆς παραδραμούσης, ὑπαρχομένης δὲ τῆς χειμερίου, καὶ ἀκμάσας αὐτὸς πρόεισι πλημμυρῶν. πληθύοντα δὲ ἄρα βορρᾶς ἐπωθεῖ αὐτόν, καὶ ἐξάπτει κατιέναι ἄγριον. καὶ ὁ μὲν καταφέρει [6] ὡς ἐς πλοῦν ἀγώγιμον τὸν καθ' ἑαυτοῦ κρύσταλλον, ὁ δὲ ἀντιπίπτει ὁ βορρᾶς αὐτῷ καταπνέων σκληρὸν ⟨καὶ⟩ μάλα γε κρυμώδες. οὔκουν αὐτῷ [7] ἐκβάλλειν ἐς τὸ πέλαγος

[1] ⟨καὶ⟩ βάθους ⟨ὅτι⟩ *add.* H.
[2] κατακολπίζουσι χρείᾳ τῶν περιοικούντων δηλονότι.
[3] καί τινες μέντοι καί.
[4] δ' ἄν.

war, gives up, and is hauled ashore. A student of Homer might say that mules were hauling tree-trunks, as Homer sings [*Il.* 23. 110] in the celebrated tale of the funeral of Patroclus.

26. There is also in the Ister a bay of immense depth and like the sea in its wide compass. Moreover that this bay attains a considerable depth is sufficiently proved by the following fact: merchant vessels which cross the sea put in to this bay and, when the bay is angered by the winds that blow and lash it into waves and drive it mad, are just as afraid of it as they are of the sea. And there are also islands in it, and even creeks along the shore into which one can run for safety. There are besides, promontories and capes running out, on which the waves in their fury dash and burst whenever the river at its very fullest is, as it were, forced into a narrow space as it presses on to the sea. This commonly occurs when the third autumnal season [a] is past and the winter season is setting in and the river is running in full flood. And as it rises the north wind urges it forward and causes it to descend in fury. And the stream carries down the ice it contains as though for an easy voyage.[b] But the north wind opposes it with its violent and icy blasts : it does not permit it to discharge into the

The Ister in winter

[a] That is, φθινόπωρον.
[b] Or 'for a voyage of commerce'?

---

[5] *Ges* : τετάρτης MSS, *del. H.*
[6] καταφέρει μάλα γε κρυμῶδες καὶ σκληρὸν ὡς εἰς . . . καταπνέων σκληρόν MSS ; ⟨καὶ⟩ *add. H, transposing* μάλα γε κρ.
[7] αὐτόν.

⟨τὴν⟩ [1] ὠδῖνα ὡς ἂν εἴποις ἐπιτρέπει, ἀλλ' ἀναχέων [2] καὶ ἀνωθούμενος ἵστησιν.[3] ὁ κρύσταλλος οὖν ἐπινηχόμενός [4] τε καὶ ἀναστελλόμενος ἐς βάθος χωρεῖ καὶ ἁδρύνεται ἐς πολύ· καὶ ἐντεῦθεν ὑπορρεῖ μὲν τοῦ Ἴστρου [5] τὸ γνήσιον ὕδωρ ὁδοῖς ὡς ἂν εἴποις κρυπταῖς, τὸ δὲ ἐπίκτητόν οἱ καὶ νόθον ἐπίκειται πεδίου δίκην, καὶ κατὰ τούτου τηνικάδε τῆς ὥρας ὁδοιποροῦσιν οἱ τῇδε ἄνθρωποι κατὰ ζεύγη καὶ μόνιπποι.[6] ὅπως [7] μὲν οὖν ἐλέγχει τε καὶ βασανίζει τὴν πῆξιν τοῦ ποταμοῦ τοῦδε καὶ τοῦ Θρακίου Στρυμόνος τὸ πονηρόν τε καὶ δολερὸν θηρίον ἡ ἀλώπηξ, ἀνωτέρω εἶπον· ὁ δ' οὖν κρύσταλλος ὁ ἐν τῷ Ἴστρῳ καὶ νηὶ φορτηγῷ κατὰ ῥοῦν φερομένῃ περιτραφεὶς εἶτα ἐπέδησεν αὐτήν, καὶ οὔτε ἱστίων ἡπλωμένων ἔτι δεῖ, οὔτε ⟨ὁ⟩ [8] πρῳράτης τὰ πρόσω βλέπει, οὔτε ὁ τῆς νεὼς ἄρχων ἐπιστρέφει τοὺς οἴακας· πεπήγασι γάρ, ἐπεὶ καὶ τὸ πᾶν σκάφος τῷ περικειμένῳ κατείληπται δεσμῷ, καὶ ἔοικεν οὐ μὰ Δία νηί,[9] οὐ γὰρ ἔτι τοῖς κύμασι τύπτεται, ἀλλὰ ἐν πολλῷ ⟨τῷ⟩ [10] πεδίῳ λόφῳ τινὶ ἀνεστῶτι ἢ καὶ νὴ Δία σκοπιᾷ ἄκρᾳ.[11] ἐνταῦθά τοι καὶ οἱ περίνεῳ καὶ οἱ ναῦται ἐκπηδῶσι καὶ κατὰ τοῦ ποταμοῦ θέουσι, καὶ ἁμάξας ἄγουσι καὶ τὸν φόρτον μετῆραν ἐπὶ τοῦ τέως ὕδατος. καὶ πάλιν μετὰ τὴν χειμέριον ὥραν τοῦ αὐτοῦ φερομένου σφοδρῶς φέρουσί [12] τε καὶ ἄγουσι τὰ ἄχθη [13]. μένει δὲ ἡ ναῦς ἔστ' ἂν

[1] ⟨τὴν⟩ *add. H.*
[2] ἀνακωχεύων *H.*
[3] *Reiske* : ἵησιν.
[4] ἀνειργόμενος *H.*
[5] ὁ Ἴστρος.
[6] *Jac* : μονίππους.
[7] *Jac* : ὁπόσα.
[8] ⟨ὁ⟩ *add. H.*
[9] *Gill* : νήσῳ.

sea what you might call its offspring, but causes it to overflow, resists it, and brings it to a halt. So the ice which is floating and checked sinks and solidifies to a great depth. In consequence the Ister's own water flows beneath, along what you might call hidden channels, while the newly acquired and alien surface resembles a plain, and at this season of the year the people thereabouts travel along it driving a pair or on horseback. Now the way in which that mischievous and crafty animal the fox tests and examines this river and the Strymon in Thrace to see if they are frozen, I have described earlier on.[a] Well, the ice on the Ister freezes hard even round a merchant vessel on its way downstream and imprisons it: it is no use to spread the sails; the man at the prow looks no more ahead; the ship's captain cannot move the rudders to and fro; they are fixed fast, for the whole vessel is caught in the surrounding fetters and looks, I declare, not like any ship, for it is no longer beaten by the waves, but like some hill rising from a wide expanse of plain or for all the world like some lofty watch-tower. Thereupon the passengers and the sailors jump out and hurry down the river and fetch wagons and transfer the cargo on to what was lately the water. Then again when the winter season is over and the river begins to flow strongly they still carry their loads. But the ship remains

Ships ice-bound

[a] See 6. 24.

---

[10] ⟨τῷ⟩ *add. H.*
[11] *Grasberger*: σκοπιᾷ ἢ ἄκρᾳ MSS, *H.*
[12] παραφέρουσι.
[13] ἄχθη καὶ τὸν Ἴστρον πατοῦσι βόες MSS; *Jac would place the words* καὶ . . . βόες *after* ὕδατος *above.*

ὑπανῇ μὲν τὰ τοῦ κρυμοῦ, τακῇ[1] δὲ ὁ κρύσταλλος καὶ λυθῇ, ἐλευθέρα δὲ τοῦ παραδόξου πείσματος ἡ ὁλκὰς ἀπολυθῇ. ἐνταῦθά τοι τοῦ καιροῦ καὶ οἱ ἁλιεῖς μακέλλας λαβόντες, ἔνθα αὐτοὺς ἄγει θυμὸς διακόπτουσι τὴν πῆξιν τοῦ ὕδατος, καὶ τάφρον κυκλοτερῆ ἐργάζονται κατιοῦσαν[2] ἐς τὸ ὕδωρ· εἴποις ἂν ἢ φρέατος εἶναι στόμα ἢ μεγίστου πίθου καὶ πάνυ γάστριδος. οὐκοῦν ἰχθύες πολλοὶ τὸν κρύσταλλον[3] διαδρᾶναι θέλοντες οἱονεὶ στέγην ἐπικείμενον[4] καὶ ποθοῦντες τὸ φῶς ἀσμένως ἐς τὸ ἀνεῳγμένον στόμιον ἐσνέουσι, καὶ γίνονται πλῆθος ἄμαχοι, καὶ ἐπωθοῦνται ἀλλήλοις, αἱροῦνται δὲ ἅτε ἐν βόθρῳ στενῷ ῥᾳδίως. καὶ πάρεστι λαβεῖν κυπρίνους τε καὶ κορακίνους ἄδην καὶ πέρκας καὶ ξιφίαν, ἀλλ' οὔπω μέγαν καὶ ἔτι τοῦ κέντρου τοῦ προμετωπιδίου ἄμοιρον· καὶ ἀντακαῖον, καὶ τοῦτον ἁπαλόν, ἐπεὶ οἵ γε μεγάλοι καὶ προήκοντες τὴν ἡλικίαν γένοιντο ἂν καὶ κατὰ τὸν θύννον τὸν μέγιστον. οὗτός τοι καὶ πιότατός ἐστι τὰς λαπάρας καὶ τὴν γαστέρα, καὶ φαίης ἂν ὑὸς οὔθατα[5] εἶναι θηλαζούσης βρέφη. δορὰν δὲ ἔχει τραχεῖαν, καὶ μέντοι καὶ τὰ δόρατα λεαίνουσι ταύτῃ δορυξόοι. ὑπὸ δὲ τῷ μυελῷ τοῦδε τοῦ ζῴου ἀρξάμενος ἐκ μέσης τῆς κεφαλῆς μέχρι[6] τῆς οὐρᾶς καθήκων ὑμὴν ὑγρὸς καὶ στενὸς ἔρχεται. τοῦτον οὖν πρὸς τὴν εἵλην αὖον ἐργασάμενος ἕξεις εἰ ἐθέλεις[7] μάστιγα[8] ὡς ἐλαύνειν ζεῦγος ἵππων· σκύτους[9] γὰρ ἤ τι[10] ἢ οὐδὲν διαφέρει. ἐς μέγεθος δὲ ἤδη προήκων, οὐκ ἂν αὐτὸν θεάσαιτό τις ὑπεκδυόμενον τοῦ κρυστάλλου καὶ ἐμπίπτοντα ἐς τὸν βόθρον, ἀλλ' ἢ πέτραν

[1] καμῇ.
[2] κατιόντες.
[3] κρυμόν.
[4] ἐπικειμένην.

stationary until the frost relaxes and the ice melts and is dissolved, and the merchant vessel, freed from its strange cable, is released.

Fishing in winter

At that season fishermen also take picks and hack through the ice wherever they feel inclined, and contrive a circular hole reaching down to the water. You would say that it was the mouth of a well or of a huge, very pot-bellied jar. Thereupon multitudes of fish wishing to escape from the ice which is pressing down upon them like a roof, and longing for the light, swim joyfully up to the opening that has been made, and come in crowds past numbering and jostle one another, and being in a confined hole are easily captured. And it is possible to catch carp and crow-fish in abundance and perch and the swordfish, though the last-named is not yet fully grown and is still without the frontal spike; sturgeon too, young and tender, for the large ones of mature age may be the size of the biggest tunny. The Sturgeon is extremely fat along the sides and the belly; you might say they were the dugs of a sow that was suckling its young. It has a rough skin and spear-makers actually polish their spear-shafts on it. Beneath the spinal marrow of this creature a supple, narrow membrane beginning at the middle of the head, runs down as far as the tail. Now if you let this dry in the sun you will obtain, should you wish it, a whip to drive a pair of horses with. For it differs hardly at all from a leather thong. When however the fish has grown to its full size one would not see it emerging from the ice and falling into the

The Sturgeon

---

[5] οὖρθρα MSS, οὖθαρ *edd.*
[6] καὶ μέχρι.
[7] θέλεις.
[8] *Reiske*: καὶ μάστιγα.
[9] *Jac*: ζεῦγος.
[10] τινι.

ὑπελθὼν πολυσκεπῆ [1] ἢ ἐν ἄμμῳ βαθείᾳ ἑαυτὸν ἐγκρύψας εἶτα ὑποθάλπει καὶ μάλα ἀγαπητῶς. δεῖται δὲ οὔτε πόας τηνικάδε οὔτε ἰχθύος ἐς βορὰν ἑτέρου, κρυμοῦ δὲ ὄντος ἀργὸς εἶναι ἐθέλει, καὶ τέρπεται [2] τῇ σχολῇ, καὶ τὴν ἑαυτοῦ πιμελὴν ἐσθίει, ὥσπερ οὖν καὶ [3] οἱ πολύποδες ἐν ἀθηρίᾳ τῶν πλεκτανῶν τῶν ἰδίων παρατραγόντες ἑαυτοὺς καὶ ἐκεῖνοι βόσκουσι. χειμῶνος δὲ λήγοντος καὶ ὑπαρχομένου ἦρος καὶ ἐλευθέρου τοῦ Ἴστρου ῥέοντος μισεῖ τὴν ἀργίαν καὶ ἀναπλεύσας ἐμφορεῖται τοῦ κατὰ τὸ ὕδωρ ἀφροῦ· πολὺς δὲ οὗτός ἐστι μορμύροντος τοῦ [4] ῥεύματος καὶ ὠθουμένου σφοδρότατα. ἐνταῦθά τοι καὶ ἁλίσκεται ῥᾳδίως, ἐλλοχώντων αὐτὸν τῶν ἁλιέων καὶ τὸ ἄγκιστρον ἐς τὸν ἀφρὸν καθιέντων σὺν τῇ ὁρμιᾷ. καὶ τὸ μὲν κρύπτεται ὑπὸ τῇ λευκότητι, καὶ ⟨ἡ⟩ [5] αἴγλη τοῦ χαλκοῦ εὐσύνοπτός [6] οἱ [7] οὐκ ἔστι, καὶ διὰ ταῦτά τοι περιχανὼν καὶ λάβρως σπῶν τοῦ προειρημένου σιτίου καταπίνει τὸν δόλον, καὶ ἀπόλωλεν ἐντεῦθεν ὅθεν τὰ πρῶτα ἐτρέφετο.

27. Ὄνομα φυτοῦ κυνόσπαστος (καλεῖται δὲ ἄρα καὶ ἀγλαόφωτις ἡ αὐτή· βούλομαι γὰρ ἐκτῖσαι χρέος ὑπομνησθείς) ὃ μεθ' ἡμέραν μὲν ἐν τοῖς ἄλλοις διαλέληθε καὶ οὐκ ἔστι πάνυ τι [8] σύνοπτον, νύκτωρ δὲ ἐκφαίνεται καὶ διαπρέπει, ὡς ἀστήρ· φλογώδης γάρ ἐστι καὶ ἔοικε πυρί. οὐκοῦν σημεῖόν τι ταῖς ῥίζαις παραπήξαντες αὐτῆς ἀπαλλάττονται, οὔτε τὴν χρόαν ἔχοντες μεθ' ἡμέραν εἰ μὴ τοῦτο δράσαιεν μνημονεῦσαι οὔτε μὴν τὸ εἶδος.

[1] πολυσκεπῆ εὗρεν.
[2] *Gill*: τρέφεται.
[3] καὶ ὥσπερ οὖν *H*.
[4] δὲ τοῦ.

hole, but either it slips beneath some all-sheltering rock or buries itself in deep sand and is only too glad to keep warm. And at that time it needs no vegetation, no other fish to eat, but prefers to remain inactive while the frost lasts, and is happy to be idle and consumes its own fat, just as octopuses also when unable to catch any prey nibble their own tentacles and feed off themselves. But when winter is over and spring is beginning and the Ister is flowing freely, it hates to be inactive and, swimming up to the surface, takes its fill of the foam on the water, and there is foam in abundance as the stream roars and boils in violent tumult. Then is the time when it is easily captured as the fishermen lie in wait for it and let down hook and line into the foam. The whiteness of the foam conceals the hook and the bright sheen of the bronze is invisible to the fish; hence, as it opens its jaws and takes a heavy draught of the aforesaid food, it swallows the bait and meets its death from the very thing that before sustained it.

The Peony, how plucked

27. There is a plant of the name of *Cynospastus* (it is also called *Aglaophotis* (peony): I have remembered and wish to fulfil my obligations [a]) which by daytime passes unnoticed among the rest and is hardly visible, but at night it becomes visible and shines out like a star, for it is of a fiery nature and like a flame. Therefore men plant some mark near the roots and then go away, for if they did not do this they would be unable by day to remember either the colour or even

[a] See above, ch. 24.

[5] ⟨ἡ⟩ *add. Jac.*
[6] *Reiske* : εὐσύνοπτον.
[7] *Jac:* οἷον.
[8] πάντῃ.

παρελθούσης δὲ τῆς νυκτὸς ἥκουσι, καὶ θεασάμενοι τὸ σημεῖον ὅπερ οὖν κατέλιπον καὶ γνωρίσαντες ἔχουσι συμβαλεῖν ὅτι ἄρα τοῦτο ἐκεῖνό ἐστιν οὗ καὶ δέονται, ἐπεί τοι τελέως [1] ὅμοιόν ἐστι τοῖς παρεστῶσι καὶ οὐδὲ ὀλίγον διαλλάττει αὐτῶν. οὐκ ἀνασπῶσι δὲ αὐτοὶ τὸ φυτὸν τόδε, ἢ οὐ χαιρήσουσι πάντως. οὔκουν οὔτε περισκάπτει τις οὔτε ἀνασπᾷ, ἐπεὶ καί, φασί, τὸν πρώτιστον ὑπ' ἀπειρίας τῆς κατ' αὐτὸ φύσεως προσαψάμενον οὐκ ἐς μακρὰν ἀπώλεσεν. ἄγουσιν οὖν κύνα νεανίαν ἡμερῶν ἀτροφήσαντα καὶ λιμώττοντα ἰσχυρῶς, καὶ τούτου σπάρτον ἐξάψαντες εὖ μάλα στερεὸν καὶ τῆς ἀγλαοφώτιδος τῷ κάτω στελέχει βρόχον τινὰ δύσλυτον προσαρτήσαντες ὡς οἷοί τέ εἰσι μακρόθεν, εἶτα τῷ κυνὶ προτιθέασι κρέα πάμπολλα ὀπτὰ κνίσης προσβάλλοντα· ὁ δὲ ὑπὸ τοῦ λιμοῦ φλεγόμενος καὶ στρεβλούμενος ὑπὸ τῆς κνίσης ἐπὶ τὰ προκείμενα ᾄττει [2] κρέα, καὶ ὑπὸ ῥύμης [3] αὐτόρριζον ἀνασπᾷ τὸ φυτόν. ἐπὰν δὲ ὁ ἥλιος ἴδῃ τὰς ῥίζας, ὁ κύων ἀποθνήσκει παραχρῆμα. θάπτουσι δὲ ἐν αὐτῷ τῷ χώρῳ [4] αὐτόν, καί τινας δράσαντες ἀπορρήτους ἱερουργίας καὶ τιμήσαντες τοῦ κυνὸς τὸν νεκρὸν ὡς ὑπὲρ αὐτῶν τεθνεῶτος εἶτα μέντοι προσάψασθαι τολμῶσι τοῦ φυτοῦ τοῦ προειρημένου, καὶ κομίζουσιν οἴκαδε. καὶ καταχρῶνταί φασιν ἐς πολλὰ καὶ λυσιτελῆ· ἐν δὲ τοῖς καὶ τὴν ἐκ τῆς σελήνης νόσον ἐνσκήπτειν τοῖς ἀνθρώποις λεγομένην ἰᾶσθαί φασιν αὐτήν, καὶ τῶν ὀφθαλμῶν τὸ πάθος, ὅπερ οὖν ὑγροῦ ἐπικλύσαντος καὶ παγέντος [5] ἀφαιρεῖ τὴν ὄψιν αὐτούς.[6]

[1] τοί γε ἄλλως.
[2] *Reiske*: ἔναντι.
[3] *Reiske*: ὑπ' ὀδύνης.
[4] χωρίῳ.

the appearance of the plant. But when the night is over they come and see the mark which they left and recognise it and are able to guess that this is the very plant that they need; for otherwise it is completely like the plants all round it, differing from them not one whit. But they themselves do not pull up this plant; if they did they would certainly regret it. Accordingly no one either digs round it or pulls it up, for (they say) the first man who in ignorance of its nature touched it, was destroyed by it shortly afterwards. And so they bring a strong dog that has not been fed for some days and is ravenously hungry and attach a strong cord to it, and round the stalk of the Peony at the bottom they fasten a noose securely from as far away as they can; then they put before the dog a large quantity of cooked meat which exhales a savoury odour. And the dog, burning with hunger and tormented by the savour, rushes at the meat that has been placed before it and with its violent movement pulls up the plant, roots and all. But when the sun sees the roots the dog immediately dies, and they bury it on the spot, and after performing some mysterious rites and paying honour to the dead body of the dog as having died on their behalf, they then make bold to touch the aforesaid plant and carry it home. It is useful, they say, for many purposes; for instance, it is said to cure the disease with which the moon is reputed to afflict men;[a] also that affliction of the eyes in which moisture floods them and then congeals and so robs them of their sight.[b]

[a] Known as σεληνιασμός, epilepsy.
[b] *I.e.* cataract, ὑπόχυσις.

[5] *ῥαγέντος.* [6] *Schn*: *αὐτοῖς.*

28. Κόχλος ἐστὶ θαλάττιος, μικρὸς μὲν τὸ μέγεθος, ἰδεῖν δὲ ὡραιότατος, καὶ ἐν θαλάττῃ τίκτεται τῇ καθαρωτάτῃ καὶ ἐν ταῖς ὑφάλοις πέτραις καὶ ἐν ταῖς καλουμέναις χοιράσιν. ὄνομα δὲ νηρίτης ἐστὶν αὐτῷ, καὶ διαρρεῖ λόγος διπλοῦς ὑπὲρ τοῦδε τοῦ ζῴου, καὶ ἐς ἐμέ γε ἀφικέσθην ἄμφω τὼ λόγω, καὶ μέντοι καὶ διαμυθολογῆσαι μικρὰ ἄττα ἐν μακρᾷ τῇ συγγραφῇ οὐδὲν ἀλλ' ἢ διαναπαῦσαί τε τὴν ἀκοὴν καὶ ἐφηδῦναι τὸν λόγον. τῷ Νηρεῖ τῷ θαλαττίῳ, ὅνπερ οὖν ἀληθῆ τε καὶ ἀψευδῆ ἀκούομεν δεῦρο ἀεί, πεντήκοντα μὲν θυγατέρας τὴν Ὠκεανοῦ Δωρίδα Ἡσίοδος ᾄδει τεκεῖν· μέμνηται δὲ αὐτῶν καὶ Ὅμηρος ἐν τοῖς ἑαυτοῦ μέτροις. ἕνα δέ οἱ γενέσθαι παῖδα ἐπὶ ταῖς τοσαύταις θυγατράσιν ἐκεῖνοι μὲν [1] οὔ φασι, λόγοι δὲ θαλάττιοι ὑμνοῦσι. καὶ Νηρίτην αὐτὸν κληθῆναι λέγουσι καὶ ὡραιότατον γενέσθαι καὶ ἀνθρώπων καὶ θεῶν, Ἀφροδίτην δὲ συνδιαιτωμένην ἐν τῇ θαλάττῃ ἡσθῆναί τε τῷ Νηρίτῃ τῷδε καὶ ἔχειν αὐτὸν φίλον. ἐπεὶ δὲ ἀφίκετο χρόνος ⟨ὁ⟩ [2] εἱμαρμένος, ⟨καθ' ὃν⟩ [3] ἔδει τοῖς Ὀλυμπίοις ἐγγραφῆναι καὶ τήνδε τὴν δαίμονα τοῦ πατρὸς παρακαλοῦντος, ἀνιοῦσαν αὐτὴν ἀκούω καὶ τὸν ἑταῖρόν τε καὶ συμπαίστην τὸν αὐτὸν ἐθέλειν ἄγειν. τὸν δὲ οὐχ ὑπακοῦσαι λόγος ἔχει τοῦ Ὀλύμπου προτιμῶντα τὴν σὺν ταῖς ἀδελφαῖς καὶ τοῖς γειναμένοις διατριβήν. παρῆν δὲ ἄρα αὐτῷ καὶ ἀναφῦσαι πτερά, καὶ τοῦτο ἐγῷμαι δῶρον τῆς Ἀφροδίτης δωρουμένης· ὁ δὲ καὶ ταύτην παρ' οὐδὲν ποιεῖται τὴν χάριν. ὀργίζεται τοίνυν ἡ Διὸς παῖς, καὶ ἐκείνῳ μὲν ἐς τὸν κόχλον τόνδε ἐκτρέπει τὴν μορφήν, αὐτὴ δὲ αἱρεῖται ὀπαδόν τε καὶ

The Nerites: two myths

28. There is in the sea a shellfish with a spiral shell, small in size but of surpassing beauty, and it is born where the water is at its purest and upon rocks beneath the sea and on what are called sunken reefs. Its name is *Nerites*: two stories are in circulation touching this creature, and both have reached me; moreover the telling of a short tale in the middle of a lengthy history is simply giving the hearer a rest and sweetening the narrative. Hesiod sings [*Th.* 233] of how Doris the daughter of Oceanus bore fifty daughters to Nereus the sea-god, whom to this day we always hear of as truthful and unlying. Homer also mentions them in his poems [*Il.* 18. 38]. But they do not state that one son was born after all that number of daughters, though he is celebrated in mariners' tales. And they say that he was named Nerites and was the most beautiful of men and gods; also that Aphrodite delighted to be with Nerites in the sea and loved him. And when the fated time arrived, at which, at the bidding of the Father of the gods, Aphrodite also had to be enrolled among the Olympians, I have heard that she ascended and wished to bring her companion and play-fellow. But the story goes that he refused, preferring life with his sisters and parents to Olympus. And then he was permitted to grow wings: this, I imagine, was a gift from Aphrodite. But even this favour he counted as nothing. And so the daughter of Zeus was moved to anger and transformed his shape into this shell, and of her own accord chose in his place for her attendant and servant Eros, who also was young

[1] *μὲν οὖν.*
[2] ⟨*ὁ*⟩ *add. H.*
[3] ⟨*καθ' ὃν*⟩ *add. H.*

θεράποντα ἀντ' ἐκείνου τὸν Ἔρωτα, νέον καὶ τοῦτον καὶ καλόν, καί οἱ τὰ πτερὰ τὰ ἐκείνου δίδωσιν. ὁ δὲ ἄλλος λόγος ἐρασθῆναι βοᾷ Νηρίτου Ποσειδῶνα, ἀντερᾶν δὲ τοῦ Ποσειδῶνος, καὶ τοῦ γε ὑμνουμένου Ἀντέρωτος ἐντεῦθεν τὴν γένεσιν ὑπάρξασθαι. συνδιατρίβειν οὖν τά τε ἄλλα τῷ ἐραστῇ τὸν ἐρώμενον ἀκούω καὶ μέντοι καὶ αὐτοῦ ἐλαύνοντος κατὰ τῶν κυμάτων τὸ ἅρμα τὰ μὲν κήτη τἄλλα καὶ τοὺς δελφῖνας καὶ προσέτι καὶ τοὺς Τρίτωνας ἀναπηδᾶν ἐκ τῶν μυχῶν καὶ περισκιρτᾶν τὸ ἅρμα καὶ περιχορεύειν,[1] ἀπολείπεσθαι δ' οὖν [2] τοῦ τάχους τῶν ἵππων πάντως [3] καὶ πάντη· μόνα δὲ ἄρα τὰ παιδικά οἱ παρομαρτεῖν καὶ μάλα πλησίον, στόρνυσθαι δὲ [4] αὐτοῖς καὶ τὸ κῦμα καὶ διίστασθαι τὴν θάλατταν αἰδοῖ Ποσειδῶνος· βούλεσθαι γὰρ τῇ τε ἄλλῃ τὸν θεὸν εὐδοκιμεῖν τὸν καλὸν ἐρώμενον καὶ οὖν καὶ τῇ νήξει [5] διαπρέπειν. τὸν δὲ Ἥλιον νεμεσῆσαι τῷ τάχει τοῦ παιδὸς ὁ μῦθος λέγει, καὶ ἀμεῖψαί οἱ τὸ σῶμα ἐς τὸν κόχλον τὸν νῦν,[6] οὐκ οἶδα εἰπεῖν ὁπόθεν ἀγριάναντα· οὐδὲ γὰρ ὁ μῦθος λέγει. εἰ δέ τι χρὴ συμβαλεῖν ὑπὲρ τῶν ἀτεκμάρτων, λέγοιντ' ἂν [7] ἀντερᾶν Ποσειδῶν καὶ Ἥλιος. καὶ ἠγανάκτει μὲν ἴσως ὁ Ἥλιος ὡς ἐν θαλάττῃ φερομένῳ,[8] ἐβούλετο δὲ [9] αὐτὸν οὐκ ἐν τοῖς κήτεσιν ἀριθμεῖσθαι, ἀλλ' ἐν ἄστροις φέρεσθαι. καὶ τὼ μὲν μύθω ἐς τοσοῦτον ἐληξάτην· ἐμοὶ δὲ τὰ ἐκ τῶν θεῶν ἵλεα ἔστω, καὶ τά γε παρ' ἐμοῦ ἔστω πρὸς αὐτοὺς εὔστομα. εἰ δέ τι θρασύτερον εἴρηται τοῖς μύθοις, ἐκείνων τὸ ἔγκλημα.

[1] περιχορεύειν, ὃ καὶ Ὅμηρος ἐν Ἰλιάδι [13.27] λέγει ἡμῖν.
[2] γοῦν.
[3] πάντας.
[4] δὲ ἄρα.
[5] τὴν ἕξιν MSS, νῆξιν *Schn.*

and beautiful, and to him she gave the wings of Nerites.

But the other account proclaims that Poseidon was the lover of Nerites, and that Nerites returned his love, and that this was the origin of the celebrated Anteros (mutual love). And so, as I am told, for the rest the favourite spent his time with his lover, and moreover when Poseidon drove his chariot over the waves, all other great fishes as well as dolphins and tritons too, sprang up from their deep haunts and gambolled and danced around the chariot, only to be left utterly and far behind by the speed of his horses; only the boy favourite was his escort close at hand, and before them the waves sank to rest and the sea parted out of reverence to Poseidon, for the god willed that his beautiful favourite should not only be highly esteemed for other reasons but should also be pre-eminent at swimming.

But the story relates that the Sun resented the boy's power of speed and transformed his body into the spiral shell as it now is: the cause of his anger I cannot tell, neither does the fable mention it. But if one may guess where there is nothing to go by, Poseidon and the Sun might be said to be rivals. And it may be that the Sun was vexed at the boy travelling about in the sea and wished that he should travel among the constellations instead of being counted among sea-monsters. Thus far the two fables; but may the gods be good to me, and for my part let me observe a religious silence regarding them. But if my fables have said anything overbold, the fault must be laid to their charge.

---

[6] νοῦν. [7] *Jac*: λέγονται.
[8] *Jac*: φερόμενος. [9] τε.

29. Ἔνθα ὁ Τάναρος [1] ποταμὸς καὶ ὁ Ἠριδανὸς συμβάλλετον (οὗτος μὲν καὶ διὰ δόξης ἰὼν καὶ κλέους, ἐκεῖνος δὲ οὐ πάνυ τι [2] γνώριμος) ἐνταῦθά τοι θῆραι ναὶ μὰ Δία ἰχθύων ἴδιαι καὶ ἐς ἐμὲ ἥκουσαι μέτροις Μυτιληναίου [3] ἀνδρός, ὃν ᾔδειν καὶ αὐτός, μηδὲ ἐξ ἡμῶν ἀγέραστοι γενέσθωσαν τῷ λόγῳ τῷδε. πεπεδημένων αὐτοῖς ὑπὸ κρυστάλλου τῶν ῥευμάτων ὅσοι περιοικοῦσιν αὐτοὺς τῇ μὲν ὥρᾳ τῇ χειμερίῳ ἀροῦσί τε καὶ σπείρουσι· καὶ γάρ πως καὶ εὔγεων χῶρον κεκλήρωνται. εἶτα ὑπαρχομένου τοῦ ἦρος, τῶν ῥευμάτων τῶν προειρημένων δι' ἣν αἰτίαν εἶπον ἔτι ἑστώτων, κολπώδη τινὰ τόπον προαιροῦνται οἱ γεωργοὶ τέως, νῦν δὲ ἁλιεῖς, καὶ περιτέμνουσι τοῦτον εὖ μάλα τεθηγμένοις πελέκεσι, καὶ τὸ ὕδωρ ἀναφαίνεται περιφερὲς κατὰ τέλμα· οὐ μὴν πλησίον ἔτι τῆς ὄχθης κόπτουσιν, ἀλλὰ ἐῶσι τὸν κρύσταλλον ὡς ἐξ ἀρχῆς ἐνετράφη. περιβάλλουσιν οὖν τῷ χώρῳ τῷ γεγυμνωμένῳ πλατὺ δίκτυον, καὶ μέντοι καὶ περιβάλλουσιν αὐτῷ [4] κάλων ἁδρότερον. καὶ τοῦτό γε τὸ δίκτυον ἐπισπῶσιν ἄνδρες ἐπὶ τῆς ὄχθης ἑστῶτες, καὶ ἁλιεῖς καὶ ἄλλοι· καὶ μέντοι ⟨καὶ⟩ [5] τὴν τῶν ἰχθύων ἅλωσιν θεῶνται πολλοὶ τῆς τέχνης οὐκ ἐπαΐοντες, ψυχαγωγία δέ τις ὕπεισιν αὐτούς. ὅταν γε μὴν ἀγόμενοι τῆς ὄχθης πλησίον ἀφίκωνται, τηνικαῦτα καὶ τὸν ἐνταῦθα τέμνουσι κρύσταλλον οἱ ἔξωθεν ὑδροθῆραι· τῇ γάρ τοι θήρᾳ ἐνέχονται, καὶ ἀναστέλλουσι [6] τοῖς ἰχθύσι τὸν ἔξω πόρον. τούτου δὲ οὕτω γενομένου πλῆρες ἰχθύων τὸ δίκτυον ἐκεῖνο τὴν περιτμηθεῖσαν ἐπωθεῖ τοῦ

[1] *Jac*: Ταίναρος.
[2] πάντη.
[3] Μιτυληναίου.
[4] *Oud*: αὐτό.

29. At the spot where the Tanarus[a] and the Eridanus meet (the latter has achieved renown and fame, whereas the former is hardly known at all) an altogether peculiar manner of fishing is in vogue; it has come to my knowledge through the poems of a man of Mytilene, an acquaintance of my own, and must not pass without a tribute in my narrative.

Winter fishing in the Eridanus

When the rivers have become ice-bound those who live in their neighbourhood plough and sow in the winter season, for it is their lot to possess a fertile land. Then at the beginning of spring while the aforesaid rivers are still immobile for the reason that I explained, the erstwhile farmers now fishermen select some spot like a bay and with well-sharpened hatchets cut round it so that a circle of water, like a pond, appears. They do not however cut close to the bank as yet but leave the ice as it froze originally. So then they throw a wide net round the space which they have laid open, and round the net a stoutish rope. This net is drawn in by men standing on the shore, fishermen and others, and there are many who though they know nothing of the art, watch the fish being caught: they feel a certain fascination in it. But as the men are drawn in[b] and approach the bank, then the fishermen on the dry land cut the ice there also, for they have an interest in the capture and try to prevent the fish from escaping. When this has been done as described, the net, full of fish, pushes the block of ice

[a] Mod. Tanaro; an important tributary of the Po, which it joins just below Valenza in Piedmont.

[b] *I.e.* the men standing on the island of ice, as explained in the following sentence.

[5] ⟨*καί*⟩ *add. H.* [6] *Reiske*: ἀναστέλλονται.

κρυστάλλου πέτραν καὶ συνεπάγει,[1] καὶ οἵ γε ἐφεστῶτες ἁλιεῖς αὐτῇ ἐοίκασιν ἐπὶ νήσου φέρεσθαι πλωτῆς. ἴδια μὲν δὴ καὶ ταῦτα ἰχθύων τῶν ἐκεῖσε καὶ θήραις ἑτέραις οὐκ ἂν εἰκασμένα. δώσει δὲ Ὅμηρος εἰπεῖν μοι ὅτι καὶ διπλοῦν αἱροῦνται μισθὸν οἵδε οἱ ἄνδρες, τὸν μὲν ἐκ τοῦ ποταμοῦ, τὸν δὲ ἐκ τῆς γῆς, ὡς τοὺς αὐτοὺς εἶναι καὶ ναύτας καὶ γεωργούς.

[1] *Ges*: συνεπάγη.

that has been cut round and draws it along with it, while the fishermen who are standing on the block look as if they were being carried along on a floating island. Such is the peculiar method of catching the fish there and quite unlike any other. And Homer will allow me to say that these men earn a double wage [*Od.* 10. 84], one from the river and another from the land, since the same men are both mariners and farmers.

BOOK XV

ΙΕ

1. Θήραν ἰχθύων Μακέτιν ἀκούσας οἶδα, καὶ ἥδε ἡ θήρα ἐστί. Βεροίας τε καὶ Θεσσαλονίκης μέσος ῥεῖ ποταμὸς ὄνομα Ἀστραῖος. εἰσὶν οὖν ἐνταῦθα ἰχθύες τὴν χρόαν κατάστικτοι· τίνας ⟨δὲ⟩[1] αὐτοὺς οἱ ἐπιχώριοι καλοῦσι, Μακεδόνας ἐρέσθαι λῷόν ἐστιν. οὐκοῦν οὗτοι ποιοῦνται τροφὴν μυίας ἐπιχωρίους ἐν τῷ ποταμῷ πετομένας οὐδέν τι ταῖς ἀλλαχόθι μυίαις προσεικασμένας οὐδὲ μὴν σφηκῶν ὄψει παραπλησίας, οὐδ' ἂν εἴποι τις ταῖς καλουμέναις ἀνθηδόσι[2] τὴν μορφὴν εἰκότως ἂν ἀντικρίνεσθαι τοῦτο τὸ ζῷον οὐδὲ ταῖς μελίτταις αὐταῖς· ἔχει[3] δέ τινα τῶν προειρημένων ἑκάστου μοῖραν ἰδίαν. ἔοικεν[4] οὖν τὸ μὲν θράσος μυίᾳ,[5] τὸ δὲ μέγεθος εἴποις ἂν ἀνθηδόνα, σφηκὸς δὲ τὴν χρόαν ἀπεμάξατο, βομβεῖ δὲ ὡς αἱ μέλιτται. καλοῦσι δὲ ἵππουρον αὐτὴν πᾶν ὅσον ἐστὶν[6] ἐπιχώριον. ἐκζητοῦσιν[7] οὖν ἐπικείμεναι[8] τῷ ῥεύματι τροφὴν τὴν ἑαυταῖς[9] φίλην, οὐ μὴν δύνανται τοὺς ὑπονηχομένους[10] λαθεῖν ἰχθύας. ὅταν οὖν[11] αὐτῶν[12] ἐπιπολάζουσαν τὴν μυῖαν θεάσηταί τις, ἡσυχῇ καὶ ὕφυδρος νέων ἔρχεται,

[1] ⟨δέ⟩ *add. H.*
[2] *Ges* : ἡμέρεσι.
[3] ἔχουσι.
[4] ἐοίκασιν.
[5] μυίαις.
[6] *Schn* : εἰσίν.
[7] *Jac* : ἐκδιαιτῶσιν.
[8] ἐπικείμενοι.
[9] ἑαυτοις.
[10] *Abresch* : ἐπινηχομένους.

BOOK XV

1. I have heard and can tell of a way of catching fish in Macedonia, and it is this. Between Beroea and Thessalonica there flows a river called the Astraeus.[a] Now there are in it fishes of a speckled hue, but what the natives call them, it is better to enquire of the Macedonians. Now these fish feed upon the flies of the country which flit about the river and which are quite unlike flies elsewhere; they do not look like wasps, nor could one fairly describe this creature as comparable in shape with what are called *Anthêdones* (bumble-bees), nor even with actual honey-bees, although they possess a distinctive feature of each of the aforesaid insects. Thus, they have the audacity of the fly; you might say they are the size of a bumble-bee, but their colour imitates that of a wasp, and they buzz like a honey-bee. All the natives call them *Hippurus*.[b] These flies settle on the stream and seek the food that they like; they cannot however escape the observation of the fishes that swim below. So when a fish observes a Hippurus on the surface it swims up noiselessly under water for fear of disturbing the surface and to

Fly-fishing in Macedonia

[a] Astraeum is the name of a town, but no river Astraeus is known; presumably the Axius is intended.

[b] This is one of the species *Stratiomys*, known as 'Soldier flies.'

---

[11] γοῦν. [12] αὐτοῦ.

κινῆσαι τὸ ἄνω δεδοικὼς ὕδωρ, ἵνα μὴ σοβήσῃ τὸ θήραμα. εἶτα ἐλθὼν πλησίον κατὰ τὴν σκιὰν αὐτῆς, ὑποχανὼν κατέπιε τὴν μυῖαν, ὡς οἶν ἐξ ἀγέλης [1] λύκος ἁρπάσας ἢ χῆνα ἐξ αὐλῆς ἀετός· καὶ τοῦτο δράσας ὑπεσῆλθε τὴν φρίκην. ἴσασιν οὖν οἱ ἁλιεῖς τὰ πραττόμενα, καὶ ταῖσδε μὲν ταῖς μυίαις ἐς δέλεαρ τῶν ἰχθύων χρῶνται οὐδὲ ἕν· ἐὰν γὰρ αὐτῶν προσάψηται χεὶρ ἀνθρωπίνη, ἀφήρηνται μὲν τὴν χρόαν τὴν συμφυῆ, μαραίνεται δὲ αὐταῖς τὰ πτερὰ καὶ ἄβρωτοι γίνονται τοῖς ἰχθύσι, καὶ διὰ ταῦτα οὐ προσίασιν αὐταῖς, ἀπορρήτῳ φύσει τὰς ᾑρημένας μεμισηκότες· σοφίᾳ δ' οὖν περιέρχονται τοὺς ἰχθῦς ὑδροθηρικῇ, δόλον αὐτοῖς ἐπινοήσαντες οἷον. τῷ ἀγκίστρῳ περιβάλλουσιν ἔριον φοινικοῦν, ἥρμοσταί τε τῷ ἐρίῳ δύο πτερὰ ἀλεκτρυόνος ὑπὸ τοῖς καλλαίοις [2] πεφυκότα καὶ κηρῷ τὴν χρόαν προσεικασμένα· [3] ὀργυιᾶς δὲ ὁ κάλαμός ἐστι, καὶ ἡ ὁρμιὰ δὲ τοσοῦτον ἔχει τὸ μῆκος. καθιᾶσιν οὖν τὸν δόλον, ἑλκόμενος δὲ ὑπὸ τῆς χρόας ὁ ἰχθὺς καὶ οἰστρώμενος ἀντίος ἔρχεται, καὶ θοίνην ὑπολαμβάνων ἐκ τοῦ κάλλους τῆς ὄψεως ἕξειν θαυμαστήν, εἶτα μέντοι περιχανὼν ἐμπαλάσσεται τῷ ἀγκίστρῳ, καὶ πικρᾶς τῆς ἑστιάσεως ἀπολέλαυκεν ᾑρημένος.

2. Οἱ θαλάττιοι κριοί, ὧνπερ οὖν ὄνομα μὲν ἐς τοὺς πολλοὺς διαρρεῖ, ἱστορία δὲ οὐ πάνυ τι [4] σαφής, εἰ μὴ [5] ὅσον χειρουργίᾳ [6] δείκνυται, χειμάζουσι μὲν περὶ τὸν Κύρνειόν τε καὶ Σαρδῷον

[1] ἀγελῶν. [2] *Reiske*: καλλέοις.
[3] παρεικασμένα. [4] πάντη.
[5] εἰ μή] ἤ. [6] γραφῇ χειρουργία καὶ πλάσματι.

avoid scaring its prey. Then when close at hand in the fly's shadow it opens its jaws and swallows the fly, just as a wolf snatches a sheep from the flock, or as an eagle seizes a goose from the farmyard. Having done this it plunges beneath the ripple. Now although fishermen know of these happenings, they do not in fact make any use of these flies as baits for fish, because if the human hand touches them it destroys the natural bloom; their wings wither and the fish refuse to eat them, and for that reason will not go near them, because by some mysterious instinct they detest flies that have been caught. And so with the skill of anglers the men circumvent the fish by the following artful contrivance. They wrap the hook in scarlet wool, and to the wool they attach two feathers that grow beneath a cock's wattles and are the colour of wax. The fishing-rod is six feet long, and so is the line. So they let down this lure, and the fish attracted and excited by the colour, comes to meet it, and fancying from the beauty of the sight that he is going to have a wonderful banquet, opens wide his mouth, is entangled with the hook, and gains a bitter feast, for he is caught.[a]

2. Ram-fishes,[b] whose name has a wide circulation, although information about them is not very definite except in so far as displayed in works of art, spend the winter near the strait between Corsica and

The Ram-fish

[a] This is the first clear mention of fishing with an artificial fly. But see 12. 43n. Martial, over a hundred years before, had referred to the use of a fly (5. 18. 8 *quis nescit | avidum uorata decipi scarum musca?*), but it need not have been artificial.

[b] ' An unknown sea-monster. . . . From the second part of the story κριὸς has been conjectured to be . . . perhaps . . . the Killer Whale ' (Thompson, *Gk. fishes*).

πορθμόν, καὶ φαίνονταί γε καὶ ἔξαλοι. περινήχονται δὲ ἄρα αὐτοὺς καὶ δελφῖνες μεγέθει μέγιστοι. ὁ τοίνυν ἄρρην κριός, λευκὴν τὸ μέτωπον ταινίαν ἔχει περιθέουσαν (εἴποις ἂν Λυσιμάχου τοῦτο διάδημα ἢ Ἀντιγόνου ἤ τινος τῶν ἐν Μακεδονίᾳ βασιλέων ἄλλου)· κριὸς δὲ θῆλυς, ὡς οἱ ἀλεκτρυόνες τὰ κάλλαια,[1] οὕτω τοι καὶ οὗτος ὑπὸ τῇ δέρῃ ἠρτημένους πλοκάμους ἔχει. ἁρπάζει δὲ ἄρα τοῖνδε τοῖν κριοῖν ἑκάτερος νεκρὰ [2] σώματα, καὶ ποιεῖται τροφὴν αὐτά. ἀλλὰ καὶ ζῶντας ἁρπάζει,[3] καὶ τῷ τῆς νήξεως κλύδωνι, πολὺς ὢν καὶ ὑπέρογκος, καὶ ναῦς περιτρέπει, χειμῶνα αὐταῖς ἐξ ἑαυτοῦ τοσοῦτον [4] ἐργασάμενος. ἁρπάζει δὲ καὶ ⟨τοὺς⟩ [5] ἀπὸ γῆς ἑστῶτας τῆς πλησίον. λέγουσι δὲ οἱ τὴν Κύρνον κατοικοῦντες, νεὼς διεφθαρμένης ἐν χειμῶνι ἄνδρα εὖ μάλα νηκτικὸν πολλὴν θάλατταν διανύσαντα λαβέσθαι τινὸς ἄκρας σφίσιν ἐπιχωρίου, καὶ ἀνελθόντα ἑστάναι καὶ μάλα ἀδεῶς, ⟨ὡς⟩ [6] ἤδη κινδύνων ἁπάντων ἐλεύθερον γενόμενον καὶ ἐν ἀδείᾳ τοῦ ζῆν καὶ ἐξουσίᾳ ὄντα. κριὸν οὖν παρανηχόμενον θεάσασθαι τὸν ἑστῶτα, καὶ ἀναφλεχθέντα ὑπὸ τοῦ λιμοῦ ἑλίξαι τε ἑαυτὸν καὶ κυρτῶσαι καὶ τῷ οὐραίῳ μέρει πολλὴν ἐλάσαι θάλατταν, εἶτα ἑαυτὸν μετεωρίσαι ἀρθέντα ὑπὸ τοῦ οἰδήσαντος κύματος, καὶ ἐπὶ τὴν ἄκραν φθάσαι ἀναταθέντα καὶ δίκην καταιγίδος ἢ στροβίλου ἁρπάσαι τὸν ἄνθρωπον. καὶ τὸ μὲν Κύρνειον ἅρπαγμά τε καὶ θήραμα τοῦ κριοῦ ἐς τοσοῦτον· μυθοποιοῦσι δὲ οἱ τὸν Ὠκεανὸν περιοικοῦντες [7]

[1] κάλλεα. [2] καὶ νεκρά.
[3] *Reiske*: καὶ ζῶντας ἁρπάζει ἀλλά. [4] *Jac*: τοῦτον.
[5] ⟨τούς⟩ *add. H.* [6] ⟨ὡς⟩ *add. Jac.* [7] οἰκοῦντες.

Sardinia and actually appear above water. And round about them swim dolphins of very great size. Now the male Ram-fish has a white band running round its forehead (you might describe it as the tiara of a Lysimachus [a] or an Antigonus or of some other king of Macedon), but the female has curls, just as cocks have wattles, attached below its neck. Male and female alike pounce upon dead bodies and feed on them, indeed they even seize living men, and with the wave caused by their swimming, since they are large and of immense bulk, they even overturn vessels, such a storm do they unaided raise against them. And they even snatch men standing on the shore close at hand. The inhabitants of Corsica tell how, when a ship was wrecked in a storm, a man who was a very strong swimmer managed to swim over a wide expanse of sea and to secure a hold on some headland in their country; he climbed out and stood there, all fear banished, for he was now free from all perils, with no anxiety for his life, his own master. Now a Ram-fish which was swimming by caught sight of him as he stood, and inflamed with hunger turned about, arched its back, and with its tail drove a great mass of water forward, and then rose as the swelling wave lifted it, and in a moment was carried up on to the headland and like a hurricane or whirlwind seized the man. So much for the Ram-fish's prey ravished off Corsica.

Those who live on the shores of Ocean tell a fable

[a] Lysimachus, c. 360–281 B.C., after the death of Alexander became ruler of Thrace and NW Asia Minor, later of Thessaly and Macedonia.—Antigonus I, 4th cent. B.C., general of Alexander, whom he aspired to succeed as ruler of his empire. Defeated and killed at the Battle of Ipsus, 301 B.C.

τοὺς πάλαι τῆς Ἀτλαντίδος βασιλέας τοὺς ἐκ τῆς Ποσειδῶνος σπορᾶς φέρειν ἐπὶ τῆς κεφαλῆς [1] τὰς τῶν κριῶν τῶν ἀρρένων ταινίας, γνώρισμα τῆς ἀρχῆς τοῦτο· καὶ τὰς ἐκείνων γαμετὰς τὰς βασιλίδας τοὺς πλοκάμους τῶν ἑτέρων καὶ ἐκείνας φορεῖν τῆς ἀρχῆς ἔλεγχον. ἔστι δὲ ἄρα τοὺς μυκτῆρας τὸ ζῷον τοῦτο καρτερὸν δεινῶς, καὶ πολὺ πνεῦμα ἐσπνεῖ, καὶ ἕλκει ἀέρα ἐφʼ ἑαυτὸν πάμπολυν, θηρᾷ δὲ καὶ τὰς φώκας τὸν τρόπον τοῦτον. αἱ μὲν συνεῖσαι πλησίον που κριὸν εἶναι καὶ φέρειν σφίσιν ὄλεθρον, ὡς ὅτι τάχιστα ἐκνήχονται καὶ παρελθοῦσαι ἐς τὴν γῆν καὶ τὰς ὑπάντρους πέτρας ὑπελθοῦσαι καταδύονται, οἱ δὲ αἰσθόμενοι τὴν φυγὴν μεταθέουσι καὶ ἀντίοι στάντες τοῦ ἄντρου κατὰ τὴν τοῦ χρωτὸς ὀσμὴν ἔνδον εἶναί σφισι τὴν ἄγραν συνιᾶσι, καὶ ὡς ἴυγγί τινι βιαιοτάτῃ ἕλκουσι ταῖς ῥισὶ τὸν μεταξὺ ἑαυτῶν καὶ τῆς φώκης ἀέρα. ἡ δὲ ὡς βέλος ἢ δόρατος αἰχμὴν ἐκκλίνει τὴν τοῦ πνεύματος προσβολήν, καὶ τὰ μὲν πρῶτα ὑπαναχωρεῖ, τελευτῶσα δὲ ὑπὸ τῆς βιαιοτάτης ἕλξεως ἐκσπᾶται τοῦ ἄντρου, καὶ ἄκουσα ἀκολουθεῖ, ὥσπερ οὖν ἱμᾶσί τισιν ἢ σχοίνοις κατατεινομένη, καὶ τέτριγε καὶ γίνεται τῷ κριῷ δεῖπνον. τάς γε μὴν ἐκπεφυκυίας τῶν μυκτήρων τοῦ κριοῦ τρίχας οἱ ταῦτα ἐξετάζειν δεινοὶ λέγουσιν ἐς πολλὰ ἀγαθάς.

3. Ἐν δὲ τῷ ῥίῳ τῷ Βιβωνικῷ θύννων ἔθνη μυρία. καὶ οἱ μὲν αὐτῶν κατὰ τοὺς σῦς εἰσι

[1] ἐπιφέρειν ἐπὶ τὰς κεφαλάς.

[a] Vibo was the Roman name for the Greek city Hipponium, on the W coast of the Bruttii. The gulf went by various

of how the ancient kings of Atlantis, sprung from the seed of Poseidon, wore upon their head the bands from the male Ram-fish, as an emblem of their authority, while their wives, the queens, wore the curls of the females as a proof of theirs. Now this creature has exceedingly powerful nostrils and inhales a great quantity of breath, drawing to itself an immense amount of air; and it hunts seals in the following manner. Directly the seals realise that a Ram-fish is somewhere close at hand, bringing destruction upon them, they swim ashore with all possible speed and pass over the land and plunge into the shelter of rocky caverns. But the Ram-fish perceive that they have fled and give chase, and as they face the cave they know from the smell of flesh that their prey is within, and, as though by some all-powerful spell, with their nostrils they draw in the air that intervenes between themselves and the seal. But the seal avoids the attack of the monster's breath, as it might an arrow or a spear-point, and at first withdraws, but is finally dragged out of the cave by the overmastering pull and follows against its will, just as though it were bound fast with thongs or cords, and shrieking provides the Ram-fish with a meal.

Those who are skilled at exploring these matters assert that the hairs which grow from the nostrils of the Ram-fish serve many purposes.

3. In the gulf of Vibo [a] there are shoals of Tunny past numbering, and some are, like hogs, solitary, and

The Tunny

names, Hipponiates sinus, Sinus Terinaeus or Napetinus or Vibonensis.

μονίαι καὶ καθ᾿ ἑαυτοὺς νήχονται μέγιστοι ὄντες, οἱ δὲ συνδυασθέντες· καί ἔστον κατὰ τοὺς λύκους συννόμω,[1] ἄλλοι δὲ κατ᾿ ἀγέλας, ὥσπερ οὖν τὰ αἰπόλια, πλατείας νομὰς νενεμημένοι. ἐπιτέλλοντος δὲ τοῦ Σειρίου καὶ τῆς ἀκτῖνος ἐνακμαζούσης ὀξύτατα, ἐπὶ τὸν Εὔξεινον στέλλονται· καὶ τοῦ κύματος αὐτοῖς ἐμπύρου δοκοῦντος, ἀλλήλοις συνυφασμένοι νήχονται, καὶ τῇ τῶν σωμάτων συναφῇ σκιᾶς τινος ἁμωσγέπως μεταλαγχάνουσιν.

4. Λέγει δὲ Δημόστρατος, ἀνὴρ ἁλιευτικῆς σοφίας ἐπιστήμων ναὶ μὰ Δία καὶ ἑρμηνεῦσαι χρηστός, εἶναί τινα ἰχθὺν ὡραῖον τὸ εἶδος, καὶ καλεῖσθαι σελήνην τοῦτον, τὸ μέγεθος βραχύν, κυανοῦν τὸ εἶδος, πλατὺν τὸ σχῆμα. τὰ νῶτα δέ οἱ λοφιὰς ἔχειν καὶ τάσδε ἀνατείνειν ὁ αὐτός φησι· μαλακὰς δὲ εἶναι αὐτὰς καὶ οὔτε ἀντιτύπους οὔτε τραχείας. ταύτας οὖν, ὅταν ὁ ἰχθὺς οὗτος ὑπονήχηται, διαιρεῖσθαι καὶ ἀποδεικνύναι κύκλου ἡμίτομον, καὶ εἶναι σελήνης ὅσα ἰδεῖν[2] τῆς διῃρημένης σχῆμα. καὶ ταῦτα μὲν Κύπριοι δὴ ἁλιεῖς φασιν· Δημοστράτου δὲ καὶ οὗτος ὁ λόγος. πληρουμένης μὲν τῆς σελήνης τὸν ἰχθὺν τόνδε ᾑρημένον πεπληρῶσθαί τε αὐτὸν καὶ πληροῦν καὶ τὰ δένδρα, ἐὰν τούτοις προσαρτήσῃς φέρων αὐτόν· ληγούσης δὲ ἄρα ὑποτετῆχθαι καὶ ἐκλείπειν, καὶ φυτοῖς προσαχθέντα αὐαίνειν αὐτά. ὀρυττομένων τε φρεάτων, ἐὰν μὲν τοῦ μηνὸς ὑποφαινομένου ἐς τὸ εὑρεθὲν ὕδωρ ἐμβάλῃ τις τὸν ἰχθὺν τοῦτον, ἀέναον ἔσται τὸ ὕδωρ καὶ οὐκ ἐπιλείψει ποτέ· εἰ δὲ ὑπολήγοντος, λήξει[3] τὸ ὕδωρ. καὶ μέντοι καὶ

[1] ἔστον . . . συννόμω] *Lorenz*: εἰς τὸν . . . σύννομον.

swim by themselves and are of very great size; others swim in couples or range together, as wolves do; others again swim in companies, just like herds of goats, ranging over wide feeding-grounds. But at the rising of the Dog-star and when the sun's rays are at their fiercest, they set out for the Euxine. And if the waves seem hot to them they swim interwoven with one another and by the contact of their bodies somehow contrive to get a certain amount of shade.

The 'Moon-fish'

4. Demostratus, a man deeply versed in fishing lore and excellent at expounding it, says that there is a certain fish of great beauty and that it is called the 'Moon-fish';[a] it is small, dark blue in colour, and flat in shape. He says too that it has dorsal fins which it raises, but that they are soft and neither unyielding nor rough. These fins, whenever the fish dives, open out and form a half-circle and present to the eye the shape of a half-moon. This is what the fishermen of Cyprus say, but Demostratus adds that if this fish is caught when the moon is at the full, it too is at the full, and causes trees to expand if one brings it and attaches it to them. But when the moon is waning the fish pines and dies, and if applied to plants they too wither. And when wells are being dug, if, as the moon is waxing, you throw this fish into the water which you have found, it will flow continually and never fail; if however you do this when the moon is waning, the flow will cease. In

[a] Unidentified.

[2] *ὅσα ἰδεῖν transposed by H*, καὶ ταῦτα μὲν ὅσα ἰδεῖν Κύπριοι.
[3] *οὐχ ἕξει.*

ἐς πηγὴν ὑπανατέλλουσαν εἰ τὸν αὐτὸν [1] ἐμβάλοις ἰχθύν, ἕξεις ἢ πεπληρωμένην αὐτὴν ἢ κενὸν τὸν χῶρον τὸ ἐντεῦθεν.

5. Ὅπως μὲν ἐσνέουσί τε ἐς τὴν Προποντίδα, καὶ ὅπως ἐκνέουσιν ἄρα οἱ θύννοι, οἶδα εἰπὼν ἄνω που τῶν λόγων τῶνδε· νοείτω δέ μοί τις ἐνταῦθα Ἡράκλειαν καὶ Τίον καὶ Ἄμαστριν,[2] πόλεις Ποντικάς. οὐκοῦν οἱ τόνδε τὸν χῶρον πάντα οἰκοῦντες τὴν τῶν θύννων ἐπιδημίαν ἴσασι κάλλιστα, καὶ μέντοι καὶ ἀφικνοῦνται τηνικάδε τοῦ ἔτους,[3] καὶ ὅπλα κατ᾽ αὐτῶν εὐτρέπισται πολλά, ναῦς καὶ δίκτυα καὶ σκοπιὰ ὑψηλή. σκοπιὰ δὲ ἄρα αὕτη ἐπί τινος αἰγιαλοῦ παγεῖσα ἀνέστηκεν ἐν περιωπῇ σφόδρα ἐλευθέρᾳ· καὶ αὐτῆς τὸ ποίημα περιηγήσασθαι ἐμοὶ μὲν οὐκ ἔστι μόχθος, σοὶ δὲ τῷ ἀκούοντι τῆς τῶν ὤτων τρυφῆς † τ ἐκειν.† [4] δύο πρέμνα ἐλάτης ὑψηλὰ δοκίσι πλατείαις διειλημμένα ἕστηκε, πυκναῖς ταύταις διυφασμέναις ⟨καὶ⟩ [5] ἀνελθεῖν τῷ σκοπῷ καὶ ἐπιβῆναι μάλα ἀγαθαῖς. αἱ δὲ ναῦς ἐρέτας ἑκάστη καὶ ἓξ ἔχει παρ᾽ ἑκάτερα νεανίας εὖ μάλα ἐρέττοντας· δίκτυα δὲ προμήκη, οὐ κοῦφα λίαν καὶ ἀνεχόμενα τοῖς φελλοῖς, μολίβῳ γε μὴν βριθόμενα μᾶλλον. ἀθρόαι δὲ ἄρα αἱ τῶνδε τῶν ἰχθύων ἀγέλαι ἐσνέουσιν. ἦρος δὲ ὑπολάμποντος καὶ τῶν ἀνέμων εἰρηναῖον ἤδη καταπνεόντων καὶ τοῦ ἀέρος φαιδροῦ τε ὄντος καὶ οἱονεὶ μειδιῶντος καὶ τοῦ κύματος κειμένου καὶ λείας οὔσης τῆς

[1] αὐτόν γε.
[2] Ἀμάστρειαν.
[3] τοῦ ἔτους τηνίκα.
[4] *Corrupt*: ἐκτικόν *Post*.
[5] ⟨καί⟩ *add. H.*

the same way if you throw this same fish into a bubbling spring, you will henceforward either find it full of water or you will find the spot empty.

5. I know that I have somewhere earlier on in this discourse[a] described how Tunny swim into and out of the Propontis. Just consider the cities along the Black Sea—Heraclea, Tium,[b] and Amastris. Now the inhabitants of the whole of that country know exactly of the coming of the Tunny, and at that season of the year[c] the fish arrive, and much gear has been got ready to deal with them, boats and nets and a high lookout-place. This lookout-place is fixed on some beach and stands where there is a wide, uninterrupted view. It is no trouble to me to explain, and you who listen should be pleased to hear, how it is constructed.[d] Two high pine-trunks held apart by wide balks of timber, are set up; the latter are interwoven in the structure at short intervals and are of great assistance to the watchman in mounting to the top. Each of the boats has six young men, strong rowers, on either side. The nets are of considerable length; they are not too light and so far from being kept floating by corks are actually weighted with lead, and these fish swim into them in shoals. And when the spring begins to shine and the breezes are blowing softly and the air is bright and as it were smiling and the waves are at rest and the sea smooth, the watcher, whose

Tunny-fishing in the Euxine

[a] See 9. 42.

[b] 'Tieum' in the atlases of Droysen, Grundy, and Perthes.

[c] About mid-July; see above, ch. 3.

[d] The text is defective and the translation provisional. Reading ἑκτικόν (conj. Post), translate 'it is capable of producing delight for the ears of you, *etc.*'

θαλάττης ὁ σκοπὸς ἰδὼν σοφίᾳ τινὶ ἀπορρήτῳ καὶ φύσει ὄψεως ὀξυωπεστάτῃ λέγει μὲν τοῖς θηραταῖς ὁπόθεν ἀφικνοῦνται· εἰ δέοι [1] γε μὴν πρὸς τὴν ἀκτὴν παρατεῖναι τὰ δίκτυα, καὶ τοῦτο ἐκδιδάσκει· εἰ δὲ ἐνδοτέρω, δίδωσιν ὥσπερ οὖν στρατηγὸς τὸ σύνθημα ἢ [2] χορολέκτης τὸ ἐνδόσιμον· ἐρεῖ [3] γε μὴν πολλάκις καὶ τὸν πάντα ἀριθμόν, καὶ οὐχ ἁμαρτήσεται τοῦ σκοποῦ. ἐκεῖνα δὲ ὁποῖα. ὅταν ἑαυτοὺς ὠθήσωσιν ἐς τὸ πέλαγος ἡ τῶν θύννων ἴλη, ὁ τὴν σκοπιὰν φυλάττων καὶ ἀκριβῶν τὴν τῶν προειρημένων ἱστορίαν καὶ μάλα ὀξὺ ἐκβοήσας λέγει διώκειν ἐκεῖθι καὶ τοῦ πελάγους ἐρέττειν εὐθύ.[4] οἱ δὲ ἐξαρτήσαντες ἐλάτης τῶν τὸν σκοπὸν ἀνεχουσῶν τῆς ἑτέρας σχοῖνον εὖ μάλα μακρὰν τῶν δικτύων ἐχομένην, εἶτα ἐπαλλήλοις [5] ταῖς ναυσὶν ἐρέττουσι κατὰ στοῖχον, ἔχονταί τε ἀλλήλων, ἐπεί τοι καὶ τὸ δίκτυον ἐφ' ἑκάστῃ διῄρηται. καὶ ἥ γε πρώτη τὴν ἑαυτῆς ἐκβαλοῦσα μοῖραν τοῦ δικτύου ἀναχωρεῖ, εἶτα ἡ δευτέρα δρᾷ τοῦτο καὶ ἡ τρίτη, καὶ δεῖ καθεῖναι τὴν τετάρτην· οἱ δὲ τὴν πέμπτην ἐρέττοντες ἔτι μέλλουσι, τοὺς δὲ ἐπὶ ταύτῃ οὐ χρὴ καθεῖναί πω· εἶτα ἐρέττουσι ἄλλοι ἄλλῃ καὶ ἄγουσι τοῦ δικτύου τὴν μοῖραν, εἶτα ἡσυχάζουσι. νωθεῖς δὲ ἄρα ὄντες οἱ θύννοι καὶ ἔργον τι τόλμης ἐχόμενον ἀδυνατοῦντες δρᾶσαι, πεπιεσμένοι μένουσί τε καὶ ἀτρεμοῦσιν· οἱ δὲ ἐρέται, ὡς ἁλούσης πόλεως, αἱροῦσιν ἰχθύων ποιητὴς ἂν εἴποι [6] δῆμον. οὐκοῦν, ὦ φίλοι Ἕλληνες, καὶ Ἐρετριεῖς ἴσασι ταῦτα καὶ Νάξιοι κατὰ κλέος, τῆς θήρας τῆς τοιαύτης μαθόντες ὅσα Ἡρόδοτός

[1] *Jac*: δὲ οἱ. [2] καί.
[3] *Jac*: αἱρεῖ. [4] ἰθύ.

mysterious skill and naturally sharp sight enable him to see the fish, announces to the fishermen the quarter from which they are coming: if on the one hand the men ought to spread their nets near the shore, he instructs them accordingly; but if closer in, like a general he gives the signal, or like a conductor, the keynote. And frequently he will tell the total number of fish and not be off the mark. And this is what happens. When the company of Tunnies makes for the open sea the man in the lookout who has an accurate knowledge of their ways shouts at the top of his voice telling the men to give chase in that direction and to row straight for the open sea. And the men after fastening to one of the pines supporting the lookout a very long rope attached to the nets, then proceed to row their boats in close order and in column, keeping near to one another, because, you see, the net is distributed between each boat. And the first boat drops its portion of the net and turns back; then the second does the same, then the third, and the fourth has to let go its portion. But the rowers in the fifth boat delay, for they must not let go yet. Then the others row in different directions and haul their part of the net, and then pause. Now the Tunny are sluggish and incapable of any action that involves daring, and they remain huddled together and quite still. So the rowers, as though it were a captured city, take captive —as a poet might say—the population of fishes. And so, my Grecian friends, the people of Eretria and Naxos know of these things by report, for they have learnt about this method of fishing all that

---

[5] *Jac*: ἐπ' ἀλλήλαις. [6] *Jac*: εἶπε.

τε καὶ ἄλλοι λέγουσι. τὰ δὲ ἔτι λοιπὰ τῆς θήρας ἀκούσεσθε ἄλλων.

6. Θύννων δὲ ἄρα ᾑρημένων τῇ θήρᾳ τῇ Ποντικῇ (ἐγὼ δ' ἂν φαίην ὅτι καὶ Σικελικῇ· ⟨ἢ⟩ [1] τί καὶ βουλόμενος ἂν τὸν ἡδὺν Θυννοθήραν ὁ Σώφρων ἔγραψε; πάντως δὲ καὶ ἀλλαχόθι ἄγραι τῶνδε τῶν θύννων εἰσί) τῷ ⟨οὖν⟩ [2] δικτύῳ ἤδη περιπλακέντων αὐτῶν Ποσειδῶνι πάντες εὔχονται ἀλεξικάκῳ [3] τηνικάδε. καὶ ὁπόθεν καὶ τόδε τοῦ δαίμονος τὸ ὄνομα, ἀξιῶ εἰπεῖν, ἐμαυτὸν καὶ μάλα γε ἀπαιτῶν τί καὶ βουλόμενοι ἐπεφήμισαν [4] τοῦτό οἱ· δέονται τοῦ Διὸς ἀδελφοῦ τοῦ θαλάττης κρατοῦντος μήτε τὸν ἰχθὺν τὸν ξιφίαν τῇδε τῇ ἴλῃ συνέμπορον ἀφικέσθαι μήτε μὴν δελφῖνα. ὁ γοῦν γενναῖος ξιφίας πολλάκις τὸ δίκτυον διέκειρε, καὶ ἀφῆκεν ἐλεύθερον διεκπαῖσαι [5] τὴν ἀγέλην. καὶ δελφὶς δὲ ἐπίβουλον δικτύῳ ζῷον· διατραγεῖν γάρ τοι δεινός ἐστιν.

7. Ὕεται ἡ Ἰνδῶν γῆ διὰ τοῦ ἦρος μέλιτι ὑγρῷ, καὶ ἔτι πλέον ἡ Πρασίων χώρα, ὅπερ οὖν ἐμπῖπτον ταῖς πόαις καὶ ταῖς τῶν ἑλείων καλάμων κόμαις, νομὰς τοῖς βουσὶ καὶ τοῖς προβάτοις

[1] ⟨ἢ⟩ *add. Jac.*
[2] ⟨οὖν⟩ *add. H.*
[3] πάντως ἀλεξικάκῳ.
[4] *Jac*: τοῦδε τοῦ δ. ἀξιῶ τὸ ὄ. . . . τε καὶ βουλόμενος ἐπευφήμασα MSS, *H.*
[5] *Jac*: διεκπέσαι.

[a] Pisistratus, driven from Athens, took refuge in Eretria, where he was joined by Lygdamis of Naxos among many

Herodotus [a] and others relate. What remains to be told of it you shall learn from others.

6. When Tunny have been caught by fishermen of the Euxine (and I might add off Sicily also, for what else had Sophron in mind when he wrote his delightful *Tunny-fisher*? Anyhow there are Tunny-fisheries in other places besides.)—when therefore they are safely enmeshed in the net, then is the time when everybody prays to Poseidon the Averter of Disaster. And as I ask myself the reason, I think it worth while to explain what induced them to attach the name 'Averter of Disaster' to the god. They pray to the brother of Zeus, the Lord of the Sea, that neither swordfish nor dolphin may come as fellow-traveller with the shoal of Tunny. At any rate your noble sword-fish has many a time cut through the net and allowed the whole company to break through and go free. The dolphin also is the net's enemy, for it is skilful at gnawing its way out.

Tunny-fishers and Poseidon

7. During the springtime in India it rains liquid honey, and especially in the country of the Prasii; [b] and it falls on the grass and on the leaves of reeds in the marshes, providing wonderful pasturage for

Honey-dew in India

others. He was induced to make a surprise attack upon the Athenians by the soothsayer Amphilytus, who delivered an oracle in which P. saw himself as a tunny-fisher waiting the moment to haul in his net and capture the fish; see Hdt. 1. 61–3.

[b] Prasiaea was reputed one of the richest and largest of the kingdoms of India. Its capital was Palibothra (mod. Patna) on the Ganges.

παρέχει θαυμαστάς, καὶ τὰ μὲν ζῷα ἑστιᾶται τὴν δαίτην ἡδίστην[1] (μάλιστα γὰρ ἐνταῦθα οἱ νομεῖς ἄγουσιν αὐτά, ἔνθα καὶ μᾶλλον ἡ δρόσος ἡ γλυκεῖα κάθηται πεσοῦσα), ἀντεφεστιᾷ[2] δὲ καὶ τὰ ζῷα τοὺς νομέας· ἀμέλγουσι γὰρ περιγλύκιστον γάλα, καὶ οὐ δέονται ἀναμίξαι αὐτῷ μέλι, ὅπερ οὖν δρῶσιν Ἕλληνες.

8. Ὁ δὲ Ἰνδὸς μάργαρος (ἄνω γὰρ εἶπον περὶ τοῦ Ἐρυθραίου) λαμβάνεται τρόπῳ τοιῷδε. πόλις ἐστὶν ἧς ἦρχε Σώρας ὄνομα, ἀνὴρ γένους βασιλικοῦ, ὅτε καὶ Βάκτρων ἦρχεν Εὐκρατίδης· ὄνομα δὲ τῇ πόλει Περίμουλα,[3] κατοικοῦσι δὲ αὐτὴν ἄνδρες Ἰχθυοφάγοι. ὅθεν ὁρμωμένους σὺν τοῖς δικτύοις φασὶ τοὺς[4] προειρημένους περιλαμβάνειν ἀγκῶσι μεγάλοις αἰγιαλοῦ κύκλον εὐμεγέθη· γίνεσθαι δὲ τὸν προειρημένον λίθον ἐκ κόγχης στρόμβῳ ἐμφεροῦς μεγάλῳ, νήχεσθαί τε κατὰ ἀγέλας τοὺς μαργάρους, καὶ ἔχειν ἡγεμόνας, ὡς ἐν τοῖς σμήνεσιν αἱ μέλιτται τοὺς καλουμένους βασιλέας· ἀκούω δὲ εἶναι καὶ τοῦτον διαπρεπῆ καὶ τὴν χρόαν καὶ τὸ μέγεθος. ἀγώνισμα δὲ ἄρα ποιοῦνται συλλαβεῖν αὐτὸν οἱ κολυμβηταὶ οἱ ὕφυδροι· τούτου γὰρ ᾑρημένου καὶ τὴν ἀγέλην αἱροῦσι πᾶσαν ἐρήμην ὡς ἂν εἴποι τις καὶ ἀπροστάτευτον οὖσαν· ἀτρεμεῖ γὰρ καὶ οὐκέτι πρόεισιν, οἷα δήπου ποίμνη τὸν νομέα ἀφῃρημένη κατά τινα τύχην ἐχθράν· ὁ δὲ διαφεύγει καὶ μάλα γε σοφῶς ἐξελίττει, καὶ προηγεῖται καὶ σώζει τὸ

[1] *Radermacher*: τήνδε τὴν ἡδίστην MSS, ἡ. τήνδε ἑστίασιν *H.*
[2] ἀνθεστιᾷ *H.*
[3] Περίμουδα.
[4] τούσδε τούς.

cattle and sheep. And the animals feast off the food with the greatest delight, for the shepherds make a point of leading them to spots where this honeyed dew falls more plentifully and settles. And they in return feast their herdsmen, for the milk which the latter draw is of the utmost sweetness and they have no need to mix honey with it as the Greeks do.

Pearl-fishing in the Indian Ocean

8. The Pearl-oyster of India (I have spoken earlier on of the one in the Red Sea) [a] is obtained in the following manner. There is a city of which one Soras by name was ruler, a man of royal lineage, at the time when Eucratides was ruler of Bactria.[b] And the name of the city is Perimula,[c] and it is inhabited by Ichthyophagi (fish-eaters). These men, it is said, set out from there with their nets and draw a ring of wide embrace round a great circle of the shore. The aforesaid stone is produced from a shell resembling a large trumpet-shell, and the Pearl-oysters swim in shoals and have leaders, just as bees in their hives have 'kings,' as they are called. And I have heard that the 'leader' too is conspicuous for his colour and his size. Now divers beneath the waters make it their special aim to capture him, for once he is caught they catch the entire shoal, since it is, so to say, left destitute and without a leader; for it remains motionless and ceases to advance, like a flock of sheep that by some mischance has lost its shepherd. But the leader makes good his escape and slips out with the utmost adroitness and takes

[a] See 10.13.
[b] 2nd cent. B.C.
[c] Island and town off the NW coast of Ceylon.

ὑπήκοον. τοὺς δὲ ληφθέντας ἐν πιθάκναις λέγονται ταριχεύειν· ὅταν δὲ ἡ σὰρξ μυδήσῃ καὶ περιρρυῇ, καταλείπεται ἡ ψῆφος. ἄριστος δὲ ἄρα ὁ Ἰνδικὸς γίνεται καὶ ὁ τῆς θαλάττης τῆς Ἐρυθρᾶς. γίνεται δὲ καὶ κατὰ τὸν Ἑσπέριον ὠκεανόν, ἔνθα ἡ Βρεττανικὴ νῆσός ἐστι· δοκεῖ δέ πως χρυσωπότερος[1] ἰδεῖν εἶναι, τάς τε αὐγὰς ἀμβλυτέρας ἔχειν[2] καὶ σκοτωδεστέρας. γίνεσθαι δέ φησιν Ἰόβας καὶ ἐν τῷ κατὰ Βόσπορον πορθμῷ, καὶ τοῦ Βρεττανικοῦ ἡττᾶσθαι αὐτόν, τῷ δὲ Ἰνδῷ καὶ τῷ Ἐρυθραίῳ μηδὲ τὴν ἀρχὴν ἀντικρίνεσθαι. ὁ δὲ ἐν Ἰνδίᾳ χερσαῖος οὐ λέγεται φύσιν ἔχειν ἰδίαν, ἀλλὰ ἀπογέννημα εἶναι κρυστάλλου, οὐ τοῦ ἐκ τῶν παγετῶν συνισταμένου, ἀλλὰ τοῦ ὀρυκτοῦ.

9. Γεράνων μὲν οὖν πέρι τῶν πτηνῶν ἐν τοῖς προτέροις[3] λόγοις εἰπεῖν ἐμαυτὸν καλῶς οἶδα, θαλάττιον δὲ γέρανον ἰχθὺν Κορινθίῳ πελάγει ἔντροφον ἀκοῦσαί φημι. ἐπικλίνει[4] δὲ ἄρα τοῦτο τὸ πέλαγος, ἔνθα ὁ γέρανος ἀνιχνεύθη οὗτος, τῷ[5] πρὸς τὰς Ἀθήνας πελάγει τοῦ ἰσθμοῦ κατὰ τὴν πλευρὰν τὴν ἐς αὐτὰς ὁρῶσαν. μῆκος μὲν οὖν ἦν προήκων ἐς πεντεκαίδεκά που πόδας μεμετρημένους δικαίῳ μέτρῳ, ἐγχέλεως[6] δὲ εἶχεν, ὡς ἀκούω, οὐ μέντοι τῆς μεγίστης ⟨τὸ⟩[7] πάχος. κεφαλὴ δὲ ἄρα ἐκείνῳ καὶ στόμα γεράνου ἐστὶ τῆς πτηνῆς,[8]

[1] *Ges*: χρυσῷ ὁπότερος.
[2] ἔχων.
[3] *Jac*: πρεσβυτέροις.
[4] ἀπέκλεινε.
[5] ἐν τῷ.
[6] ἐγχέλυος.
[7] ⟨τό⟩ *add. Ges.*
[8] τοῦ πτηνοῦ.

[a] The Pearl-mussel, *Unio margaritiferus*, of the British Isles is found in fresh water, but the pearl it produces is smaller than the Orient pearl.

the lead and rescues those that obey him. Those however that are caught the Ichthyophagi are said to pickle in jars. And when the flesh turns clammy and falls away, the precious stone is left behind. The best ones are those from India and from the Red Sea; but they are also found in the western ocean where the island of Britain is, though this kind has a more golden appearance, and a duller, duskier sheen.[a] Juba asserts that they occur also in the strait leading to the Bosporus and are inferior to the British kind, and are not for a moment to be compared with those from India and the Red Sea. But the land-pearl[b] of India is said not to have an independent origin but to be generated not from the ice formed by frost but from excavated rock-crystal.

The Crane-fish

9. I am well aware that earlier on in my discourse I have spoken of cranes, the birds, but I claim to have heard of a sea-crane,[c] a fish that lives in the sea of Corinth. Now this stretch of sea, where the Crane-fish has been tracked down, lies near the sea which approaches Attica on that side of the Isthmus that faces Athens.[d] The fish reaches a length of perhaps fifteen feet reckoned accurately, but it is not (so I learn) as bulky as the largest eel. It has the head and mouth of the bird, and its scales[e] you

[b] The 'ground-pearl' is the outer pearly covering of *Margarodes*, one of the Coccidae; see A. D. Imms, *Gen. textbook of Entomology*[5] (1942), 389; D. Sharp, *Insects*, 598 (Camb. Nat. Hist. 6). For other views see *RE* 14. 1682, art. 'Margarita.'

[c] Perhaps the 'Oar-fish,' *Regalecus banksi*; but Gossen suggests *Nemicthys scolopaceus*.

[d] In other words 'in the Saronic gulf.'

[e] Or, if λόφια (Thompson, *Gk. fishes*, s.v. Γέρανος) is read, 'crest.'

λεπίδες δὲ αὐτῷ, γεράνου πτερὰ καὶ ταύτας [1] εἴποις [2] ἄν. ἔρχεται δὲ οὐχ ἑλικτὴν τὴν νῆξιν, ὥσπερ οὖν οἱ τῶν ἰχθύων κατὰ τὰς ἐγχέλεις [3] στενοὶ καὶ μακροί. ἔχει δὲ ῥώμην καὶ μάλα ἁλτικήν· πηδᾷ γοῦν ὥσπερ ἀπὸ νευρᾶς οἰστὸς ἀφεθείς. . . .[4] λέγουσιν οὖν οἱ λόγοι οἱ μὲν Ἐπιδαύριοι τοῦτο τὸ ζῷον οὐδενὸς ἰχθύος ἔκγονον, ἀλλὰ τὰς πτηνὰς γεράνους φευγούσας τὸν Θρᾴκιον κρυμὸν καὶ τὸν ἑσπέριον τὸν ἄλλον ἐμπίπτειν μὲν τῷ πνεύματι, τάς γε μὴν θηλείας ἐς μίξιν οἰστρᾶσθαι, τοὺς δὲ ἄρρενας αὐταῖς ἐπιφλέγεσθαι, καὶ μέντοι καὶ ἐς τὴν πρὸς αὐτὰς ὁμιλίαν κυμαίνειν, καὶ διὰ ταῦτά τοι καὶ ἀναβαίνειν [5] αὐτὰς ἐθέλειν, τὰς δὲ οὐχ ὑπομένειν· μὴ γὰρ οἵας [6] τε εἶναι μετέωρον μίξιν φέρειν· [7] τοὺς δὲ ἐγκρατεῖς οὐκ ἔχοντας γενέσθαι τοῦ πόθου, ἐκβάλλειν τὴν γονήν. καὶ εἰ μὲν τύχοιεν ὑπὲρ γῆς φερόμενοι, τὴν δὲ ἐκπίπτειν ἐς οὐδὲ ἕν, ἀλλ' ἀπόλλυσθαι ἄλλως· εἰ δὲ ὑπὲρ τοῦ πελάγους πέτοιντο, ἐνταῦθά τοι τὴν θάλατταν ὥσπερ οὖν θησαύρισμα παραλαβοῦσαν φυλάττειν ἔμβρυον, καὶ γεννᾶν τὸ ζῷον τοῦτο, ἀλλ' οὐ διαφθείρειν ὥσπερ ἔς τινα ἄγονον καὶ στερίφην γαστέρα ἐμπεσόν. καὶ τούτων μὲν τῶν λόγων ἅτερος καὶ δὴ διηνύσθη ὁ Ἐπιδαύριος. λέγει δὲ ἄλλος, οὗ τὸ γένος οὐκ οἶδα, ἑτέραν ὁδὸν τραπόμενος, εἶτα μέντοι οὐ ταὐτὰ ὁμολογεῖ, ὡς δ' ἂν [8] μὴ δοκοίην [9] ἀμαθὴς εἶναι αὐτοῦ, λελέξεται μέντοι καὶ ἐκεῖνος. Δημόστρατος, οὗπερ οὖν καὶ ἀνωτέρω μνήμην ἐποιησάμην, 'εἶδον τὸν ἰχθὺν' ἦ δ' ὅς, 'καί μ' ἐσῆλθεν αὐτοῦ θαῦμα, καὶ ἐβουλήθην αὐτὸν ποιῆσαι τάριχον, ἵνα εἴη [10] καὶ ἄλλῳ

might say were the feathers of a crane. But it does not swim in the sinuous fashion of those fishes which are slim and long like eels. It is an exceedingly powerful jumper; at any rate it springs forward like an arrow shot from a bowstring . . . Now the accounts from Epidaurus state that this creature is not the offspring of any fish, but that cranes fleeing from the frosts of Thrace and of the west generally, encounter the wind, and that the female birds are stimulated to mate, while the male birds are inflamed with desire and agitated with a passion to couple, which makes them want to mount the females. They however will not permit it, for they cannot bear the burden of coupling in mid air, and so the males frustrated in their desire ejaculate semen. If they happen to be flying over land, the semen is spent to no purpose but is lost and wasted. If however they are flying over the ocean, then the sea takes up and preserves the embryo as though it were a treasure, and generates this creature, not destroying it as though it had entered some unfruitful, sterile womb. Here then of the two versions is the Epidaurian one fully set out. But the other version, whose origin I cannot tell, takes a different direction and does not agree with the former, but I shall mention it as well so that I may not appear to be ignorant of it. Demostratus, whom I also mentioned earlier on, says, 'I saw the fish and was filled with astonishment, and I was anxious to pickle it so that others might be able

Demostratus quoted

[1] *Schn*: ταῦτα.
[2] *Ges*: εἶπες.
[3] ἐγχέλυς.
[4] *Lacuna.*
[5] *Jac*: διαβαίνειν.
[6] οἷον.
[7] φέρειν, οὐδὲ ἑτέραν.
[8] *Schn*: ὡς ἂν δέ.
[9] *Jac*: δοκοῖμι or δοκῇ.
[10] *Jac*: ᾗ.

βλέπειν. οὐκοῦν ἐνεργῶν ὄντων καὶ ἀνοιγνύντων τῶν μαγείρων αὐτόν,[1] ἐπεσκόπουν τὰ σπλάγχνα αὐτός. ἀκάνθας τε εἶδον ἐξ ἑκατέρας τῆς πλευρᾶς συνιούσας τε καὶ ἐγκλινούσας τὰ πέρατα ἐς ἀλλήλας, τρίγωνοι δέ' φησιν 'ἦσαν ὥσπερ οὖν καὶ αἱ κύρβεις, ἧπάρ τε ἐνέκειτό οἱ προῆκον ἐς μῆκος, ὑπέκειτο δὲ αὐτῷ καὶ χολὴ μακρὰν ἔχουσα τὴν φῦσαν κατὰ τὰ φασκώλια· εἶπες δ' ἂν ἰδὼν τὴν χολὴν κύαμον ὑγρὸν εἶναι. ἐξαιρεθέντα οὖν ἄμφω, καὶ ἡ χολὴ καὶ τὸ ἧπαρ, τὸ μὲν ἕτερον [2] διωγκώθη [3] καὶ ἐῴκει ἰχθύος ἥπατι μεγίστου, διατήξασα δ' ἡ χολὴ τὸν λίθον (καὶ γάρ πως ἔτυχε τεθεῖσα ἐπὶ λίθου) εἶτα ἠφανίσθη.' ἄμφω δὲ τὼ λόγω ἐνταῦθα ὁρίζομεν.

10. Θήραν δὲ πηλαμύδων εἰπεῖν μὴ πάνυ τι [4] συνειθισμένην οὐκ ἔστιν ἔξω τῆσδε τῆς σπουδῆς. δέκα νεανίαι τὸ ἀκμαιότατον ἀνθοῦντες ἀναβαίνουσι ναῦν ἐλαφρὰν καὶ διὰ ταῦτά τοι [5] καὶ ταχυτάτην· διανέμοντες δὲ ἑαυτοὺς ἐς ἑκατέραν ἴσους τὴν πλευρὰν καὶ κορεσθέντες εὖ μάλα τροφῆς, εἶτα μέντοι τοῖς ἐρετμοῖς ἕκαστος ἐπιχειροῦσι, πλανώμενοι δεῦρο καὶ ἐκεῖσε. κάθηται δὲ εἷς ἐπὶ τῆς πρύμνης, καὶ ἐντεῦθέν τε καὶ ἐκεῖθεν παρασείρους καθίησιν ὁρμιάς· ἤρτηνται δὲ τούτων καὶ ἄλλαι, καὶ συνῆπται πάσαις τὰ ἄγκιστρα, καὶ ἕκαστον ἄγκιστρον δέλεαρ φέρει Λακαίνης πορφύρας μαλλῷ [6] κατειλημένον,[7] καὶ πτερὸν μέντοι

[1] αὐτὸν καὶ τεμνόντων τὴν γαστέρα.
[2] *Reiske* : οὐ μεθ' ὕστερον.
[3] *Ges* : μέντοι διογκωθῆναι MSS, μέντοι *del. Schn, H.*
[4] πάντη οὖν.
[5] μέντοι.
[6] *Ges* : μαλῷ *or* μᾶλλον.
[7] *Schn* : κατειλημμένον.

to see it. And so when the cooks got to work and opened it up, with my own eyes I inspected its internal organs and observed spines on both sides which met and turned their points towards one another; they were,' he says, 'triangular like the three-sided law-tablets, and imbedded in them was a liver of considerable length, and below that was a gall-bladder, with a long tube as in skin-bags. You would have said on seeing it that it was a damp bean-pod. So both gall-bladder and liver were extracted, and the latter swelled up till it equalled the liver of the largest fish, whereas the gall-bladder, which happened somehow to have been placed on a stone, caused the stone to melt away and disappeared from sight.'

Here I conclude the two accounts.

The Pelamyd

10. It is not irrelevant to our present study to describe the altogether singular manner in which Pelamyds[a] are caught. Ten young men in the prime of strength embark in a boat, light and therefore capable of great speed, arranging themselves in equal numbers on either side; and after satisfying themselves with a good meal they each lay hold of an oar and roam this way and that. And one youth sits in the stern and lets down horse-hair lines on either side of the vessel. To these other lines are fastened, and to all of them hooks are attached, and each hook carries a bait wrapped round with wool of Laconian purple, and further, to each hook is

[a] 'Usually a small Tunny; and then either the young of the common tunny, or one of the lesser species. . . . [The word] seems to be used especially of the tunny of the Black Sea' (Thompson, *Gk. fishes*).

λάρου ἑκάστῳ ἀγκίστρῳ προσήρτηται, ὥστε ἡσυχῇ[1] διασείεσθαι ὑπὸ τοῦ προσπίπτοντος ὕδατος. τούτων οὖν ἱμέρῳ προσνέουσιν αἱ πηλαμύδες· μία δὲ ἡ προτένθης[2] ὅταν τὸ στόμα ἐναπερείσῃ, προσίασι καὶ αἱ λοιπαί, καὶ δονεῖται τὰ ἄγκιστρα ὑπὸ τὸν αὐτὸν καιρὸν περιπαρέντα τοῖς ἰχθύσιν. οἱ ἄνδρες οὖν τοῦ μὲν ἐρέττειν ἔτι ἀπέστησαν, παρῆκαν δὲ τὰς κώπας, ἐξαναστάντες δὲ ἀνασπῶσι τὰς μηρίνθους εὐαγρούσας καὶ μέντοι καὶ βριθομένας τοῖς ἰχθύσιν· ὅταν δὲ ἐς τὴν ναῦν ἐμπέσωσι, διαφαίνεται τῆς εὐθηρίας τὸ μαρτύριον ἐκ τοῦ πλήθους τῶν ἰχθύων τῶν ἑαλωκότων.

11. Ἡ χερσαία γαλῆ ὅτι ἦν ἄνθρωπος ἤκουσα· καὶ ὅτι τοῦτο ἐκαλεῖτο, καὶ ὅτι ἦν γόης καὶ φαρμακίς, καὶ ὅτι δεινῶς ἀκόλαστος ἦν καὶ ἀφροδίτην παράνομον ἐνόσει, καὶ ταῦτα ἐς ἀκοὴν τὴν ἐμὴν ἀφίκετο· καὶ ὡς ἐς τοῦτο τὸ ζῷον τὸ κακὸν ἔτρεψεν αὐτὴν Ἑκάτης τῆς θεοῦ μῆνις οὐδὲ τοῦτό με λέληθεν. ἡ μὲν οὖν θεὸς ἵλεως ἔστω· μύθους δὲ ἐῶ καὶ μυθολογίαν ἄλλοις. ὅτι δέ ἐστι θηρίον ἐπιβουλότατον, καὶ νεκροῖς ἀνθρώποις ἐπιτίθενται γαλαῖ, καὶ μὴ φυλαττομένοις[3] ἐπιπηδῶσι, καὶ συλῶσι τοὺς ὀφθαλμοὺς καὶ ἐκροφοῦσι,[4] δῆλόν ἐστι. φασὶ δὲ καὶ ὄρχεις γαλῆς γυναικὶ κατ' ἐπιβουλὴν ἢ ἑκούσῃ περιαφθέντας ἐπίσχειν τοῦ[5] ἔτι μητέρα[6] γίνεσθαι, καὶ ἀνάστέλλειν

[1] ἡσυχῇ ὥστε.
[2] ἡ ⟨μάλιστα⟩ προ- *add.* *H.*
[3] φυλαττόμεναι.
[4] *Ges.*: ἐκροφῶσι.
[5] τό.

attached the feather of a sea-mew so as to be gently fluttered by the impact of the water. Now the Pelamyds in their eagerness for these objects come swimming up, and when the 'foretaster'[a] has applied its mouth to them the rest approach and at the same moment the hooks are agitated as they pierce the fish. Meanwhile the men have stopped rowing and laid aside their oars and standing up draw up the lines with their plentiful catch, indeed even laden with fish. And when they tumble into the boat the evidence of a successful day's sport is manifest in the great number captured.

The Marten

11. I have heard that the land-Marten was once a human being. It has also reached my hearing that 'Marten' was its name then; that it was a dealer in spells and a sorcerer; that it was extremely incontinent, and that it was afflicted with abnormal sexual desires. Nor has it escaped my notice that the anger of the goddess Hecate transformed it into this evil creature.—May the goddess be gracious to me: fables and their telling I leave to others. But it is clearly a most malicious animal: Martens set upon human corpses, leap upon them if they are unprotected, pluck out their eyes and swallow them. They say too that if the testicles of a Marten are hung on a woman either by treachery or with her consent, they prevent her from becoming a mother and make her refrain from the sexual act. If the

[a] The title of an official at Athens who on the eve of the Apaturia tasted the food provided for the public feast to see if it was satisfactory.

[6] μητέρας.

μίξεως.[1] σπλάγχνα δὲ γαλῆς σκευασίαν τινὰ προσλαβόντα, ἣν ἴστωσαν οἱ σοφοὶ ταῦτα, καὶ[2] ἐς οἶνον ἐμβληθέντα κατ' ἐπιβουλήν, φιλίαν ὡς λόγος διίστησι, καὶ ἡνωμένην τέως εὔνοιαν διακόπτει. καὶ ὑπὲρ μὲν τούτων τοὺς γόητάς τε καὶ φαρμακέας Ἄρει φίλῳ κολάζειν καὶ δικαιοῦν[3] καταλείπωμεν. εἴη δ' ἂν καὶ ἰχθὺς γαλῆ, σμικρὸς οὗτος, καὶ οὐδέν τι κοινὸν πρὸς τοὺς καλουμένους γαλεοὺς ἔχων. οἱ μὲν γάρ εἰσι σελάχιοι[4] καὶ πελάγιοι, καὶ ⟨ἐς⟩[5] μέγεθος προήκοντες εἶτα μέντοι κυνὶ ἐοίκασιν· ἡ γαλῆ δέ, φαίης[6] ἂν αὐτὴν εἶναι τὸν καλούμενον ἥπατον. ἰχθὺς δὲ ἔστιν αὕτη βραχύς, καὶ τὼ ὀφθαλμὼ ἐπιμέμυκε· κόρας δὲ ἔχει κυάνου χρόᾳ προσεικασμένας. καὶ τὸ μὲν γένειον ἔχει τοῦ ἡπάτου μεῖζον, ἡττᾶται δὲ αὖ πάλιν τοῦ χρέμητος κατά γε τοῦτο. πετραίαν δὲ οὖσαν τὴν γαλῆν καὶ νεμομένην φυκία ἀκούω πάντων σωμάτων οἷς ἂν νεκροῖς ἐντύχῃ τοὺς ὀφθαλμοὺς καὶ ταύτην ὡς τὴν χερσαίαν ἐσθίειν. χρῶνται δὲ αὐτῇ ἐς τὰ ὅμοια ἁλιεῖς ὅσοι κατὰ τοὺς Ἠπειρώτας φαρμακεύουσι πονηροὶ καὶ οὗτοι σοφισταὶ κακῶν. ἐπεὶ δὲ ὠμοβόρον ἐστὶ τὸ τῶν ἰχθύων ⟨τῶνδε⟩[7] φῦλον, πᾶν τὸ ταῖς ὑδροθηρίαις γένος συμβιοῦν καὶ τὰς ὑποδύσεις[8] τὰς κατωτάτω μετιὸν μελαίνουσι τὰς ἑαυτῶν βάσεις καὶ τὰ κοῖλα τῶν χειρῶν, ἀφανίζειν πειρώμενοι τὴν ἐξ αὐτῶν αὐγήν· τὰ γάρ τοι τῶν ἀνθρώπων μέλη,[9] ὡς ὅτι μάλιστα ἐκλάμποντα ἐν τῷ ὕδατι, ἐφολκὰ τῶν ἰχθύων τούτων[10] ἐστίν.

[1] μίξεως] *H, comp.* 4. 34 *and* 9. 54; αὐτῶν MSS, ἀνδρῶν *Jac.*
[2] καὶ ταῦτα.
[3] *Abresch*: διακαίειν.
[4] σελάχη V, *H.*
[5] ⟨ἐς⟩ *add. Schn.*

inwards of a Marten are dressed in a certain way, which I leave to those skilled in these matters, and dropped with evil intent into wine, they break up (so they say) a friendship, and sunder relations hitherto harmonious. In reward for these activities let us leave spell-binders and sorcerers to our friend Ares [a] to punish and judge.

The Marten-fish

There is also a fish called Marten (*galê*): it is small and has nothing in common with those known as dog-fish (*galeus*), for the latter are cartilaginous, live in the sea, attain to a considerable length, and resemble a dog. But the Marten-fish one might identify with the Hepatus,[b] as it is called. This is a small fish and blinks its eyes; the pupils are a dark blue colour. Its barbel is larger than that of the Hepatus; on the other hand it yields to the Chromis in this respect. I am told that the Marten lives among rocks, feeds on sea-weed, and that it too like the land Marten eats the eyes of all bodies that it finds dead. Fishermen who practise sorcery after the manner of those that dwell on the continent of Asia, being evilly disposed and skilled in mischief, use it for the same purpose as men use the land Marten. And since this species of fish is carnivorous, all men who spend their lives fishing and who explore the deepest recesses black their feet and the palms of their hands in an attempt to nullify the light that radiates from them, for men's limbs appear extremely bright in water and so attract these fish.

[a] Cp. Ael. *VH* 5. 18: cases of poisoning came before the court of the Areopagus.
[b] Unidentified; see 9, 38 n.

---

[6] *φαίην.* [7] ⟨*τῶνδε*⟩ *add. H.* [8] *Abresch*: *ἀπο-*.
[9] *ἀνθρωπίνων μελῶν.* [10] *πάντων.*

12. Χῆμαι δὲ θαλάττιαι ζῷόν εἰσι καὶ αὗται διάφορον· αἱ μὲν γὰρ αὐτῶν τραχεῖαι πεφύκασιν, αἱ δὲ λεῖαι πάνυ· καὶ τὰς μὲν τοῖς δακτύλοις πιέσας συνθλάσεις, τὰς δὲ συντρίψεις λίθῳ καὶ μόλις·[1] καὶ αἱ μὲν αὐτῶν μελάνταται τὴν χρόαν εἰσίν, αἱ δέ, ἀργύρῳ φαίης ἂν αὐτὰς προσεοικέναι, αἱ δὲ ἀνακραθείσας[2] περίκεινται τὰς χρόας τὰς[3] προειρημένας. γένη δὲ αὐτῶν[4] διάφορα καὶ εὐναὶ πάνυ ποικίλαι· αἱ μὲν γὰρ ἐν ταῖς ψάμμοις κεῖνται διεσπαρμέναι ταῖς τῶν αἰγιαλῶν, διαναπαύονταί τε κατὰ τῆς ἰλύος, αἱ δὲ ὑπόκεινται τῷ βρύῳ, αἱ δὲ εἰλημμέναι τῶν σπιλάδων εἶτα αὐταῖς προσέχονται μάλα ἐγκρατῶς. ἐν δὲ τῇ καλουμένῃ Ἰστριάδι θαλάττῃ αἵδε αἱ χῆμαι κατὰ τὴν ὥραν τὴν θέρειον, ὑπαρχομένου τοῦ ἀμήτου, δίκην ἀγέλης ἀλλήλαις συμφέρονται, καὶ ἀναπλέουσι κούφως, τά γε πρῶτα βαρεῖαί[5] τε καὶ ἐπαχθεῖς οὖσαι καὶ οὐκ ἀναπλεύσασαι, ἀλλὰ[6] τηνικάδε οὐκέτι τοιαῦται. ἀποδιδράσκουσι δὲ τὸν νότον, καὶ φεύγουσι τὸν βορρᾶν, καὶ οὐδὲ τὸν εὗρον[7] ἀνέχονται. χαίρουσι δὲ ἀκύμονι θαλάττῃ, καὶ ζεφύρου καταπνεούσαις αὔραις ἡδείαις τε καὶ μαλακαῖς. ὑπὸ ταύταις οὖν τοὺς ἑαυτῶν εἰλυοὺς[8] ἐκλιποῦσαι, μεμυκυῖαί τε καὶ κατάκλειστοι ἔτι, ἀνίασιν ἐκ τῶν μυχῶν, καὶ ἀκύμονος οὔσης τῆς θαλάττης νέουσι· καὶ τότε ἀνοίξασαι τὰς ἑαυτῶν στέγας ἐκκύπτουσιν, ὡς ἐκ τῶν ἰδίων θαλάμων αἱ νύμφαι ἢ τὰ ῥόδα πρὸς τὴν εἵλην ὑπαλεανθέντα καὶ ἐκκύψαντα τῶν καλύκων. οὐκοῦν[9] κατὰ μικρὰ

[1] καὶ μόλις *del.* H.
[2] ἀνακραθεῖσαι.
[3] *Jac*: ἁπάσας τάς.
[4] αὐταῖς.
[5] ἑαυταῖς βαρεῖαι.
[6] *Jac*: ἀλλ' αἱ.

12. Clams of the sea are of different kinds, for some of them are rough, others perfectly smooth; some you can crush by the mere pressure of the fingers, others you will hardly smash with a stone; some are of a deep black colour, others you might compare with silver, others again are clothed in a blend of the aforesaid colours. Their species differ and their habitats are very various, for some lie scattered in the sands of the sea-shore or rest at times in the mud, others lie low beneath the sea-moss, while others lay hold of reefs and cling to them with might and main. In the Istrian Sea,[a] as it is called, these Clams in summer time at the beginning of the harvesting season swim along together like a herd of cattle, floating lightly to the surface, although up to this time they have been too heavy and weighty to float upwards, but now they are no longer so. And they avoid the South wind and flee before the North, and cannot endure even the East wind, but their delight is in a waveless sea and when the pleasant and gentle breezes of the West wind blow. And so beneath their influence they quit their burrows, with their shells still closed and fast shut, and mount upwards from their recesses and, when the sea is waveless, swim around. And then they open their coverings and peep forth, like brides looking down from their private chambers or like rosebuds that, warmed a little, have peeped out of their flower-cups towards the sun's heat. And so little by little they

The Clam

[a] That part of the Euxine that lies off Istrus, S of the mouths of the Danube.

[7] τὸν εὖρον οὐδέ. [8] εἰλέους.
[9] οὐκοῦν καὶ αἱ κόγχαι.

ὑποθαρροῦσαι,[1] ⟨καὶ⟩ [2] μάλα γε ἀσμένως ἡσυχάζουσι καὶ ἀτρεμοῦσι τὸν ἑταῖρον ἄνεμον προσδεχόμεναι, καὶ τὸν μὲν ὑπεστόρεσαν χιτῶνα, τὸν δὲ ὤρθωσαν, καὶ πλέουσι τῷ μὲν ἱστίῳ αἱ χῆμαι, τῷ δὲ σκάφει χρώμεναι. καὶ προΐασι μὲν τὸν τρόπον τοῦτον, ἡσυχίας οὔσης καὶ εὐδίας (οὐδὲν φαίης ἂν μακρόθεν ἰδὼν ἢ νηΐτην στόλον εἶναι)· ἐὰν δὲ αἴσθωνται νεὼς ἐπίπλουν ἢ ἔφοδον θηρίου ἢ νῆξιν ἰχθύος ἁδροῦ, ἑαυτὰς ὑφ' ἑνὶ κρότῳ τῶν ὀστράκων πτύξασαι, κατώλισθόν τε ἀθρόαι καὶ ἠφανίσθησαν.

13. Ὁ δὲ αἱμόρρους (εἴη δ' ἂν γένος ἔχεως οὗτος) μάλιστα ἐν τοῖς πετρώδεσι χηραμοῖς ἤθη τε ἔχει καὶ διατριβάς. μῆκός τε σώματος εἴληχε πόδα, πλάτος δὲ ἐξ εὐρείας τῆς κεφαλῆς μείουρος κάτεισιν ἔστε ἐπὶ τὴν οὐράν· καὶ πῇ μὲν φλογώδης ἰδεῖν ἐστι, πῇ δὲ δεινῶς μέλας· φρίττει δὲ τὴν κεφαλὴν οἱονεὶ κέρασί τισιν. ἕρπει δὲ ἥσυχος ἐπιθλίβων τὰς τῆς νηδύος φολίδας, λοξὸν δὲ οἶμον πρόεισιν. ἠρέμα οὖν [3] ὑπηχεῖ, ὡς καταγνῶναι νωθείαν αὐτοῦ καὶ οὐδένειαν. δακὼν δὲ νύγμα ἐργάζεται, καὶ τοῦτό γε ἰδεῖν ἐστι παραχρῆμα κυανοῦν, καρδιώττει γε μὴν ὁ πληγεὶς μάλα [4] οἴκτιστα,[5] ἐκκρίνει δὲ ἡ γαστὴρ ὀχετούς. νὺξ δὲ ἀφίκετο ἡ πρώτη, καὶ αἷμα ἐκρεῖ διά τε ῥινῶν καὶ αὐχένος καὶ μέντοι καὶ δι' ὤτων σὺν ἰῷ χολώδει, οὖρα δὲ ἀφίησιν ὕφαιμα ἡ κύστις. εἰ δὲ καὶ ὠτειλαί εἰσί τινες παλαιαὶ περὶ τὸ σῶμα, ῥήγνυνται καὶ αὗται. εἰ δὲ θῆλυς αἱμόρρους κρούσει τινὶ ⟨ἰὸν⟩ [6] μεθίησιν, καὶ ἐς τὰ οὖλα ὁ

[1] ὑποθαρσοῦσαι.
[2] ⟨καὶ⟩ *add.* H.
[3] γοῦν.
[4] ἀλλά.

gather courage and are glad to rest quietly while waiting for the friendly breeze; and one of their coverings the Clams spread beneath them, the other they raise, and with the latter for sail and the former for skiff they float along. And in this way they move forward when the sea is calm and the weather fine. To see them from a distance you would say that it was a fleet of ships. If however they perceive some vessel approaching or some savage creature advancing or some monstrous fish swimming by, with one clash of their shells they fold up, sink in a mass, and are gone.

The 'Haemorrhous'

13. The *Haemorrhous* or 'Blood-letter' is a species of snake which lives and has its haunts chiefly among rocky hollows. Its body is one foot long, and its width tapers downwards from its broad head to its tail. At one time it has a fiery hue, at another pitch-black, and on its head there bristle what look like horns. It crawls softly as it scrapes the scales of its belly along the ground, and its course is crooked. And so it makes a gentle rustling, which shows how sluggish and how feeble it is. But when it bites it makes a puncture which immediately appears dark blue, and the victim suffers agonising pains in his stomach, while the belly discharges copious fluid. On the first night after, blood streams from the nose and throat and even from the ears together with a bile-like poison, and the bladder emits blood-stained water. Also if there are any old scars on the body they break open. But if a female Blood-letter darts poison as it strikes, the poison mounts to the gums, blood streams copiously

---

[5] *Jac*: ὤκιστα. [6] ⟨ἰόν⟩ *add. OSchn.*

ἰὸς ἀναθεῖ, καὶ ἐκ τῶν ὀνύχων ἄκρων [1] αἷμα ἐκχεῖται πάμπολυ, καὶ ἐκθλίβονται τῶν οὔλων οἱ ὀδόντες. τούτῳ φασὶ τῷ θηρίῳ περιπεσεῖν ἐν Αἰγύπτῳ τὸν τοῦ Μενέλεω κυβερνήτην Κάνωβον Θώνιδος βασιλεύοντος, καὶ συνεῖσαν τὴν Ἑλένην τοῦ δακετοῦ τὴν ἰσχὺν κατάξαι μὲν αὐτοῦ τὴν ῥάχιν, ἐξελεῖν δὲ τὸ φάρμακον. ἐς τίνα δὲ ἄρα χρείαν ἔσπευσε λαβεῖν τὸ θησαύρισμα τοῦτο, οὐκ οἶδα.

14. Κομίζουσι δὲ ἄρα τῷ σφετέρῳ βασιλεῖ οἱ Ἰνδοὶ τίγρεις πεπωλευμένους καὶ τιθασοὺς πάνθηρας καὶ ὄρυγας τετράκερως, βοῶν δὲ γένη δύο, δρομικούς τε καὶ ἄλλους ἀγρίους δεινῶς. ἐκ τούτων γε τῶν βοῶν καὶ τὰς μυιοσόβας [2] ποιοῦνται, καὶ τὸ μὲν ⟨ἄλλο⟩ [3] σῶμα παμμέλανές εἰσιν οἵδε, τὰς δὲ οὐρὰς ἔχουσι λευκὰς ἰσχυρῶς. καὶ περιστερὰς ὠχρὰς κομίζουσιν, ἅσπερ [4] οὖν καὶ λέγουσι μήτε ἡμεροῦσθαι μήτε ποτὲ πραΰνεσθαι, καὶ ὄρνιθας δέ, οὓς κερκορώνους [5] φιλοῦσιν ὀνομάζειν, καὶ κύνας γενναίους, ὑπὲρ ὧν ἄνω μοι λέλεκται, καὶ πιθήκους λευκοὺς καὶ μελαντάτους ἄλλους· [6] τοὺς γάρ τοι πυρροὺς ὡς γυναιμανεῖς ἐς τὰς πόλεις οὐκ ἄγουσιν, ἀλλὰ καί ποθεν ἐπιπηδήσαντες ἀναιροῦσιν, ὡς μοιχοὺς μεμισηκότες.

[1] ἐκ τῶν ὀνύχων ἄκρων, *after* μεθίησιν *in the* MSS, *transposed by OSchn, comp.* Nic. *Th.* 305.

[2] *Ges* : τοὺς (τὰς) μυιοσόβους.

[3] ⟨ἄλλο⟩ *add. H.*

[4] οἵασπερ.

[5] κερκίωνας *Ges.*

[6] ἄλλους καὶ τοῦτους πιθήκους.

from the finger-nails, and the teeth are forced out from the gums. This, they say, was the savage creature that Canobus, the helmsman of Menelaus, encountered in Egypt during the reign of Thonis; and when Helen realised how strong this venomous beast was she broke its spine and extracted the poison. But for what purpose she was eager to obtain this precious stuff I am unable to say.[a]

The tale of Canobus and Helen

14. The people of India bring to their king tigers that they have trained, tame panthers,[b] four-horned antelopes, two kinds of oxen, the one swift of foot, the other exceedingly wild. From these oxen they contrive fly-whisks, and whereas the rest of their body is entirely black, their tails are dazzlingly white. They bring also pale-yellow doves which are said never to become domesticated, never to be tamed; those birds too which they are accustomed to call *Cercorônoi* (mynahs);[c] and hounds of good pedigree (I have spoken of these above);[d] and apes, some white, some the deepest black: the reddish ones,[e] which are too fond of women, they do not introduce into their towns, but if they can contrive somehow to spring upon them, they put them to death, because they detest them as adulterers.

Animals presented to the Indian King

[a] It seems impossible to identify this snake; see Gow-Scholfield on Nicander, *Th.* 282–319.

[b] 'Panther' and 'leopard' are synonymous terms, although in 7. 47 Ael. appears to distinguish them. Perh. render 'snow-leopard' or 'ounce.'

[c] κερκορώνος conjecturally identified with κερκίων, the Indian mynah; though κερκο- 'would suggest one of the handsome long-tailed Jays' (Thompson, *Gk. birds*).

[d] See 4. 19; 8. 1.

[e] The Orang-utan (Gossen § 241).

15. Ἰνδῶν δὲ ὁ μέγας βασιλεὺς μιᾶς ἡμέρας ἀνὰ πᾶν ἔτος ἀγωνίας προτίθησι τοῖς τε ἄλλοις ὅσοις εἶπον ἑτέρωθι, ἐν δὲ[1] τοῖς καὶ ζῴοις ἀλόγοις, ἀλλὰ ἐκείνοις ⟨γε⟩[2] ὧν ἐκπέφυκε κέρατα. κυρίττει δὲ ταῦτα ἄλληλα, καὶ φύσει τινὶ θαυμαστῇ μέχρι νίκης ἁμιλλᾶται, ὥσπερ οὖν ἀθληταὶ ἢ ὑπὲρ ἄθλων μεγίστων ἰσχυριζόμενοι ἢ ὑπὲρ κλέους σεμνοῦ καὶ φήμης τινὸς ἀγαθῆς. εἰσὶ δὲ οἱ ἀγωνισταὶ οἵδε οἱ ἄλογοι ταῦροί τε ἄγριοι καὶ κριοὶ ἥμεροι καὶ οἱ καλούμενοι † μέσοι †[3] καὶ ὄνοι μονόκερῳ καὶ † ὕαιναι.†[4] φασὶ δὲ εἶναι τοῦτο τὸ ζῷον δορκάδος μὲν ἧττον, ἐλάφου δὲ πολλῷ θρασύτερον καὶ θυμούμενον ἐς κέρας. εἶτα ἐπὶ πᾶσιν οἱ ἐλέφαντες ἀγωνισταὶ παρίασιν· προχωροῦσι δὲ οὗτοι καὶ μέχρι θανάτου τιτρώσκοντες ἀλλήλους τοῖς κέρασιν, καὶ πολλάκις μὲν ὁ ἕτερος κρατεῖ καὶ ἀποκτείνει τὸν ἀντίπαλον, πολλάκις δὲ καὶ συναποθνήσκουσιν.

16. Θεόφραστος οὔ φησι τοῦ ἔχεως τὰ βρέφη διεσθίειν τῆς μητρὸς τὴν γαστέρα, ὥσπερ οὖν θυροκοποῦντα, ἵνα τι καὶ παίσω,[5] καὶ ἐξαράττοντα πεφραγμένην ἔξοδον, ἀλλὰ τοῦ θήλεος θλιβομένου[6] καὶ τῆς γαστρός οἱ σ τ ε ι ν ο μ έ ν η ς (Ὁμηρείως δὲ εἶπον), τὴν δὲ οὐκ ἀντέχειν ἀλλὰ διαρρήγνυσθαι. καί με πείθει λέγων, ἐπεί τοι καὶ θαλάττιαι βελόναι ἄκολποί τε οὖσαι καὶ λεπταὶ ὅτι τὰ αὐτὰ πάσχουσιν ὑπὸ τῶν σφετέρων βρεφῶν καὶ ἐκεῖναι ἄνω που τῶν λόγων εἶπον. Ἡρόδοτον δὲ ἀξιῶ μή μοι

[1] δὲ δή.
[2] ⟨γε⟩ *add.* H.
[3] *Corrupt.*
[4] *Corrupt.*
[5] παίξω.
[6] τὸν θῆλυν θλιβόμενον.

15. In India the Great King on one day in every year arranges contests not only for various creatures, as I have said elsewhere,[a] but among them between dumb animals also, or at any rate for those which are born with horns. And these butt each other and struggle with an instinct truly astonishing until one is victorious, as in fact athletes do, using all their strength to win the highest prizes or to achieve glorious renown and a noble fame. But these dumb combatants are wild bulls, tame rams, and what are called *mesoi* [b] and one-horned asses and *hyainai*. They say that this animal is smaller than a gazelle but far more spirited than a stag and that it vents its fury with its horns. And last of all there come forward elephants to the fight: they advance and wound one another to the death with their tusks, and frequently one comes off victor and kills its adversary; frequently also both die together.

Animal contests in India

16. Theophrastus [c] denies that the young of a Viper eat through their mother's belly, as though they were breaking open a door (if I may be allowed the jest) or forcing an exit that had been blocked; but as the female is subjected to pressure and as its belly is (to use the language of Homer) 'straitened,' [d] it is unable to hold out and so bursts. And his statement convinces me, for, you see, Pipe-fish too having no womb and being slim, go through the same process with their young, as I have explained somewhere earlier on.[e] But I trust that Herodotus will

The Viper and its young

The Pipe-fish

[a] See ch. 24.

[b] *Mesoi* and *hyainai* have not been identified, and edd. regard the words as corrupt.

[c] Not in any extant work. [d] E.g. *Il.* 14. 34.

[e] See 9. 60.

μηνίειν, εἰ μύθοις ἐγγράφω ὅσα ὑπὲρ τῆς τῶν ἔχεων ᾠδῖνος ᾄδει.

17. Φυσικὴ δὲ ἄρα ἦν τις κοινωνία καὶ συγγένεια λέοντι καὶ δελφῖνι ἀπόρρητος· οὐ γὰρ ὅτι βασιλεύουσιν ὁ μὲν τῶν χερσαίων ὁ δὲ τῶν ἐναλίων, τοῦτο ἀπόχρη, ἀλλὰ γάρ τοι κἂν τήκωνται [1] προϊόντες ἐς γῆρας,[2] ὁ μὲν τὸν χερσαῖον πίθηκον ἔχει φάρμακον, ὁ δὲ ἀναζητεῖ τὸν συμφυῆ. ὡς γάρ ἐστι καὶ ἐν θαλάττῃ πίθηκος, εἶπόν που· καὶ ἔστι καὶ τῷδε οὗτος ἀγαθόν, ὡς ἐκείνῳ ἐκεῖνος.

18. Ἔστι δὲ ἄρα ἐν τοῖς ἀδιηγήτοις καὶ ἀριθμοῦ περιττοτέροις καὶ σηπεδών, κακὸν ἑρπετόν· ὁμόχρουν τε εἶναι τῷ αἱμόρρῳ καὶ τήνδε φησὶ Νίκανδρος καὶ ἀδελφὴν κατὰ σχῆμα. καὶ τοῦτο ἐκεῖνος λέγει· ὠκυτέρα τε [3] εἶναι δοκεῖ, παρίστησι δὲ καί τινα σμικρότητος [4] φαντασίαν· γυρὸν γὰρ [5] καὶ ἑλικτὸν πρόεισι τὸν οἶμον, καὶ μάλιστα ἐν τούτῳ διαψεύδεται τοὺς ὁρῶντας ὅση τὸ μέγεθός ἐστιν. δεινὸν δὲ ἄρα τὸ ἐξ αὐτῆς τραῦμα· πρόεισι γοῦν καὶ ὑποσήπει, καὶ τήν γε θῆρα [6] τὴν προειρημένην ἀποδείκνυσι φερώνυμον. ὁ γοῦν ἰὸς ἐπὶ πᾶν ὠθεῖται τὸ σῶμα τάχει ἀμάχῳ, καὶ μέντοι καὶ ἡ θρὶξ καὶ ἐκείνη μυδῶσα ἀφανίζεται, λείβονται δὲ αἱ ὀφρῦς καὶ αἱ βλεφαρίδες, καὶ τοὺς ὀφθαλμοὺς ἀχλὺς κατέχει, καὶ ἔφηλοι γίνονται.

[1] καὶ τήκονται.
[2] γῆρας καὶ ἄλλως νοσήσαντες.
[3] δέ *Ges.*
[4] ἀφίησι . . . σμικράν.
[5] γὰρ καὶ πέριξ.

not be angry with me if I reckon as fables all that he says [3. 109] regarding the birth of Vipers.

17. It seems that there is a certain natural association and kinship of a mysterious kind between the Lion and the Dolphin. It is not merely that one is king of land-animals and the other of fishes of the sea, but that when they advance to old age and begin to waste away, the Lion takes a land-monkey by way of medicine while the Dolphin searches for its equivalent in the sea: I have stated somewhere [a] that the sea also contains a 'monkey,' and this is beneficial to the Dolphin, just as the land-monkey is to the Lion.

Lion and Dolphin compared

18. Among the creatures which I have not described and which are past numbering, is the *Sêpedon*, an evil reptile. Nicander says [*Th.* 320–33] that it is the same colour as the Blood-letter and is akin to it in appearance. This also he says: it seems to move more quickly, but conveys the impression of being smaller, for its path is crooked and tortuous, and it is chiefly for this reason that it deceives the spectator as to its real size. Now the wound which it inflicts is terrible: at any rate it spreads and festers and proves that the aforesaid creature is true to its name. At any rate the poison forces its way over the entire body with irresistible speed, and what is more, the hair turns clammy and perishes; the eyebrows and eyelashes fall away; darkness comes over the eyes and they are covered with white spots.

The 'Sepedon'

[a] See 12. 27.

[6] *Ges*: θήραν.

19. Χερσαία χελώνη ζῷον λαγνίστατον, ἀλλὰ ὅ γε ἄρρην· ὁμιλεῖ δὲ ἡ θήλεια ἄκουσα. καὶ λέγει Δημόστρατος, ἀνήρ, ὡς λέγω [1] καὶ τοῦτο, τῶν ἐκ τῆς Ῥωμαίων βουλῆς γενόμενος (καὶ οὔ τί που διὰ τοῦτο ἤδη τεκμηριῶσαι ἱκανός, δοκεῖ δέ μοι ἐπιστήμης τῆς ἁλιευτικῆς ἐς ἄκρον ἐλάσαι καὶ ὅσα ἔγνω εἰπεῖν κάλλιστα· εἰ δέ τί οἱ καὶ ἄλλο ἐσπούδασται τοῦδε σοβαρώτερον, καὶ σοφίας τῆς περὶ τὴν ψυχὴν προσέψαυσεν, οὐκ ἂν θαυμάσαιμι [2]) λέγει δὲ ὅδε ὁ ἀνήρ, ὑπὲρ ὅτου μὲν [3] ἑτέρου τὴν ὁμιλίαν ἀναίνεται ἡ θήλεια οὐκ ἔχειν σαφῶς εἰδέναι, τεκμαίρεσθαι δὲ ἐκεῖνό φησιν. ἡ θήλεια οὐκ ἄλλως ὁμιλεῖ ἢ πρὸς τὸν ἄρρενα ὁρῶσα· καὶ ὁ μὲν ἐξέπλησε τὴν ἐπιθυμίαν κᾆτα ἀπηλλάγη, ἡ δὲ ἑαυτὴν ἐπιστρέψαι ἡκίστη ἐστὶ τῷ τε ὄγκῳ τοῦ χελωνίου καὶ ἐρεισθεῖσα ἐς τὴν γῆν. δεῖπνον οὖν ἕτοιμον ὑπὸ τοῦ γαμέτου καταλέλειπται τοῖς τε ἄλλοις ζῴοις καὶ οὖν καὶ τοῖς ἀετοῖς. αἱ μὲν οὖν ταῦτα ὀρρωδοῦσιν, ὡς ἐκεῖνος λέγει, οἵ γε μὴν ἄρρενες [4] σωφρονούσας αὐτὰς καὶ τιθεμένας πρὸ τοῦ ἡδέος τὸ σωτήριον οὐκ ἔχουσιν ἀναπεῖσαι.[5] οἱ δὲ φύσει τινὶ ἀπορρήτῳ ἴυγγα [6] προσείουσιν [7] ἐρωτικὴν καὶ δέους ἐπίληθον ἅπαντος.[8] ἦσαν δὲ ἄρα ἐρωτικῶς ἐχούσης χελώνης ἴυγγες οὐκ ᾠδαὶ μὰ Δία, ὁποίας Θεόκριτος ὁ τῶν νομευτικῶν παιγνίων συνθέτης ληρεῖ, ἀλλ᾽ ἀπόρρητος πόα, ἥσπερ οὖν οὔτε ἐκεῖνος ὄνομα εἰδέναι φησίν, οὔτε ἄλλον ἐγνωκέναι ὁμολογεῖ· ἐοίκασι δὲ τῇ πόᾳ

[1] *Jac*: λέγει.
[2] οὐκ ἂν θ.] θαυμάσια αὐτοῦ.
[3] μὲν καί.
[4] ἄρρενες οἵδε.

The Tortoise, male and female

19. The land-Tortoise is a most lustful creature, at least the male is; the female however mates unwillingly. And Demostratus, a member, I may add, of the Roman Senate—not that this makes him a sufficient voucher, though in my opinion he attained the summit of knowledge in matters of fishing and was an admirable expounder of his knowledge; nor should I be surprised if he had made a study of some weightier subject and had dealt with the science of the soul.—This Demostratus admits that he does not know precisely whether there is any other reason for the female declining to copulate, but he claims to vouch for the following fact. The female couples only when looking towards the male, and when he has satisfied his desire he goes away, while the female is quite unable to turn over again owing to the bulk of her shell and because she has been pressed into the ground. And so she is abandoned by her mate to provide a meal for other animals and especially for eagles. This then, according to Demostratus, is what the females dread, and since their desires are moderate and they prefer life to pleasurable indulgence, the males are unable to coax them to the act. And so by some mysterious instinct the males cast an amorous spell 'that brings forgetfulness of all' fear [Hom. *Od.* 4. 221]. It seems that the spells of a Tortoise in loving mood are by no means songs, like the trifles which Theocritus, the composer of sportive pastoral poems, wrote, but a mysterious herb of which Demostratus admits that neither he nor anyone else knows the name. Apparently the males

---

[5] *ἀναπείθειν.*
[6] *ἀμάχῳ ἴυγγα ἀπορρήτως.*
[7] *Schn*: *προϊᾶσιν.*
[8] *Jac*: *παντός.*

καλλωπίζεσθαι καί τινας ἀπορρήτους . . . † παλιώρας †.[1] εἰ γοῦν ἐκείνην διὰ στόματος ἔχοιεν, τὰ ἔμπαλιν γίνεται [2] τῶν προειρημένων· θρύπτεται μὲν γὰρ ὁ ἄρρην,[3] μεταθεῖ δὲ ἡ θήλεια ἡ τέως φεύγουσα νῦν φλεγομένη, καὶ ἐξοιστρᾶται καὶ ἱμείρει τῆς συνόδου· δέος δὲ ἐκείναις [4] φροῦδόν ἐστι, καὶ ὑπὲρ ἑαυτῶν ὀρρωδοῦσιν ἥκιστα.

20. Θεσσαλονίκῃ τῇ Μακεδονίτιδι χῶρός ἐστι γειτνιῶν καὶ καλεῖται Νίβας. οὐκοῦν οἱ ἐνταῦθα ἀλεκτρυόνες ᾠδῆς τῆς συμφυοῦς ἀμοιροῦσι καὶ σιωπῶσι πάντα πάντη. καὶ διαρρεῖ λόγος παροιμιώδης ἐπὶ τῶν ἀδυνάτων, ὃς λέγει 'τότε ἂν ἔχοιτε [5] τόδε τι, ὅταν Νίβας κοκκύσῃ.'

21. Ὅτε Ἀλέξανδρος τὰ μὲν ἐδόνει τῆς Ἰνδῶν γῆς τὰ δὲ ᾕρει, πολλοῖς μὲν καὶ ἄλλοις ζῴοις ἐνέτυχεν, ἐν δὲ τοῖς καὶ δράκοντι, ὅνπερ οὖν ἐν ἄντρῳ τινὶ νομίζοντες ἱερὸν Ἰνδοὶ μετὰ πολλοῦ τοῦ θειασμοῦ προσετρέποντο,[6] οὐκοῦν παντοῖοι ἐγένοντο οἱ Ἰνδοὶ δεόμενοι τοῦ Ἀλεξάνδρου μηδένα ἐπιθέσθαι τῷ ζῴῳ· ὁ δὲ κατένευσε. παριούσης οὖν τὸ ἄντρον τῆς στρατιᾶς καὶ κτύπου γενομένου, εἶτα ὁ δράκων ᾔσθετο· ὀξυηκοώτατον δὲ ἄρα ζῴων ἐστὶ καὶ ὀξυωπέστατον. συριγμὸν μὲν οὖν ἀφῆκε μέγιστον καὶ φύσημα, ὡς ἐκπλῆξαί τε πάντας καὶ ἐκταράξαι. ἐλέγετο δὲ ἄρα πήχεων ἑβδομήκοντα εἶναι, ἐφάνη γε μὴν [7] οὐ πᾶς·

[1] *Lacuna* : παλιώρας '*vox nihili*'.
[2] *Schn* : ἔμπαλιν γίνεται τά.
[3] ἄρρην ἐρῶν ὡς οὐκ ἐρῶν.
[4] ἐκείνῳ.

adorn themselves with this herb, and some mysterious. . . . At any rate if they hold this herb in their mouth there ensues the exact opposite to what I have described: the male becomes coy, but the female hitherto reluctant is now full of ardour and pursues him in a frenzied desire to mate; fear is banished and the females are not in the least afraid for their own safety.

The Cock in Nibas

20. There is a region near to Thessalonica in Macedon which goes by the name of Nibas. Now the Cocks there lack their natural faculty of crowing and are absolutely silent. There is current a proverbial saying applied to things that are impossible, it is to this effect: 'You shall have such-and-such when Nibas crows.'

A monstrous Snake

21. When Alexander threw some parts of India into a commotion and took possession of others he encountered among many other animals a Serpent which lived in a cavern and was regarded as sacred by the Indians who paid it great and superstitious reverence. Accordingly Indians went to all lengths imploring Alexander to permit nobody to attack the Serpent; and he assented to their wish. Now as the army passed by the cavern and caused a noise, the Serpent was aware of it. (It has, you know, the sharpest hearing and the keenest sight of all animals.) And it hissed and snorted so violently that all were terrified and confounded. It was reported to measure 70 cubits although it was not visible in all its length,

---

[5] *Bernhardy*: ἔχητε. [6] *Ges*: προετρέποντο.
[7] καὶ ἐφάνη μέν.

μόνην γὰρ ἐξέκυψε τὴν κεφαλήν.[1] καὶ οἵ γε ὀφθαλμοὶ ᾄδονται αὐτοῦ τὸ μέγεθος ἔχειν Μακεδονικῆς περιφεροῦς ἀσπίδος [2] μεγάλης.

22. Ταῖς κορώναις ἔργον τοὺς ἀετοὺς ἐρεσχελεῖν ἐστιν. οἱ δὲ ὑπερφρονοῦσιν αὐτῶν, καὶ ἐκείναις [3] μὲν ἀπολείπουσι τὴν κάτω φέρεσθαι πτῆσιν, αὐτοὶ δὲ τὸν αἰθέρα [ὑψηλότερον ὄντα] [4] ὠκίστοις [5] τέμνουσιν πτεροῖς, οὐ δήπου δεδιότες (πῶς γὰρ ἂν τοῦτο εἴποι τις, τὴν τῶν ἀετῶν ἀλκὴν καλῶς ἐπιστάμενος;) ἀλλὰ ἰδίᾳ τινὶ μεγαλονοίᾳ ἐῶσιν ἔρρειν ἐκείνας κάτω.

23. Τὸν ἰχθὺν τὸν πομπίλον οὐ μόνον Ποσειδῶνος λέγουσιν ἱερὸν εἶναι, ἀλλὰ καὶ τῶν ἐν Σαμοθρᾴκῃ θεῶν φίλον. ἁλιέα γοῦν τινα ἐν τοῖς ἄνω τοῦ χρόνου τιμωρίαν ὑποσχεῖν τῷδε τῷ ἰχθύι. καὶ τὸ μὲν ὄνομα ἦν ὡς λόγος τοῦ ἁλιέως Ἐπωπεύς,[6] ἦν δὲ ἐξ Ἰκάρου τῆς νήσου, καὶ υἱὸς αὐτῷ ἦν. ἀθηρίας οὖν ποτε γενομένης ἰχθύων, ἀνήγαγε τὸν βόλον μόνους θηράσαντα πομπίλους,[7] οὕσπερ οὖν καὶ δεῖπνον σὺν τῷ παιδὶ ὁ Ἐπωπεὺς ἔθετο. οὐκ ἐς μακρὰν δὲ δίκη τιμωρὸς [8] μετῆλθεν αὐτόν· τῇ γὰρ ἁλιάδι αὐτοῦ κῆτος ἐπελθὸν ἐν ὄψει τοῦ παιδὸς τὸν Ἐπωπέα κατέπιε. λέγουσι δὲ καὶ τοὺς δελφῖνας πολεμίους τῷ πομπίλῳ εἶναι, οὐ μὴν οὐδὲ ἐκείνους καλῶς ἀπαλλάττειν ὅταν αὐτοῦ γεύσωνται· σφαδάζουσι γὰρ παραχρῆμα καὶ ἐκμαί-

[1] μόνη . . . ἡ κεφαλή *H*.
[2] περιφεροῦς μεγάλης *del. H*.
[3] *Ges*: ἐκείνας.
[4] [ὑψ. ὄντα] *gloss, H*.
[5] τοῖς ὠκίστοις.
[6] *Gill*: Ὀπωπεύς *here and below*.
[7] τοὺς πομπίλους.
[8] τιμωρὸς αὐτῶν.

for it only put its head out. At any rate its eyes are said to have been the size of a large, round Macedonian shield.

22. Crows make it their business to worry Eagles, but they despise the Crows and leave them to fly at a lower level, while they themselves cleave the upper air on the swiftest of wings, not of course because they are afraid (how could anyone knowing well what the might of Eagles is say such a thing!): it is rather from what I may call their own magnanimity that they allow those birds to go their miserable way down below.

Crow and Eagle

23. They say that the Pilot-fish is sacred not only to Poseidon but is also beloved of the gods of Samothrace.[a] At any rate a certain fisherman in the olden days was punished by this fish. The name of the fisherman was, according to the story, Epopeus, and he came from the island of Icarus[b] and had a son. Now on one occasion after they had failed to find any fish Epopeus drew up his net with a catch consisting entirely of Pilot-fish, off which he and his son made a meal. But not long after, avenging justice overtook him, for a sea-monster attacked his boat and swallowed Epopeus before the very eyes of his son.

The Pilot-fish

And they also say that Dolphins are the enemies of the Pilot-fish, and they again do not escape unharmed when they eat one, for they immediately begin to writhe and go quite mad, and being

[a] The Cabiri, who were later confused with the Dioscuri.

[b] Icaria, an island of the Sporades off the SW coast of Asia Minor.

νονται, καὶ ἀτρεμεῖν ἀδυνατοῦντες ἐπὶ τοὺς αἰγιαλοὺς ἐκφέρονται, καὶ ἅπαξ ἐκβρασθέντες ὑπὸ τοῦ κύματος κορώναις τε εἰναλίαις[1] καὶ λάροις δεῖπνόν εἰσιν. λέγει δὲ Ἀπολλώνιος ὁ Ῥόδιος ἢ Ναυκρατίτης ὅτι καὶ ἄνθρωπός ποτε οὗτος ἦν, καὶ ἐπόρθμευεν. ὁ δὲ Ἀπόλλων ἠράσθη κόρης, καὶ ἐπειρᾶτο αὐτῇ ὁμιλῆσαι· ἡ δὲ ἀποδιδράσκουσα ἦλθεν ἐς Μίλητον καὶ ἐδεήθη Πομπίλου τινὸς θαλαττουργοῦ, ἵνα αὐτὴν διαγάγοι τὸν πορθμόν· ὁ δὲ ὑπήκουσεν. ἐπιφανεὶς δὲ ὁ Ἀπόλλων τὴν μὲν κόρην ἁρπάζει, τὴν δὲ ναῦν λίθον ἐργάζεται, τὸν δὲ Πομπίλον ἐς τὸν ἰχθὺν τοῦτον μετέβαλεν.

24. Ἰνδοὶ δὲ ἄρα καὶ περὶ τοὺς βοῦς τοὺς δρομικοὺς τίθενται σπουδήν. καὶ ὑπὲρ τῆς ὠκύτητος τῆς ἐκείνων ἁμιλλῶνται βασιλεύς τε αὐτὸς καὶ τῶν ἀρίστων πολλοί, καὶ ποιοῦνται ῥήτρας ἐπὶ χρυσίῳ παμπόλλῳ καὶ ἀργυρίῳ, καὶ οὐχ ἡγοῦνται αἰσχρὸν εἶναι ἐρίζεσθαι ὑπὲρ τῶνδε τῶν ζῴων, συνωρίζουσι δὲ αὐτοὺς ἄρα καὶ ὑπὲρ τῆς νίκης κυβεύουσιν. οἱ μὲν οὖν ἵπποι ζύγιοι θέουσιν, οἱ δὲ βοῦς παράσειροι, καὶ ἐγχρίμπτει τῇ νύσσῃ ὁ ἕτερος, καὶ δεῖ δραμεῖν σταδίους τριάκοντα. ἴσοι δὲ τοῖς ἵπποις οἱ βόες συνθέουσι, καὶ οὐκ ἂν ἀποκρίνειας τὸν ὠκύτερον οὔτε βοῦν οὔτε ἵππον. ἐὰν δέ ποτε ὁ βασιλεὺς πρός τινα ὑπὲρ τῶν ἑαυτοῦ βοῶν σύνθηται, ἐς τοσαύτην προχωρεῖ φιλονικίαν, ὡς αὐτὸς ἐφ' ἅρματος ἕπεσθαι, καὶ

[1] ἐναλίαις.

[a] The 'Little Manx Shearwater.' Wellmann sees in these words a reminiscence of Pancrates, epic poet, 2nd cent. A.D.,

incapable of remaining still are carried on to beaches, and when once they are cast ashore by the wave they furnish a meal to 'sea-crows'[a] [Hom. *Od.* 5. 66] and sea-mews. And Apollonius of Rhodes or of Naucratis says[b] that the Pilot-fish was once actually a human being and a ferryman. And Apollo fell in love with a maiden and attempted to lie with her, but she escaped and came to Miletus and implored one Pompilus, a seaman, to conduct her across the strait. He agreed to do so, but Apollo appeared and seized the maiden, turned the ship into stone, and transformed Pompilus into this fish.

Racing Oxen

24. The Indians devote much attention to fast-running Oxen. And the King himself and many of the nobles make the speed of their oxen the subject of contest, and lay wagers in immense sums of gold and silver, and think no shame to compete with one another respecting these animals, indeed they couple them together and gamble on the race for victory. Now the horses run yoked together, while the Oxen are harnessed alongside and one of them almost grazes the turning-post; they have to run 30 *stades*. The Oxen run as fast as the horses and you could not tell which is the faster of the two, the Ox or the horse. If, as sometimes happens, the King makes a wager with someone over his own Oxen, so full of emulous zeal does he become that he himself follows in a chariot and urges on the

whom Athenaeus (7. 283), cites as his authority for this same story; see *Hermes* 26. 523.

[b] See Powell, *Coll. Alex.* p. 6. The story was related by Apollonius in his poem Κτίσις Ναυκράτεως, but it is thought unlikely that he was born or lived at Naucratis.

παρορμᾶν τὸν ἡνίοχον. ὁ δὲ ἄρα τοὺς μὲν ἵππους ἐξαιμάττει τῷ κέντρῳ, τῶν δὲ βοῶν τὴν χεῖρα ἀνέχει· ἀκέντητοι γὰρ θέουσι. τοσαύτη δέ ἐστι περὶ τὴν βοεικὴν ἅμιλλαν ἡ φιλοτιμία, ὡς μὴ μόνους τοὺς πλουσίους ὑπὲρ αὐτῶν ἐπὶ πολλῷ φιλονικεῖν μηδὲ τοὺς δεσπότας ἀλλὰ καὶ τοὺς θεωμένους, οἷα δήπου καὶ ὁ Ἰδομενεὺς ὁ Κρὴς καὶ ὁ Λοκρὸς Αἴας παρ᾽ [1] Ὁμήρῳ φιλονικοῦντε [2] ἀποδείκνυσθον. εἰσὶ δὲ καὶ ἕτεροι παρ᾽ αὐτοῖς βόες, ἰδεῖν κατὰ τοὺς μεγίστους τράγους· ⟨καὶ⟩ [3] αὐτοὶ δὲ καθ᾽ ἑαυτοὺς ζεύγνυνται, καὶ τρέχουσιν ὤκιστα, καὶ τῶν ἵππων γε τῶν Γετικῶν οὐκ εἰσὶ νωθέστεροι.

25. Λόγος ἔχει τοὺς ἵππους τοὺς πίνοντας ἐκ τοῦ Κοσσινίτου ποταμοῦ (ἔστι δὲ οὗτος ἐν Θρᾴκῃ) δεινῶς ἐκθηριοῦσθαι· ἐκδίδωσι δὲ ὁ ποταμὸς οὗτος ἐς τὴν Ἀβδηριτῶν, καὶ ἀναλίσκεται ἐς τὴν Βιστονικὴν λίμνην. ἐνταῦθά τοι καὶ τὰ βασίλεια γενέσθαι ποτὲ Διομήδους τοῦ Θρᾳκός, ᾧ καὶ αἱ ἀνήμεροι ἐκεῖναι ἵπποι κτῆμα ἦσαν ὁ Ἡράκλειος ἆθλος. τὸ δὲ αὐτό φασι πάσχειν καὶ τοὺς ἵππους τοὺς ἐκ τῆς Ποτνιάδος κρήνης πίνοντας. αἱ δὲ Ποτνιαὶ τὸ χωρίον, ἔνθα ἡ κρήνη, οὐ μακρὰν ἀπὸ Θηβῶν ἐστιν. Ὠρείτας [4] δὲ λέγουσι καὶ Γεδρωσίους [5] ἰχθῦς παραβάλλειν τοῖς ἵπποις χόρτον. Κελτοὺς δὲ ἀκούω καὶ τοῖς βουσὶ καὶ

[1] παρὰ τῷ.
[2] φιλονεικοῦντες.
[3] ⟨καί⟩ *add.* *H.*
[4] *Jac*: Ὠραείτας.
[5] *Gill*: Ἀδρασίους.

---

[a] The Compsantus of Hdt. 7. 109.
[b] The capture of the mares of Diomedes, King of the

driver. And the latter makes the horses quite bloody with his goad, but withholds his hand from the Oxen, for they run without any goading. And feeling runs so high over this ox-racing that not only the rich and the owners but the spectators also contend for large stakes, just as in Homer [*Il.* 23. 473–93] Idomeneus of Crete and Ajax of Locris are represented contending.

There are also in India other Oxen the size of the largest he-goats. These also are yoked together and run extremely fast, at any rate they are no less spirited than the horses of the Getae.

Horses affected by certain waters

25. It is reported that Horses which drink from the river Cossinitus [a] (it is in Thrace) become terribly savage. This river empties itself into the territory of Abdera and is swallowed up in the Lake of the Bistones. Here, you know, was once the palace of Diomedes the Thracian who owned those famous wild mares, one of the 'Labours' of Heracles.[b] And they say that the same fate befalls horses that drink from the spring at Potniae.[c] The place called Potniae, where the spring is, lies not far from Thebes. They say that the inhabitants of Oraea and Gedrosia [d] give their Horses fish for fodder, and I am told that the Celts feed both their cattle and

Bistones, was the 8th Labour imposed by Eurystheus upon Heracles. They ate human flesh, but after eating their master, whom Heracles had slain, became tame.

[c] Village in Boeotia, famed as the home of the mythical Glaucus, who was torn to pieces by his mares. It lay about 1 mi. SW of Thebes.

[d] Oraea (or Orae), a town on the eastern border of Gedrosia, a region corresponding more or less to the modern Makran and extending from the Gulf of Oman to the River Indus.

τοῖς ἵπποις ἰχθῦς διδόναι δεῖπνον. ἐνταῦθά τοι λέγουσι καὶ τοὺς ἵππους τὴν ἀποπνοὴν τὴν ἐκ τῶν ἀνθρώπων φεύγοντας ἐς τὰ νοτιώτερα τῆς Εὐρώπης φέρεσθαι, μάλιστα ὅταν οἱ νότοι καταπνέωσι. καὶ Μακεδόνας δὲ καὶ Λυδοὺς ὁμολογοῦσί τινες καὶ αὐτοὺς ἰχθύσι τοὺς ἑαυτῶν ἵππους τρέφειν, καὶ τὰ πρόβατα δὲ τὰ Λύδια καὶ τὰ Μακεδονικὰ ἐκ τῶν αὐτῶν πιαίνεσθαι λέγουσιν. ἐν Μυσοῖς δὲ τῶν θηλειῶν ἵππων ἀναβαινομένων ἐπηύλουν τινές, οἷον ὑμέναιόν τινα τοῦτον τοῖς τῶν ἵππων γάμοις ἐπᾴδοντες· τάς τε ἵππους ὑπὸ τοῦ μέλους θελγομένας τάχιστα ἐγκύμονας γίνεσθαι, καὶ οὖν καὶ καλοὺς τοὺς πώλους ἀποτίκτειν. καὶ ἐκεῖνο δὲ περὶ ἵππων ἤκουσα. τοὺς πρεσβυτέρους ἤδη φασὶ καὶ προήκοντας τὴν ἡλικίαν ἀσθενῆ γεννᾶν τὰ ἐξ αὐτῶν ἔκγονα· τά τε γὰρ ἄλλα καὶ τοὺς πόδας ἀγεννεῖς ἔχειν. βίον δὲ ἵππων καὶ χρόνον ἀριθμοῦσιν ἐς τοσάδε ἔτη· τῶν μὲν [1] ἀρρένων ἐς πέντε καὶ τριάκοντα [2] . . . Ἀριστοτέλης δ' ὁ Νικομάχου λέγει πέντε καὶ ἑβδομήκοντα ἔτη διαβιῶναι ἵππον.

26. Ἐκ Σούσων τῶν Περσικῶν ἐς Μηδίαν ἀπιόντι [3] ἐν τῷ δευτέρῳ σταθμῷ πάμπολύ τι λέγεται σκορπίων πλῆθος γίνεσθαι, ὥστε τὸν τῶν Περσῶν βασιλέα, ὁπότε διίοι,[4] πρὸ τριῶν ἡμερῶν προστάττειν πᾶσι θηρεύειν αὐτούς, καὶ τῷ πλείστους θηράσαντι δῶρα διδόναι. εἰ γὰρ τοῦτο μὴ γένοιτο, ὁ χῶρος ἄβατός ἐστιν· ὑπὸ παντὶ γὰρ

[1] μὲν γάρ.
[2] *Lacuna.*
[3] *Schn* : ἀπιόντων.
[4] *Schn* ; δὴ ἴοι.

their horses on fish. In their country, it is said, the Horses actually flee from the scent of human beings and hasten to the more southerly parts of Europe, especially when the South Wind blows. And there are those who bear witness to the fact that the inhabitants of Macedonia and of Lydia also feed their horses on fish, and who assert that the sheep of Lydia and of Macedonia are fattened on the same diet. In Moesia while Mares are in process of being covered some people play the pipe, accompanying the marriage of Horses with nuptial music, as it were; and the Mares are so enchanted by the melody that they very soon become pregnant and, what is more, produce beautiful foals. This too I have heard concerning Horses. They say that when Horses are older and advanced in years the offspring which they beget is feeble, having besides other defects poor legs. The age and life of Horses men reckon as so many years: in the case of Stallions, five and thirty . . .[a] But Aristotle the son of Nicomachus states [*HA* 545 b 20] that a Horse lived for five and seventy years.

fed on fish

affected by music

their age

26. In the second stage of a journey from Susa in Persia to Media there are said to be Scorpions in multitudes, so that when the Persian King is going to pass that way he issues orders three days in advance that everybody is to hunt them, and bestows presents on the man who has caught the greatest number. For if this were not done, the region would be impassable, for 'beneath every stone'

Scorpions in Persia

[a] Some words must have been lost here, corresponding to Aristotle's ἡ δὲ θήλεια πλείω τῶν τετταράκοντα, 'in the case of Mares, more than forty.'

λίθῳ καὶ βώλῳ πάσῃ σκορπίος ἐστί. λέγουσι δὲ καὶ ὑπὸ σκολοπενδρῶν ἐξαναστῆναι Ῥοιτιεῖς· τοσοῦτο πλῆθος αὐτοῖς ἐπεφοίτησε τούτων. φασὶ δὲ καὶ ἐν Κυρήνῃ μυῶν διάφορα γίνεσθαι [1] γένη οὐ μόνον ταῖς χρόαις, ἀλλὰ καὶ ταῖς μορφαῖς· ἐνίους γὰρ αὐτῶν πλατυπροσώπους εἶναι καθάπερ τὰς γαλᾶς, καὶ αὖ πάλιν ἄλλους ἐχινώδεις,[2] οὕσπερ οὖν καὶ οἱ ἐπιχώριοι καλοῦσι ἐχινέας. ἐν Αἰγύπτῳ δὲ ἀκούω δίποδας εἶναι μῦς, καὶ μεγίστους μεγέθει φύεσθαι, τοῖς γε μὴν ἐμπροσθίοις ποσὶν ὡς χερσὶ χρῆσθαι· εἶναι γὰρ αὐτοὺς τῶν ὄπισθεν βραχυτέρους.[3] βαδίζουσι δὲ ὀρθοὶ ἐπὶ τοῖν δυοῖν ποδοῖν· ὅταν δὲ διώκωνται, πηδῶσι. Θεόφραστος λέγει ταῦτα.

27. Λέγει τις λόγος [4] τοὺς ὄρνιθας τοὺς ἀτταγᾶς μετακομισθέντας ἐς Αἴγυπτον ἐκ Λυδίας καὶ ἀφεθέντας ἐς τὰς ὕλας τὰ μὲν πρῶτα ὄρτυγος φωνὴν ἀφιέναι· χρόνῳ δὲ ὕστερον τοῦ ποταμοῦ κοίλου ῥυέντος λιμὸς ἐγένετο, καὶ πολλοὶ τῶν κατὰ τὴν χώραν ἀπώλλυντο. οὐ διέλιπον οὖν οἱ ὄρνιθες οὗτοι πολλῷ σαφέστερον καὶ ἐναρθρότερον παιδίου φθέγμα [5] ἀφιέντες καὶ λέγοντες 'τρὶς τοῖς κακοῖς τὰ κακά.' λέγει δὲ ὁ αὐτὸς λόγος ὅτι συλληφθέντες καὶ ἀγρευθέντες οὐ μόνον οὐ τιθασεύονται, ἀλλὰ οὐδὲ φωνὴν ἔτι ἀφιᾶσιν ἣν πρότερον ἠφίεσαν· ἡ δουλεία γὰρ αὐτῶν καὶ ἡ κάθειρξις [6] καταψηφίζεται σιωπήν. ἐὰν δὲ ἀφεθῶσι καὶ ἐλεύθερον ἁπλώσωσι

[1] *Jac* : γένεσθαι.
[2] *Ges* : ἐχεώδεις MSS; ἐ. [ὀξείας ἀκάνθας ἔχοντας] *del. H.*
[3] βραχυτέρους. [εἶδον τούτους, Λιβυκοί εἰσιν] *del. Jac, H.*
[4] λόγος τις λέγει.
[5] *Jac* : μεῖζον φθέγμα.

and every clod 'there lurks a scorpion.' And they say that the inhabitants of Rhoeteum [a] were driven out by centipedes, so great was the multitude that invaded them. They say too that in Cyrene there are species of mice which differ not only in colour but in form: some for instance have flat faces like martens, others again look like hedgehogs (*echinoi*), and these the natives call 'prickly mice' (*echinees*).[b] And I have heard that in Egypt there are mice [c] with only two legs, and that they grow to a great size, but their front legs they use as hands, for they are shorter than their hind legs. And they walk erect on their two legs, but when pursued they jump. This is what Theophrastus says [*fr.* 174. 8].

The Acomys

The Jerboa

27. There is a story that the birds known as Francolins when transported from Lydia to Egypt and let loose in the woods, at first uttered the note of a quail. Later on, owing to the river being confined in its hollow bed, a famine broke out and many of the inhabitants perished, whereupon these same birds never ceased to utter with a sound far clearer and more articulate than any child words meaning 'Three curses on the accursed.' And the same story tells how if they are captured and snared they not only refuse to be tamed but no longer even utter the notes which they did before: their servitude and confinement decree silence against them. If however they are let go and can unfold their wings at liberty

The Francolin

[a] Town in the Troad on the Hellespont.

[b] This is the *Mus cahirinus* of the genus Acomys, allied both to the rat and the mouse.

[c] Ael. is referring to the Jerboa.

[6] καὶ ἡ κάθειρξις *del. H.*

τὸ πτερόν, καὶ ἐς ἤθη τὰ ἑαυτῶν ἀφίκωνται πάλιν γίνονται ἔμφωνοι, ὁμοῦ καὶ τὸ φθέγμα καὶ τὴν παρρησίαν ἀναλαβόντες.

28. Λέγουσι δὲ καὶ τοὺς σκῶπας (ὧν καὶ Ὅμηρος ἐν Ὀδυσσείᾳ μέμνηται λέγων πολλοὺς αὐτοὺς περὶ τὸ ἄντρον τὸ τῆς Καλυψοῦς εὐνάζεσθαι) καὶ ἐκείνους ἁλίσκεσθαι ὀρχήσει. ἄνδρες ⟨δὲ⟩ [1] ὀρχηστικοί φασι καὶ ὀρχήσεως εἶδός τι ἐξ αὐτῶν κεκλῆσθαι, καὶ εἴ γε αὐτοῖς χρὴ πιστεύειν, ἡ ὄρχησις αὕτη σκὼψ κέκληται. καὶ τὸ μιμεῖσθαι δέ τινα [2] ἐπὶ τὸ γελοιότερον καὶ διαπαίζειν ἥδιστον δοκεῖ τοῖσδε τοῖς ὄρνισιν· ἔνθεν τοι ⟨καὶ⟩ [3] ἐτράπη ὁ λόγος, καὶ ἡμεῖς τὸ σκώπτειν οὕτω καλοῦμεν. λέγεται δὲ ὁ σκὼψ οὗτος μικρότερος εἶναι γλαυκὸς καὶ τὴν χρόαν ἔχειν μολίβῳ προσεοικυῖαν τῷ βαθυτάτῳ,[4] ἔχειν δὲ τὰ πτερὰ αὐτοῦ φασι στίγματα [5] ὑπόλευκα. ἀναφαίνει [6] τε δύο ἀπὸ τῶν ὀφρύων παρ᾽ ἑκάτερον τὸν κρόταφον πτερά. Καλλίμαχος δὲ δύο φησὶν εἶναι γένη σκωπῶν, καὶ τοὺς μὲν φθέγγεσθαι, τοὺς δὲ συγκεκληρῶσθαι σιωπῇ· καὶ τοὺς μὲν αὐτῶν λέγεσθαι σκῶπας, τοὺς δὲ ἀείσκωπας. λέγει δὲ Ἀριστοτέλης τοὺς παρ᾽ Ὁμήρῳ διὰ τοῦ σίγμα μὴ λέγεσθαι, ἀλλὰ ἁπλῶς ὀνομάζεσθαι κῶπας. τοὺς οὖν τιθέντας τὸ σίγμα ἁμαρτάνειν τῆς κατὰ τὸ ὄνομα ἀληθείας καὶ τῆς Ὁμήρου περὶ τὸν ὄρνιν κρίσεώς τε καὶ

[1] ⟨δέ⟩ *add. H.*
[2] τινας.
[3] ⟨καί⟩ *add. H.*
[4] βαθύτατα.
[5] καὶ στίγματα.

and return to their own haunts, they again become vocal and recover both their voice and their freedom of speech together.

The Little Horned Owl

28. They say that men catch the Little Horned Owl also[a] (mentioned in the *Odyssey* [5. 66] by Homer who says that it nests in great numbers round about the cavern of Calypso) by dancing. And dancers assert that a certain kind of dance is called after this bird, and if we are to believe them this dance has been called 'the Little Horned Owl.' And that anyone should caricature and imitate them in a playful way affords these birds the greatest pleasure. This is the origin of the word *skôptein* which we use, meaning 'to mock.' It is said that the Little Horned Owl is smaller than the Little Owl and that its colour resembles lead of the deepest hue, but its wings are said to have whitish speckles. And it displays two feathers rising from the brows on either temple. Callimachus [*fr.* 418 P] maintains that there are two kinds of Little Horned Owl, one kind is vocal, the other doomed to silence; the latter is called *skôps*, the former *aeiskôps*.[b] But Aristotle asserts that in Homer the word does not begin with a *sigma* (*skôps*), but that the birds are called simply *kôpes*. So those who prefix a *sigma* mistake the true spelling of the word and are mistaken as to Homer's judgment and knowledge of the

[a] 'Also,' *i.e.* as well as the Sting-ray; cp. 1. 39.

[b] 'All-the-year-round owl'; see Arist. *HA* 617 b 31, and D. W. Thompson's note in his Eng. transl. The σκώψ is a migrant.

[6] ἀναφέρει.

γνώσεως. καὶ ταῖς μὲν ἄλλαις ὥραις τοῦ ἔτους μὴ ἐσθίεσθαι αὐτούς, ἐν δὲ τῷ μετοπώρῳ δύο ἡμέραις ἢ μιᾷ τοὺς θηρωμένους, ἀλλὰ τούτους γε ἐδωδίμους εἶναι. τῶν δὲ ἀεισκώπων διαφέρουσιν οἱ σκῶπες τῷ πάχει, παραπλήσιοι δέ εἰσι τὴν ἰδέαν τρυγόνι τε καὶ φάττῃ.

29. Ἀλλὰ τό γε τῶν Πυγμαίων ἔθνος ἀκούω καὶ ἐκεῖνο καθ' ἑαυτὸ βασιλεύεσθαι, καὶ οὖν καὶ γενέσθαι παρ' αὐτοῖς ἐκλείποντος ἄρρενος βασιλέως βασιλίδα τινὰ καὶ κρατῆσαι τῶν Πυγμαίων, Γεράναν ὄνομα, ἥνπερ οὖν ἐκθεοῦντες οἱ Πυγμαῖοι σεμνοτέραις ἢ κατ' ἄνθρωπον ἐτίμων τιμαῖς. ἐκ τούτων οὖν ἐκείνη φασὶ τὴν διάνοιαν ἐξηνεμώθη, καὶ τὰς θεὰς παρ' οὐδὲν ἐτίθετο. μάλιστα δὲ τὴν Ἥραν καὶ τὴν Ἀθηνᾶν καὶ τὴν Ἄρτεμιν καὶ τὴν Ἀφροδίτην οὐδὲ ἴκταρ ἔλεγε βάλλειν πρὸς τὸ αὐτῆς κάλλος. οὔκουν ἔμελλεν ἁμαρτήσεσθαι κακοῦ νοσοῦσα τοιαῦτα· κατὰ γὰρ τὸν τῆς Ἥρας χόλον ἐς ὄρνιν αἰσχίστην τὸ εἶδος τὸ ἐξ ἀρχῆς ἤμειψε, καὶ ἔστιν ἡ νῦν γέρανος, καὶ πολεμεῖ τοῖς Πυγμαίοις, ὅτι αὐτὴν ἐξέμηναν τῇ πέρα τιμῇ καὶ ἀπώλεσαν.

bird.[a] At all other seasons of the year the Little Horned Owl is not edible, but only when caught on one or two days in the late autumn, and then it is edible. These *Skôpes* differ from the *Aeiskôpes* in bulk, and bear some resemblance to a turtle-dove or a ring-dove.

29. As to the race of Pygmies I have heard that they are governed in a manner peculiar to themselves, and that in fact owing to the failure of the male line a certain woman became queen and ruled over the Pygmies; her name was Gerana, and the Pygmies worshipped her as a god, paying her honours too august for a human being. The result was, they say, that she became so puffed up in her mind that she held the goddesses of no account. It was especially Hera, Athena, Artemis, and Aphrodite that, she said, came nowhere near her in beauty. But she was not destined to escape the evil consequences of her diseased imagination. For in consequence of the anger of Hera she changed her original form into that of a most hideous bird and became the crane of today and wages war on the Pygmies[b] because with their excessive honours they drove her to madness and to her destruction.

The Pygmies and their Queen

[a] The statement does not occur in any surviving work of Aristotle, nor is the form κῶπες found in our MSS. of Homer, though Eustathius (1523. 59, 1524. 6) says that at *Od.* 5. 66 τινὲς κῶπας γράφουσι δίχα τοῦ ς. On this passage see Wellmann in *Hermes* 51. 2.

[b] Cp. Milton *PL* 1. 575 That small infantry | Warred on by cranes.

BOOK XVI

Iϛ

1. Ἀνὴρ πορφυρεὺς ὅταν θηράσῃ πορφύραν, οὐκ ἐς ἀνθρώπων τροφήν, ἀλλ' ἐς ἐρίων βαφήν, εἰ μέλλοι μένειν ἡ ἐκ τοῦ ζῴου χρόα δευσοποιὸς καὶ δυσέκνιπτος καὶ οἵα τὴν βαφὴν ἐργάσασθαι γνησίαν ἀλλ' οὐ δεδολωμένην, μιᾷ λίθου καταφορᾷ διαφθείρει τὴν πορφύραν αὐτοῖς ὀστράκοις. ἐὰν δὲ κουφοτέρα ἡ πληγὴ γένηται, καταλειφθῇ δὲ τὸ ζῷον ἔτι ἔμπνουν, ἀχρεῖός ἐστιν ἐς τὴν βαφὴν ἡ δεύτερον βληθεῖσα τῷ λίθῳ πορφύρα· ὑπὸ γὰρ τῆς ὀδύνης ἐξανάλωσε τὴν βαφὴν ἀναποθεῖσαν ἐς τὸν τῆς σαρκὸς ὄγκον ἢ ἄλλως ἐκρυεῖσαν. τοῦτό τοι καὶ Ὅμηρος οἶδέ φασι, καὶ τοὺς ἀποθνήσκοντας ἀθρόως [1] τῷ τῆς πορφύρας θανάτῳ καταλαμβάνεσθαί φησι, τὸ ᾀδόμενον ἐν τοῖς ἑαυτοῦ μέτροις ἀναμέλπων ἐκεῖνος

ἔλλαβε πορφύρεος θάνατος καὶ μοῖρα κραταιή.

2. Ἐν Ἰνδοῖς μανθάνω σιττακοὺς ὄρνεις [2] γίνεσθαι, ὧνπερ οὖν καὶ ἀνωτέρω μνήμην ἐποιησάμην· ἃ δὲ πρότερον ὑπὲρ αὐτῶν οὐκ εἶπον, ταῦτά μοι λεχθῆναι νῦν δοκεῖ πρεπωδέστατα. γένη τρία αὐτῶν ἀκούω· οἱ πάντες δὲ οὗτοι μαθόντες ὡς παῖδες, οὕτως καὶ αὐτοὶ γίνονται λάλοι καὶ φθέγγονται φθέγμα ἀνθρωπικόν. ἐν δὲ

[1] ἀθρόως μιᾷ πληγῇ. [2] ὄρνις.

BOOK XVI

1. When a fisherman after Purple Shellfish catches one, not for human consumption but for dyeing wool, if the colour from it is to remain fast, indelible, and capable of producing the genuine tint unadulterated, then he smashes it, shell and all, with one blow of a stone. But if the blow is too light and the creature is left still alive, a second blow with the stone renders it useless for dyeing purposes. For the pain causes the fish to spend the dye which is absorbed into the mass of flesh or escapes in some other way. And this, they say, was known to Homer who says of those who die all at once that they are overtaken by the death of the Purple Shellfish: in his poem he sings in the well-known passage how

The Purple Shellfish

'Empurpled[a] death and violent fate laid hold on him' [*Il.* 5. 83].

2. I learn that in India there are Parrots, and I have also mentioned them earlier on,[b] but this seems a most fitting place to relate what I did not relate on the former occasion. I am told that there are three kinds, and all learn like children and become talkative in the same way and speak like human beings. In the forests however they utter the notes

Birds of India

[a] So Ael. understood πορφύρεος; the proper meaning is 'onrushing.'

[b] See 13. 18.

ταῖς ὕλαις ὀρνίθων μὲν ἀφιᾶσιν ἦχον, φωνὴν δὲ εὔσημόν τε καὶ εὔστομον οὐ προΐενται, ἀλλ' εἰσὶν ἀμαθεῖς καὶ οὔπω λάλοι. γίνονται δὲ καὶ ταῶς ἐν Ἰνδοῖς τῶν πανταχόθεν μέγιστοι, καὶ πελειάδες χλωρόπτιλοι· φαίη τις ἂν πρῶτον θεασάμενος καὶ οὐκ ἔχων ἐπιστήμην ὀρνιθογνώμονα, σιττακὸν εἶναι καὶ οὐ πελειάδα. χείλη δὲ ἔχουσι καὶ σκέλη τοῖς ἐν Ἕλλησι πέρδιξι τὴν χρόαν προσεοικότα. ἀλεκτρυόνες δὲ γίνονται μεγέθει μέγιστοι, καὶ ἔχουσι λόφον οὐκ ἐρυθρὸν κατά γε τοὺς ἡμεδαπούς, ἀλλὰ ποικίλον κατὰ τοὺς ἀνθινοὺς στεφάνους. τὰ δὲ πτερὰ τὰ πυγαῖα ἔχουσιν οὐ κυρτὰ οὐδὲ ἐς ἕλικα ἐπικαμφθέντα ἀλλὰ πλατέα, καὶ ἐπισύρουσιν αὐτά, ὥσπερ οὖν καὶ οἱ ταῶς, ὅταν μὴ ὀρθώσωσί τε καὶ ἀναστήσωσιν αὐτά. χρόαν δὲ ἔχει τὰ πτερὰ τῶν Ἰνδῶν ἀλεκτρυόνων χρυσωπόν τε καὶ κυαναυγῆ κατὰ τὴν σμάραγδον λίθον.

3. Γίνεται δὲ ἐν Ἰνδοῖς καὶ ἄλλο ὄρνεον, καὶ ἔχει τὸ μέγεθος κατὰ τοὺς ψᾶρας, καὶ ἔστι ποικίλον, καὶ μουσωθὲν ἀνθρώπου φωνὴν εἶτα μέντοι τῶν σιττακῶν ἐστι λαλίστερόν τε καὶ θυμοσοφώτερον. οὐ μὴν τὴν ἐξ ἀνθρώπων τροφὴν ἡδέως ὑπομένει,[1] ἀλλὰ ἐλευθερίας πόθῳ καὶ παρρησίας τῆς κατὰ τὴν συντροφίαν ἐπιθυμίᾳ ἀσπάζεται λιμὸν μᾶλλον ἢ δουλείαν μετὰ τρυφῆς. καλοῦσι δὲ αὐτὸ οἱ Μακεδόνων Ἰνδοῖς ἐποικήσαντες ἔν τε Βουκεφάλοις πόλει καὶ τῇ περὶ ταύτην καὶ τῇ καλουμένῃ Κύρου πόλει [2] καὶ ταῖς ἄλλαις, ἃς

of birds, and do not produce intelligible and distinct speech, but are unlearned and cannot talk as yet. There are also Peacocks in India, larger than anywhere else, and Doves with green plumage;[a] anyone seeing them for the first time and not possessing a knowledge of birds would say that they were parrots not doves. But they have beaks and legs the same colour as those of partridges in Greece. And the Cocks there are of immense size, and their combs are not scarlet like those of our country, but of variegated hue like flower-garlands. And their tail-feathers are not arched or curved in a circle but flat, and they trail them, just as peacocks do when not raising them aloft. And the wings of Indian Cocks are golden with the dark gleam of an emerald.

3. There is also in India another bird, the size of a starling, and it is of varied colouring and if taught to utter human speech is more talkative and by nature more intelligent than the parrot. Yet it does not willingly endure to be kept by man, but in its yearning for liberty and its desire for its natural freedom it welcomes starvation in preference to captivity with its luxuries. And the Macedonians who settled in India in the cities founded by Alexander, the son of Philip, in Bucephala[b] and the surrounding country,

The Mynah

[a] ' An Indian Green Fruit-pigeon, such as *Crocopus chlorogaster* ' (Thompson, *Gk. birds*, s.v. Πελειάς).

[b] Founded by Alexander 326 B.C. on the river Jhelum (Hydaspes) after his victory over Porus and named after his horse Bucephalus.

[1] *Ges*: ὑπομένοι.

[2] Κυροπόλει.

ἀνέστησεν Ἀλέξανδρος ὁ Φιλίππου, κερκίωνα· ἔσχε δὲ ἄρα τὸ ὄνομα τήνδε τὴν γένεσιν, ἐπειδὴ καὶ αὐτὸ διασείει τὸν ὄρρον, ὥσπερ οὖν καὶ οἱ κίγκλοι.

4. Γίνεσθαι δὲ ἐν Ἰνδοῖς καὶ κήλαν ἀκούω ὄρνιν· καὶ τὸ μέγεθος τριπλασίων[1] ὠτίδος ἐστί, καὶ τὸ στόμα ἔχει γενναῖον δεινῶς καὶ μακρὰ τὰ σκέλη· φέρει δὲ καὶ πρηγορεῶνα καὶ ἐκεῖνον μέγιστον προσεμφερῆ κωρύκῳ, φθέγμα δὲ ἔχει καὶ μάλα ἀπηχές. καὶ τὴν μὲν ἄλλην πτίλωσίν ἐστι τεφρός, τὰς δὲ πτέρυγας ἄκρας ὠχρός ἐστιν.

5. Ἀκούω δὲ ἔγωγε καὶ Ἰνδὸν ἔποπα διπλασίονα τοῦ παρ' ἡμῖν καὶ ὡραιότερον ἰδεῖν. καὶ Ὅμηρος μὲν λέγει βασιλεῖ κεῖσθαι ἄγαλμα Ἕλληνι χαλινὸν καὶ κόσμον ἵππου, ὁ δὲ ἔποψ οὗτος Ἰνδῶν βασιλεῖ ἄθυρμά ἐστι, καὶ διὰ χειρῶν αὐτὸν φέρει, καὶ ἥδεται αὐτῷ, καὶ συνεχὲς ἐνορᾷ τὴν ἀγλαΐαν τεθηπὼς τοῦ ὄρνιθος καὶ τὸ κάλλος τὸ αὐτοφυές. ἐπᾴδουσι δὲ ἄρα τῷδε τῷ ὀρνέῳ καὶ μῦθον Βραχμᾶνες, καὶ ὅ γε μῦθος ὁ ᾀδόμενος οὗτός ἐστιν. παῖς ἐγένετο Ἰνδῶν βασιλεῖ, καὶ ἀδελφοὺς εἶχεν, οἵπερ οὖν ἀνδρωθέντες ἐκδικώτατοί τε γίνονται καὶ λεωργότατοι. καὶ τούτου μὲν ὡς νεωτάτου καταφρονοῦσι, τὸν δὲ πατέρα ἐκερτόμουν καὶ τὴν μητέρα, τὸ γῆρας αὐτῶν ἐκφαυλίσαντες. ἀναίνονται οὖν ἐκεῖνοι τὴν σὺν τούτοις διατριβήν, καὶ ᾤχοντο φεύγοντες ὅ τε παῖς καὶ οἱ γέροντες. συντόνου δὲ ἄρα αὐτοὺς πορείας διαδε-

[1] τριπλάσιον.

in Cyropolis [a] and the rest, call the bird *Cercion* (mynah). The name has its origin in the fact that it too wags its rump (*cercos*) as the wagtail does.

The Adjutant stork

4. I have heard that there is also in India a bird called the 'Adjutant.' It is three times the size of a bustard, and has a mouth of astonishing size and long legs. It also has an enormous crop resembling a wallet and an extremely harsh cry. While the rest of its plumage is of an ashen colour, the wing-tips are pale.

The Hoopoe of India

5. I have heard also that the Indian Hoopoe is twice as big as the bird of our country and more beautiful in appearance. And as Homer says [*Il.* 4. 144] that the bit and trappings of a horse are laid up to be a Greek king's glory, so the Hoopoe is the joy of the Indian King: he carries it on his hand and delights in it, gazing continually in wonder at its splendour and its natural beauty.

A Brahmani myth

Now the Brahmins also relate a legend regarding this bird, and the legend they relate is as follows. A son was born to an Indian king and he had brothers who, when they were grown to manhood, became extremely lawless and violent. And they looked down upon their brother, as being the youngest, jeered at their father and mother, and showed no respect for their old age. Accordingly the parents refused to live with them and departed into exile, the aged couple with their young son. There ensued a laborious journey for them; the parents' strength

[a] Cyropolis, more commonly known as Cyreschata, was in Sogdiana. It was stormed and destroyed by Alexander in 329 B.C. The name is probably the Graecised form of some Oriental name.

ξαμένης, οἱ μὲν ἄπειπον καὶ ἀποθνήσκουσιν, ὁ δὲ παῖς οὐκ ὠλιγώρησεν αὐτῶν, ἀλλ' ἔθαψεν αὐτοὺς ἐν ἑαυτῷ, ξίφει τὴν κεφαλὴν διατεμών. ἀγασθέντα δὲ τὸν πάντ' ἐφορῶντα Ἥλιον οἱ αὐτοί φασι τῆς εὐσεβείας τὴν ὑπερβολήν, ὄρνιν αὐτὸν ἀποφῆναι, κάλλιστον μὲν ὄψει, μακραίωνα δὲ τὸν βίον· ὑπανέστηκε δέ οἱ καὶ λόφος ἐκ τῆς κορυφῆς, οἱονεὶ μνημεῖον τοῦτο τῶν πεπραγμένων ὅτε ἔφευγεν. τοιαῦτα ἄττα καὶ Ἀθηναῖοι ὑπὲρ τοῦ κορύδου τερατευόμενοι προσεῖχον μύθῳ τινί, ᾧπερ οὖν ἀκολουθῆσαί μοι δοκεῖ καὶ Ἀριστοφάνης ὁ τῆς κωμῳδίας ποιητὴς ἐν Ὄρνισι λέγων

ἀμαθὴς γὰρ ἔφυς κοὺ πολυπράγμων, οὐδ' Αἴσωπον πεπάτηκας,
ὃς ἔφασκε λέγων κορυδὸν πάντων πρώτην ὄρνιθα γενέσθαι,
προτέραν τῆς γῆς, κἄπειτα νόσῳ τὸν πατέρ' αὐτῆς ἀποθνήσκειν·
γῆν δ' οὐκ εἶναι, τὸν δὲ προκεῖσθαι πεμπταῖον. τὴν δ' ἀποροῦσαν
ὑπ' ἀμηχανίας τὸν πατέρ' αὐτῆς ἐν τῇ κεφαλῇ κατορύξαι.

ἔοικεν οὖν ἐξ Ἰνδῶν τὸ μυθολόγημα ἐπ' ἄλλου μὲν ὄρνιθος, ἐπιρρεῦσαι δ' οὖν καὶ τοῖς Ἕλλησιν. ὠγύγιον γάρ τι μῆκος χρόνου λέγουσι Βραχμᾶνες, ἐξ οὗ ταῦτα τῷ ἔποπι τῷ Ἰνδῷ ἔτι ἀνθρώπῳ ὄντι καὶ παιδὶ τήν γε ἡλικίαν ἐς τοὺς γειναμένους πέπρακται.

6. Ἐν Ἰνδοῖς γίνεται ζῷον κροκοδείλῳ χερσαίῳ παραπλήσιον ἰδεῖν· μέγεθος δὲ αὐτῷ κυνιδίου

failed, and they died. The son however did not neglect them but split his head with a sword and buried them in himself. The Brahmins assert that the all-seeing Sun was so filled with admiration for this surpassing act of piety that he transformed the boy into a bird most beautiful to behold and endowed with length of days. And from his crown there sprang up a crest, as it were in commemoration of the events of his exile. The Athenians too tell some such wondrous tale in a myth regarding the Lark, which Aristophanes, the writer of comedies, appears to me to have followed in his *Birds* [471–5] when he says

> 'No, for you were unlearned and no busybody and had not thumbed your Aesop, who used to say that the Lark was the first of all birds to be born, before the earth, and that then its father fell sick and died. But there was no earth, and the corpse was laid out for five days, and the Lark in straits and at its wits' end buried its father in its own head.'

So it seems that this fable from India, about a different bird indeed, yet spread to the Greeks as well. For the Brahmins maintain that it is long ages since the Indian Hoopoe, while still a human being and a child in years, did this to its parents.

The Pangolin

6. In India there is an animal somewhat like the land-crocodile [a] in appearance. It is the size of a

[a] See 1. 58, note a.

Μελιταίου εἴη ἄν. περίκειται δὲ ἄρα φολίδα τραχεῖαν [1] οὕτω καὶ πυκνήν, ὥστε ὅταν δαρῇ ῥίνης αὐτοῖς ἔργα παρέχει. διατέμνει δὲ καὶ χαλκόν, καὶ τὸν σίδηρον διεσθίει. καλοῦσι δὲ φαττάγην αὐτό.

7. Συροπέρδιξ γίνεται περὶ τὴν Ἀντιόχειαν τὴν Πισιδίας, καὶ σιτεῖται καὶ λίθους· μικρότερος δέ ἐστι τοῦ πέρδικος καὶ μέλας [2] τὴν χρόαν, πυρρὸς ⟨δὲ⟩ [3] τὸ ῥάμφος. οὐχ ἡμεροῦται δὲ κατὰ τὸν ἄλλον, οὐδὲ γίνεται τιθασός, ἀλλ᾽ ἄγριος ἐς τὸ ἀεὶ διαμένει. ἔστι δὲ οὐ μέγας, βρωθῆναί τε ἡδίων τοῦ ἑτέρου, καὶ τὴν σάρκα πως δοκεῖ πυκνότερος.

8. Ἡ δὲ Ἰνδῶν θάλαττα ὕδρους θαλαττίους τίκτει πλατεῖς τὰ οὐράς· τίκτουσι δὲ καὶ λίμναι μεγίστους ὕδρους. οἱ δὲ θαλάττιοι ὄφεις οἵδε κάρχαρον ἐοίκασι μᾶλλον ἔχειν τὸ δῆγμα ἤπερ οὖν ἰῶδες.

9. Ἐν Ἰνδοῖς ἵππων τε ἀγρίων καὶ ὄνων τοιούτων εἰσὶν ἀγέλαι. οὐκοῦν ἀναβαινόντων ⟨τῶν⟩ [4] ὄνων τὰς ἵππους, ὑπομένειν ἐκείνας λέγουσι, καὶ ἥδεσθαι τῇ μίξει, καὶ τίκτειν ἡμιόνους πυρροὺς [5] τὴν χρόαν καὶ ἄγαν δρομικούς, δυσλόφους δὲ καὶ δυσγαργάλεις [6] ἄλλως. ποδάγραις δὲ τούτους αἱροῦσιν, εἶτα ἀνάγεσθαι τῷ τῶν Πρασίων βασιλεῖ φασι· καὶ διετεῖς μὲν ἑαλωκότας μὴ ἀναίνεσθαι

[1] φολίδα τραχεῖαν ἄρα.
[2] πέλας *Thompson*.

Melitean [a] lapdog. The scales that cover it are so rough and of such close texture, that when flayed they perform the functions of a file. They will even cut through bronze and eat their way through iron. They call the creature *Phattagê* (pangolin).

7. The Sand-partridge occurs in the neighbourhood of Antioch in Pisidia and feeds on stones. It is smaller than the partridge and black in colour, but its beak is red. It is not to be domesticated like the partridge, nor does it grow tame, but continues wild all the time. It is not large, but is pleasanter to eat than the other, and its flesh seems somewhat firmer.

The Sand-partridge

8. The Indian Ocean produces Sea-snakes with broad tails; the lakes also produce Water-snakes of immense size. But apparently these snakes in the Ocean bite with teeth that are saw-like rather than poisonous.

Water-snakes of India

9. In India there are herds of wild horses and wild asses. Now they say that when the asses mount the mares, the latter remain passive and take pleasure in the act and produce Mules of a red colour and extremely swift of foot, but that these Mules are impatient of the yoke and generally skittish. The people are said to catch them with foot-traps and then to take them to the King of the Prasii. If they are caught as two-year-olds they do not refuse to be

The Indian Mule

[a] Melita, island off the coast of Dalmatia.

---

[3] ⟨δέ⟩ *add. H.*
[4] ⟨τῶν⟩ *add. Jac.*
[5] πυρσούς.
[6] *Toup* : γαργαλεῖς.

τὴν πώλευσιν, πρεσβυτέρους δὲ μὴ διαφέρειν τῶν καρχάρων θηρίων καὶ σαρκοφάγων μηδὲ ἕν.

10. Ἐν Πρασίοις δὲ τοῖς Ἰνδικοῖς εἶναι γένος πιθήκων φασὶν ἀνθρωπόνουν, ἰδεῖν [1] δέ εἰσι κατὰ τοὺς Ὑρκανοὺς κύνας τὸ μέγεθος, προκομία τε αὐτῶν ὁρᾶται συμφυής· εἴποι δ' ἂν ὁ μὴ τὸ ἀληθὲς εἰδὼς ἀσκητὰς εἶναι αὐτάς. γένειον δὲ αὐτοῖς ὑποπέφυκε σατυρῶδες, ἡ δὲ οὐρὰ κατὰ τὴν τῶν λεόντων ἀλκαίαν ἐστί. καὶ τὸ μὲν ἄλλο πᾶν σῶμα πεφύκασι λευκοί, τὴν δὲ κεφαλὴν καὶ τὴν οὐρὰν ἄκραν εἰσὶ πυρροί.[2] σώφρονες δὲ καὶ φύσει τιθασοί· εἰσὶ δὲ ὑλαῖοι τὴν δίαιταν,[3] καὶ σιτοῦνται τῶν ὡραίων [4] τὰ ἄγρια. φοιτῶσι δὲ ἀθρόοι ἐς τὰ τῆς Λατάγης προάστεια (πόλις δέ ἐστιν Ἰνδῶν ἡ Λατάγη), καὶ τὴν προτεθειμένην αὐτοῖς ἐκ βασιλέως ἑφθὴν ὄρυζαν σιτοῦνται· ἀνὰ πᾶσαν δὲ ἡμέραν ἥδε ἡ δαὶς αὐτοῖς εὐτρεπὴς πρόκειται. ἐμφορηθέντας δὲ ἄρα αὐτοὺς ἀναχωρεῖν αὖθις ἐς ⟨τὰ⟩ [5] ἤθη τὰ ὑλαῖά φασι σὺν κόσμῳ, καὶ σίνεσθαι τῶν ἐν ποσὶν οὐδὲ ἕν.

11. Ποηφάγον ἐν Ἰνδοῖς ζῷόν ἐστι, καὶ πέφυκέ γε διπλάσιον ἵππου τὸ μέγεθος. οὐρὰν δὲ ἔχει δασυτάτην καὶ μελαίνης ἀκράτως χρόας, καὶ εἶεν [6] αὗται αἱ τρίχες καὶ τῶν ἀνθρωπείων λεπτότεραι ἄν, καὶ ἐν μεγάλῳ τίθενται ταύτας ἔχειν Ἰνδῶν αἱ γυναῖκες· καὶ γάρ τοι παραπλέκονται ἐξ αὐτῶν καὶ κοσμοῦνται μάλα ὡραίως, ταῖς πλοκαμῖσι ταῖς

[1] *καὶ ἰδεῖν.*
[2] *πυρσοί.*
[3] *Schn*: δίαιταν καὶ τὸ γένος.
[4] *Bernard* : ὀρέων.
[5] ⟨τά⟩ *add. H.*
[6] *Jac* : εἰσιν.

broken in, but when older they are just as savage as fanged and carnivorous beasts.

10. They say that among the Prasii in India there is a race of Monkeys with human intelligence;[a] in appearance they are as large as Hyrcanian hounds, and they are seen to possess a natural forelock; anyone who did not know the facts would say that these forelocks were artificial. The beard that grows beneath their chin is like that of a satyr, while the tail is as long as a lion's. The whole of their body is white except for the head and the tip of the tail, which are red. They are sober and naturally tame. They live in the forests and feed on wild produce. They visit the suburbs of Latage (this is a city in India) in great numbers and feed on the boiled rice which the king has served out to them, and this meal is prepared and laid out for them every day. And when they have eaten their fill, it is said that they withdraw again to their haunts in the forest in an orderly fashion without damaging anything that they come across.

Monkeys of Prasiaea

11. In India there is a herbivorous animal[b] and it is twice the size of a horse. It has a very bushy tail, pitch-black in colour; the hairs of it are finer than those of man, and Indian women set great store by obtaining them, and in fact they braid them in and adorn themselves most beautifully,

The Yak

[a] Keller (*Ant. Tierw.* 1. 9) identifies this monkey with the 'Hunuman,' *Semnopithecus entellus.*

[b] The Yak, *Bos poëphagus grunniens*, is to be found on the Rupshu plateau in the SE corner of Kashmir and in Sikkim; elsewhere only in Tibet.

συμφύτοις καὶ ταύτας ὑποδέουσαι. προήκει δὲ καὶ ἐς δύο πήχεις ἑκάστης τὸ μῆκος τριχός, ἐκ μιᾶς δὲ ῥίζης ὁμοῦ τι καὶ τριάκοντα θυσανηδὸν ἐκπεφύκασι. ζῴων δὲ ἄρα ἁπάντων τοῦτο δειλότατον ἦν· ἐὰν γὰρ ὑπό τινος ὀφθῇ καὶ αἴσθηται βλεπόμενον, ᾗ ποδῶν ἔχει φεύγει,[1] καὶ κέχρηται προθυμίᾳ μᾶλλον ἢ σκελῶν ὠκύτητι. καὶ διώκεται μὲν ὑπὸ ἱππέων καὶ κυνῶν ἀγαθῶν δραμεῖν· ἐὰν μέντοι συνίδῃ ὅτι ἄρα ἁλίσκεσθαι μέλλει, τὴν οὐρὰν ἀπέκρυψεν ἔν τινι δάσει, αὐτὸ δὲ ἀντιπρόσωπον ἕστηκε, καὶ δοκεύει τοὺς θηρατάς, καὶ ὑποθαρρεῖ πως, καὶ οἴεται μηκέτι φανεῖσθαι περισπούδαστον, τῆς οὐρᾶς μὴ βλεπομένης· ἐκείνην γὰρ οἶδέν ⟨οἱ⟩ [2] εἶναι τὸ κάλλος. κενὴν δὲ ἄρα ἴσχει τὴν ὑπὲρ τοῦδε φαντασίαν· βάλλει γάρ τις αὐτὸ βέλει πεφαρμαγμένῳ, καὶ ἀποκτείνας ἀποκόψει τὴν οὐράν, τὸ ἆθλον τῆς ἄγρας. καὶ δείρας τὸ πᾶν σῶμα (ἀγαθὸν γὰρ καὶ ἡ δορά) ἀφῆκε τὸν νεκρόν· σαρκῶν γὰρ τῶν ἐκείνου δέονται Ἰνδοὶ οὐδὲ ἕν.

12. Κήτη δὲ ἦν ἄρα ἐν τῇ τῶν Ἰνδῶν θαλάττῃ πενταπλασίονα ⟨τὸ⟩ [3] μέγεθος ἐλέφαντος τοῦ μεγίστου. πλευρὰ γοῦν μία κήτους καὶ ἐς τοὺς εἴκοσι πήχεις πρόεισι, χελύνην δὲ πήχεων πεντεκαίδεκα ἔχει, τὸ δὲ πτέρωμα βραγχίου ἑκατέρου πήχεων τὸ εὖρος καὶ ἑπτά. κήρυκες δὲ καὶ πορφύραι . . .[4] ὡς καὶ χοῦν ῥᾷστα δέξασθαι· καὶ μέντοι καὶ τῶν ἐχίνων τὰ χελώνια δύναιτο ἂν τοσοῦτον στέγειν. μεγέθη δ' ἰχθύων ἄπειρα, λαβράκων μάλιστα, καὶ ἀμίαι καὶ χρυσόφρυες.

[1] φεύγει καὶ πρόεισι.

plaiting them in with their own hair. Each hair attains a length of two cubits, and there spring perhaps as many as thirty from one root, like a tassel. Now this is of all animals the most timid, for if it is seen by somebody and realises that it is being looked at, it flees as fast as it can, the pace of its legs only exceeded by its eagerness to escape. It is hunted by horsemen with swift-footed hounds. But if it realises that it is going to be caught, it hides its tail in some thicket, faces about, and stands waiting for its pursuers and plucks up its courage, fancying that, since its tail is not visible, it will no longer seem worth pursuing. For it knows that its beauty resides in its tail. And yet on this point its fancies are idle, for a man shoots it with a poisoned arrow and having killed it will cut off its tail, the reward of the chase. And after flaying the body (for the hide also is serviceable) he leaves the dead carcase, because the Indians have no use for the flesh of these animals.

Fishes of India

12. It seems that in the Indian Ocean there are sea-monsters five times the size of the largest elephant. At any rate a single rib of a Sea-monster measures as much as twenty cubits; it has a jaw of fifteen cubits; the fin beside each of the gills is seven cubits in width. The Trumpet-shells and Purple-shellfish of the Indian Ocean ⟨are large enough⟩ to contain easily six pints; further, the shells of Sea-urchins have the same capacity. As for Fishes, they are gigantic, especially the Basse, the Pelamyd,

[2] ⟨οἱ⟩ *add. Jac.*
[3] ⟨τό⟩ *add. H.*
[4] *Lacuna*: ⟨τοσοῦτοι⟩ *ex. gr. H.*

ἀκούω δὲ τούτους κατὰ τὴν ὥραν, ὅταν ἐπιρρέωσιν οἱ ποταμοὶ λάβροι [1] κατιόντες ἐκ τῆς πλημμύρας καὶ ἐς τὴν γῆν ἀναχέωνται, καὶ αὐτοὺς ὑπερχεῖσθαι κατὰ τὰς ἀρούρας καὶ ἐν ὕδατι λεπτῷ φέρεσθαί τε καὶ ἀλᾶσθαι. παυσαμένων δὲ τῶν ὑπερπιμπλάντων τοὺς ποταμοὺς ὑετῶν καὶ ἀναχωρούντων ὀπίσω τῶν ῥευμάτων καὶ ἐς τὰς ὁδοὺς τὰς κατὰ φύσιν ὑποστρεφόντων, ἐν τοῖς καθημένοις χωρίοις καὶ τοῖς τεναγώδεσι καὶ ἀπέδοις, ἔνθα δήπου φιλοῦσι καὶ αἱ νεαὶ [2] καλούμεναι κόλπους τινὰς ἔχειν, ἰχθῦς ὑπομένουσι [3] καὶ ὀκτὼ πήχεων. καὶ αἱροῦσιν οἱ γεωργοῦντες αὐτοὺς ἀσθενεῖ τῇ νήξει χρωμένους, ἅτε μὴ ἐν βυθῷ φερομένους ἀλλὰ ἐπιπολῆς, καὶ ἐκ τοῦ ὀλίγου ὕδατος ἀγαπητῶς καὶ μόλις ἀποζῶντας.

13. Ἰνδῶν δὲ ἰχθύων ἴδια καὶ ἐκεῖνα. βατίδες γίνονται παρ' αὐτοῖς οὐδέν τι μείους Ἀργολικῆς ἀσπίδος ἑκάστη, καρίδες δὲ [4] καὶ μείζους καράβων αἱ Ἰνδῶν εἰσίν. αἱ μὲν οὖν ἐκ τῆς θαλάττης ἀναθέουσαι διὰ τοῦ ποταμοῦ τοῦ Γάγγου χηλὰς μεγίστας ἔχουσι καὶ τραχείας θιγεῖν,[5] τάς γε μὴν ἐκ τῆς Ἐρυθρᾶς ἐκπιπτούσας ἐς τὸν Ἰνδὸν λείας ἔχειν πέπυσμαι τὰς ἀκάνθας, προμήκεις γε μὴν καὶ βοστρυχώδεις τὰς ἀπηρτημένας ἕλικας. χηλὰς δὲ οὐκ ἔχειν ταύτας.

14. Χελώνη δὲ ἐν Ἰνδοῖς ποταμία [6] τὸ χελώνιον [7] ἔχει σκάφης οὐ μεῖον τελείας. χωρεῖ γοῦν

[1] καὶ λάβροι.
[2] *Schn*: αἱ ἐννέαι.
[3] *Schn*: ἀπονέμουσι.
[4] τε.
[5] θιγεῖν αὐτῶν.
[6] ποταμία μεγίστη τε αὕτη καί.

and the Gilthead. And I have heard that at the season when the rivers descend in violence owing to floods and spill themselves upon the land, the Fish also are emptied over the fields and are borne hither and thither in shallow water. But when the rains which have over-filled the rivers cease, and the streams withdraw again and return to their natural courses, then Fishes of as much as eight cubits long remain in low-lying, marshy, level spots, where what is known as 'fallow land' commonly has depressions. And the cultivators catch the Fish which can only swim feebly, since they are not moving in deep water but on the surface, glad to snatch a bare existence from the shallow water.

The Skate and the Prawn of India

13. Indian fish have the following peculiarities. The Skate there is as large as an Argolic shield;[a] the Prawns[b] of India are even larger than crayfish. Now these Prawns ascend the river Ganges from the sea and have claws of immense size and rough to the touch, whereas I learn that those that quit the Red Sea for the Indus have smooth spines, and the feelers attached to them are long and curly, but they have no claws.

The Turtle and the Tortoise of India

14. The river-Turtle of India[c] has a shell as large as a full-sized skiff. At any rate each one has a

[a] The Argolic shield was circular and about 3 ft. across.

[b] The *Palaemon carcinus* of the E Indies attains the size of a lobster.

[c] The Turtles described here, in ch. 17, and in 17. 3, cannot be certainly identified.

---

[7] χελώνειον.

ἕκαστον μεδίμνους δέκα ὀσπρίων. γίνονται δὲ καὶ χερσαῖαι χελῶναι, καὶ εἶεν ἂν τὸ μέγεθος κατὰ τὰς βώλους τὰς μεγίστας, αἵπερ οὖν ἐπανίστανται ἐν τοῖς βαθέσιν ἀρώμασιν, εὐπειθοῦς μὲν οὔσης τῆς γῆς, ἐς πολὺ δὲ κατιόντος τοῦ ἀρότρου καὶ τὴν αὔλακα σχίζοντος ῥᾷστα καὶ ἐγείροντος τὰς βώλους ὑψοῦ. ταύτας δὲ καὶ ἀποδύεσθαι τὸ ἔλυτρόν φασιν. οἱ τοίνυν ἀρόται καὶ πᾶν τὸ περὶ τοὺς ἀγροὺς ἐργατικὸν ταῖς μακέλλαις ἀνασπῶσιν αὐτάς, καὶ ἐξαίρουσιν [1] ὥσπερ οὖν ἐκ τῶν θριπηδέστων φυτῶν τὰς εὐλάς. εἰσὶ δὲ γλυκεῖαι τὴν σάρκα καὶ πίονες, οὐ μὴν κατὰ τὰς θαλαττίας πικραὶ καὶ αὗται.

15. Θυμόσοφα δὲ καὶ παρ' ἡμῖν ζῷά ἐστιν, οὐ μὴν ὅσα ἐν Ἰνδοῖς [2] ἀλλὰ ὀλίγα. ἐκεῖ δὲ ὅ τε ἐλέφας τοιοῦτός ἐστι καὶ ὁ σιττακὸς καὶ αἱ σφίγγες καὶ οἱ καλούμενοι σάτυροι· σοφὸν δὲ ἄρα ἦν καὶ ὁ μύρμηξ ὁ Ἰνδός. οἱ μὲν οὖν ἡμεδαποὶ τὰς ἑαυτῶν χειὰς καὶ ὑποδρομὰς ὑπὸ τὴν γῆν ὀρύττουσι, καὶ φωλεούς τινας κρυπτοὺς ἀποφαίνουσι γεωρυχοῦντες, καὶ μεταλλείαις ὡς εἰπεῖν τισιν ἀπορρήτοις καὶ λανθανούσαις καταξαίνονται· ἀλλὰ οἵ γε Ἰνδοὶ μύρμηκες οἰκίσκους τινὰς συμφορητοὺς ἐργάζονται, καὶ τούτους γε οὐκ ἐν χωρίοις ὑπτίοις καὶ λείοις καὶ ἐπικλυζομένοις ῥᾷστα, ἀλλὰ μετεώροις καὶ ὑψηλοῖς. ἐν αὐτοῖς δὲ περιόδους τινὰς καὶ ὡς εἰπεῖν σύριγγας Αἰγυπτίας [3] ἢ λαβυρίνθους Κρητικοὺς σοφίᾳ τινὶ ἀπορρήτῳ διατρήσαντες οἰκεῖα ἑαυτοῖς ἀπέφηναν, οὐκ εὐθυτενῆ καὶ ῥᾴδια παρελθεῖν [4] ἀλλ' ἑλιγμοῖς καὶ

[1] ἐξαιροῦσιν.
[2] *Jac*: Ἰνδοῖς ἐστιν.
[3] *Gron*: Αἰγυπτίους.
[4] παρελθεῖν ἢ εἰσρεῦσαί τι.

capacity of ten *medimni* [a] of pulse. There are also land-Tortoises, and these may be the size of the largest clods of earth which are turned up in deep ploughing, provided the soil is yielding and the plough goes deep and cuts a furrow without difficulty and brings up the clods. And they say that these Tortoises shed their covering. Now the ploughmen and all who work in the fields dig them out with mattocks and extract them as we extract caterpillars from plants which are worm-eaten. The flesh of Tortoises is sweet and they are fat and by no means bitter like the Turtles.

15. In our country also there are intelligent animals, but they are few and not so numerous as in India. In that land, for example, are the Elephant, the Parrot, the Sphinx-ape, and the Satyrs,[b] as they are called. The Indian Ant [c] too, it seems, is a clever creature. True, the Ants of our country excavate their holes and burrow below ground and construct hidden lairs, as it were, by digging in the earth, and wear themselves out with their mysterious and secret mining operations, so to speak. But the Ants of India construct little houses of material brought together, and these are not in low-lying, level country, which is easily flooded, but high up on rising ground. And there with indescribable skill they bore passages and what you might call Egyptian galleries or Cretan labyrinths and make a place for themselves, not straight ahead or easy to penetrate but out of the way past a maze of tunnels;

The Ants of India

[a] *Medimnus* = about 12 gallons.
[b] A kind of ape, perh. the ‘Gibbon.’
[c] The Termite.

διατρήσεσι λοξά· καὶ ἀπολείπουσί γε ἐπιπολῆς μίαν ὀπήν, δι' ἧς εἰσίασί τε αὐτοὶ καὶ τὰ σπέρματα ὅσα ἐκλέγουσι,[1] εἶτα ἐς τοὺς ἑαυτῶν θησαυροὺς ἐσκομίζουσι. παλαμῶνται δὲ ἄρα τὰς ἐν ὕψει φωλεύσεις ὑπὲρ τοῦ τὰς ἐκ τῶν ποταμῶν ἀναχύσεις τε καὶ ἐπικλύσεις διαδιδράσκειν. καὶ αὐτοῖς ὑπὲρ τῆσδε τῆς σοφίας περιγίνεται ὥσπερ ἐν σκοπιαῖς τισιν ἢ νήσοις κατοικεῖν, ὅταν τῶν λοφιδίων ἐκείνων τὰ κύκλῳ περιλιμνάσῃ.[2] τὰ δ' οὖν χώματα ἐκεῖνα, καίτοι συμπεφορημένα, τοσοῦτον ἀποδεῖ τοῦ λύεσθαί τε καὶ διαξαίνεσθαι ὑπὸ τῆς περικλύσεως, ὡς καὶ κρατύνεσθαι αὐτά, πρῶτον μὲν ὑπὸ τῆς ἑῴας δρόσου· ὑπαμφιέννυται[3] γὰρ ὡς εἰπεῖν ἐκ ταύτης πάγου τινὰ χιτῶνα ὑπόλεπτον, πλὴν καρτερόν· εἶτα μέντοι δεσμεύεται κάτω βρυώδει τῆς ποταμίας ἰλύος φλοιῷ. καὶ μυρμήκων μὲν Ἰνδῶν πέρι Ἰόβᾳ πάλαι, ἐμοὶ δὲ νῦν ἐς τοσοῦτον λελέχθω.

16. Παρὰ τοῖς Ἀριανοῖς τοῖς Ἰνδικοῖς χάσμα Πλούτωνός ἐστι, καὶ κάτω τινὲς ἀπόρρητοι σύριγγες καὶ ὁδοὶ κρυπταὶ καὶ διαδρομαὶ ἀνθρώποις ⟨μὲν⟩[4] ἀθέατοι, βαθεῖαι δ' οὖν καὶ ἐπὶ μήκιστον προήκουσαι· γενόμεναι δὲ πῶς[5] καὶ ὀρωρυγμέναι τρόπῳ τῷ, οὔτε Ἰνδοὶ λέγουσιν, οὔτε ἐγὼ μαθεῖν πολυπραγμονῶ. ἄγουσιν οὖν[6] Ἰνδοὶ καὶ ὑπὲρ τὰ τρισμύρια ἐνταῦθα κτήνη[7] προβάτων τε καὶ αἰγῶν καὶ βοῶν καὶ ἵππων· καὶ ἕκαστος τῶν ἢ δεισάντων[8] ἐνύπνιον ἢ ὄτταν τινὰ ἢ φήμην ἢ ὄρνιν

[1] ἐκλέγονται.
[2] περιλιμνάζηται H.
[3] ἐπ- H.
[4] ⟨μέν⟩ add. H.
[5] Jac: πως.
[6] ἄγουσι γοῦν.

and on the top they leave a single hole through which they themselves enter and bring into their storehouses all the seeds which they select. You see, they construct their caves high up in order to escape from inundations and floods from rivers. The result of this clever move is that they are living as it were in watch-towers or on islands at a time when all the land around their hillocks becomes a lake. Now these mounds, although merely heaped up, are so far from being dissolved and eaten away by an inundation that they are actually strengthened, primarily by the morning dew, for they are, so to say, clothed beneath with a fine but strong coating of frost resulting from the dew; then at the base they are bound round with a bark-like coating of weeds from the river mud.

Juba long ago wrote about the Ants of India; but this is all I have to say at present.

16. In the country of the Ariani[a] of India there is a Chasm of Pluto, and at the bottom there are certain mysterious galleries, hidden paths, and passages unseen of man, though they are in fact deep and extend a very long way. But how they came to be and how they were dug, neither the Indians can say nor have I been at the pains to discover. Now the Indians bring to the spot over thirty thousand beasts—sheep, goats, cattle, and horses. And everyone who has been scared by some dream or has encountered some omen divine

The Chasm of Pluto

[a] Ariana comprehended, roughly speaking, most of the modern Persia, Afghanistan, and India as far as the river Indus.

[7] κτήνη διάφορα. [8] δεισάντων *del. H.*

οὐκ εὔεδρον ὑφορωμένων ἀντὶ τῆς ἑαυτοῦ ζωῆς ἐμβάλλει κατὰ τὴν οἴκοθεν[1] δύναμιν, ἑαυτὸν λυτρούμενος καὶ διδοὺς ὑπὲρ τῆς ἑαυτοῦ ψυχῆς τὴν τοῦ ζῴου. τὰ δὲ ἄγεται οὔτε ἐδεσμοῖς ἐπαγόμενα οὔτε ἐλαυνόμενα ἄλλως, ἑκόντα δὲ τὴν ὁδὸν τήνδε ἀνύτει ἕλξει τινὶ καὶ ἴυγγι ἀπορρήτῳ. εἶτα ἐπιστάντα τῷ στομίῳ ἑκόντα ἐμπηδᾷ, καὶ ὄψει μὲν ἀνθρωπίνῃ οὐκ ἔστιν οὐκέτι σύνοπτα ἐς γῆς χάσμα ἀπόρρητόν τε καὶ ἀχανὲς ἐμπεσόντα, ἀκούονται δ' οὖν[2] ἄνω βοῶν μὲν μυκηθμοί, τῶν δὲ οἰῶν βληχή, χρεμετισμὸς δὲ τῶν ἵππων καὶ μηκὴ τῶν αἰγῶν. καὶ εἴ τις ἐπιπολῆς βαδίζοι καὶ προχωροίη[3] τὸ οὖς παραβάλλων, ἀκούσεται ἐπὶ μήκιστον τῶν προειρημένων. οὐδὲ ἐκλείπει ποτὲ ὁ συμμιγὴς ἦχος, ἐπιπεμπόντων ὁσημέραι τὰ ὑπὲρ ἑαυτῶν ζῷα. εἰ μὲν οὖν τὰ πρόσφατα ἐξακούεται μόνα ἢ καὶ τῶν πρώτων τινά, οὐκ οἶδα, ἀκούεται δ' οὖν. καὶ εἴρηταί μοι ζῴων τῶν ἐκεῖ καὶ τοῦτο ἴδιον.

17. Ἐν δὲ τῇ καλουμένῃ Μεγάλῃ θαλάττῃ καὶ νῆσον ᾄδουσι μεγίστην, καὶ ὄνομα αὐτῆς ἀκούω Ταπροβάνην· πάνυ δὲ δολιχὴν πυνθάνομαι καὶ ὑψηλὴν τὴν νῆσον εἶναι, καὶ μῆκος μὲν ἔχειν σταδίων ἑπτακισχιλίων, πλάτος δὲ πεντακισχιλίων, καὶ ἔχειν οὐ πόλεις, ἀλλὰ κώμας πεντήκοντα καὶ ἑπτακοσίας· στέγας δὲ ἔχουσιν ἔνθα[4] κατάγονται οἱ ἐπιχώριοι ἐκ ξύλων πεποιημένας, ἤδη δὲ καὶ

[1] οἴκοθεν αὐτοῦ. [2] γοῦν.
[3] προσχωροῖ. [4] ὅθεν.

[a] The Indian Ocean.
[b] Ceylon.
[c] 7000 stades = about 789 mi., 5000 = about 568 mi.

or human, or who has seen some bird in an unfavourable quarter, casts into the Chasm what his personal means can afford by way of ransom for himself, sacrificing the life of an animal for his own life. And the victims are brought there without being hauled with ropes or otherwise compelled, and make the journey of their own free will owing to some mysterious attraction or spell. Then, as they stand on the brink, of their own accord they leap into the Chasm and are no more seen of the human eye once they have fallen into this mysterious and yawning Chasm of earth, while above are heard the lowing of cattle, the baa of sheep, the neighing of horses, and the bleating of goats. And anyone who walks over the surface of the land and comes to the spot and listens will hear the aforesaid animals for a very long while. And the confused sounds never cease, since every day the Indians send in animals for their own redemption. Now whether it is only the recent victims that are audible or some of the earlier ones also, I cannot say, but audible they are. So much for this singular trait in the animals of that country.

17. It is commonly reported that in the Great Sea,[a] as it is called, there is an island of immense area, and I have heard that its name is Taprobane.[b] And I learn that this island is very long and high: its length is seven thousand *stades* and its width five thousand;[c] it has no cities, only seven-hundred-and-fifty villages, and the dwellings where the inhabitants lodge are made of wood and even of reeds.

The island of Taprobane

The actual length of Ceylon from N to S is 271½ mi. and the width 137½ mi.

δονάκων. τίκτονται δὲ ἄρα ἐν ταύτῃ τῇ θαλάττῃ καὶ χελῶναι μέγισται, ὧνπερ οὖν τὰ ἔλυτρα ὄροφοι γίνονται· καὶ γάρ ἐστι καὶ πεντεκαίδεκα πήχεων ἓν χελώνιον, ὡς ὑποικεῖν οὐκ ὀλίγους· καὶ ἡλίους πυρωδεστάτους ἀποστέγει, καὶ σκιὰν ἀσμένοις παρέχει, πρός γε μὴν τῶν ὄμβρων τὰς καταφορὰς ἀντίτυπόν ἐστι, καὶ κεράμου παντὸς καρτερώτερον,[1] τάς τε ἐμβολὰς τῶν ὑετῶν ἀποσείεται, καὶ κροτούμενον ἀκούουσιν οἱ ὑποικοῦντες, ὡς ἔς τι τέγος ἐμπιπτόντων τῶν ὑδάτων. οὐ δέονταί γε μὴν ὡς κέραμον ῥαγέντα ἀμεῖψαι· σκληρὸν γὰρ τὸ χελώνιον,[2] καὶ ἔοικεν ὑπορωρυγμένῃ πέτρᾳ καὶ [3] ὑπάντρῳ τε καὶ αὐτορόφῳ στέγῃ.

18. Ἡ τοίνυν νῆσος ἡ ἐν τῇ Μεγάλῃ θαλάττῃ, ἣν καλοῦσι Ταπροβάνην, ἔχει φοινικῶνας μὲν θαυμαστῶς πεφυτευμένους ἐς στοῖχον, ὥσπερ οὖν ἐν τοῖς ἁβροῖς τῶν παραδείσων οἱ τούτων μελεδωνοὶ φυτεύουσι τὰ δένδρα τὰ σκιαδηφόρα, ἔχει δὲ καὶ νομὰς ἐλεφάντων πολλῶν καὶ μεγίστων. καὶ οἵ γε νησιῶται ἐλέφαντες τῶν ἠπειρωτῶν ἀλκιμώτεροί τε τὴν ῥώμην καὶ μείζους ἰδεῖν εἰσί, καὶ θυμοσοφώτεροι δὲ πάντα πάντῃ κρίνοιντο ἄν. κομίζουσί τε οὖν αὐτοὺς ἐς τὴν ἀντιπέρας[4] ἤπειρον ναῦς μεγάλας τεκτηνάμενοι (ἔχει γὰρ δήπου καὶ δάση ἡ νῆσος), πιπράσκουσί τε διαπλεύσαντες τῷ βασιλεῖ τῷ ἐν Καλίγγαις. διὰ μέγεθος δὲ ἄρα τῆς νήσου οὐδὲ ἴσασιν οἱ τὰ μέσα αὐτῆς οἰκοῦντες τὴν θάλατταν, ἀλλὰ ἠπειρώτην μὲν βίον

[1] κρατερώτερον.
[2] χελώνειον.
[3] ⟨ἢ⟩ καί *Jac.*

Now in this sea Turtles of immense size are hatched, and their shells are made into roofs, for a single shell measures fifteen cubits across, so that quite a number of persons can live underneath; and it keeps out the most fiery sun and affords a welcome shade; moreover it resists a downpour of rain, and being stronger than any tiles, it shakes off pelting showers, while the inmates beneath listen to it being pounded, as though the water were descending upon a tiled roof. Yet they have no need to exchange old for new as you must with a broken tile, for the Turtle's shell is hard and resembles a rock that has been hollowed out or the roof of a cavern vaulted by nature.

The Turtle of the Indian Ocean

18. Now this island which they call Taprobane in the Great Sea has groves of palm-trees wonderfully planted in lines, just as in luxurious parks shady trees are planted by those in charge; it has also pasturing grounds for numerous Elephants of the largest size. And these Elephants of the island are more powerful and bigger than those of the mainland, and may be judged naturally cleverer in every way. And so the people build huge ships (for the island of course has dense forests) and transport the Elephants to the mainland opposite, and having crossed, sell them to the King of the Calingae.[a] But owing to the size of the island those who live in the middle of it do not even know the sea but live as though

Taprobane, its elephants

[a] Their territory lay along the E coast of India between the mouths of the Mahanadi and Godavari rivers, far N of Ceylon; but Ael. appears to regard it as in the same latitude as the island.

[4] ἀντιπέραν.

τρίβουσι, περιερχομένην [1] δὲ αὐτοὺς καὶ κυκλουμένην πυνθάνονται θάλατταν. οἱ δὲ τῇ θαλάττῃ πρόσοικοι τῆς μὲν ἄγρας τῆς τῶν ἐλεφάντων ἀμαθῶς ἔχουσιν, ἀκοῇ δὲ αὐτὴν ἴσασι μόνῃ· περί γε μὴν τὰς τῶν ἰχθύων καὶ τὰς τῶν κητῶν ἄγρας τίθενται τὴν σπουδήν. τὴν γάρ τοι θάλατταν τὴν περιερχομένην τὸν τῆς νήσου κύκλον ἄμαχόν τι πλῆθος καὶ ἰχθύων καὶ κητῶν τρέφειν [2] φασί, καὶ ταῦτα μέντοι καὶ λεόντων ἔχειν κεφαλὰς καὶ παρδάλεων καὶ λύκων [3] καὶ κριῶν δέ, καὶ τὸ ἔτι θαῦμα σατύρων μορφὰς κήτη ἔστιν ἃ περιφέρει καὶ γυναικῶν ὄψιν,[4] αἷσπερ ἀντὶ πλοκάμων ἄκανθαι προσήρτηνται. ἔχειν δὲ καὶ ἄλλας τινὰς ὑμνοῦσιν ἐκτόπους μορφάς, ὧν τὰ εἴδη μηδ᾽ ἂν τοὺς δεινοὺς γράφειν καὶ κράσεις σωμάτων συμπλέκειν ἐς τερατείαν ὄψεων ἀκριβῶσαί ποτε καὶ σοφίᾳ γραφικῇ παραστῆσαι δύνασθαι ἄν· προμήκη δὲ ἔχει τὰ οὐραῖα καὶ ἑλικτά, πόδας γε μὴν χηλὰς ἢ πτερύγια. πυνθάνομαι δὲ αὐτὰ καὶ ἀμφίβια εἶναι, καὶ νύκτωρ μὲν ἐπινέμεσθαι τὰς ἀρούρας· πόαν μὲν γὰρ [5] ἐσθίειν τῶν ἀγελαίων τε καὶ σπερμολόγων δίκην, χαίρειν δὲ καὶ τῷ φοίνικι τῷ δρυπεπεῖ, διασείειν τε ἐκ τούτου τὰ δένδρα ταῖς σπείραις περιβάλλοντα αὐτὰς ὑγρὰς οὔσας καὶ οἵας περιπλέκεσθαι. τοῦτον οὖν τὸν φοίνικα ἐκ τοῦ σεισμοῦ τοῦ βιαίου καταρρέοντα ἐπινέμεσθαι. ὑπολήγει δὲ ἄρα νύξ,[6] καὶ σαφὴς οὔπω ἡμέρα, καὶ ἐκεῖνα ἠφανίσθη καταδύντα ἐς τὸ πέλαγος, ἑῴου [7] μέλλοντος ὑπο-

[1] τὴν περιερχομένην.
[2] ἐκτρέφειν.
[3] ἄλλων.
[4] ὄψιν ἔχουσιν.
[5] μὲν γὰρ οὖν.

they were of the mainland and only learn by report of the sea that surrounds and encircles them. Whereas those that live near to the sea are ignorant of the way in which Elephants are hunted and only know of it by hearsay: they devote themselves to catching fish and sea-monsters. For they assert that the sea which surrounds the circuit of their island breeds a multitude past numbering of fishes and monsters, and moreover that they have the heads of lions and leopards and wolves and rams, and, still more wonderful to relate, that there are some which have the forms of satyrs with the faces of women, and these have spines attached in place of hair. They tell of others too which have strange forms whose appearance not even men skilled in painting and in combining bodies of diverse shapes to make one marvel at the sight, could portray with accuracy or represent for all their artistic skill; for these creatures have immense and coiling tails, while for feet they have claws or fins. I learn too that they are amphibious [a] and that at night they graze the fields, for they eat the grass as cattle and rooks do; they enjoy the ripe fruit of the date-palm and therefore shake the trees with their coils, which being supple and capable of embracing, they fling round them. So when the shower of dates has fallen because of this violent shaking, they feed upon it. And then as the night wanes and before it is clear daylight these creatures plunge into the ocean and disappear as the dawn

its sea-monsters

[a] Ael. is apparently describing the Dugong, *Halicore dugong*, a large, herbivorous, seal-like mammal of the Indian Ocean; see O. Keller, *Ant. Tierwelt* 1. 414.

[6] ἡ νύξ.   [7] Ἑωσφόρου? *H*, ἑ. ⟨ἀστέρος⟩? *Jac.*

λάμπειν.[1] εἶναι δὲ καὶ φαλλαίνας φασὶ πολλάς, οὐ μὴν ἐς τὴν γῆν προϊέναι αὐτάς, τοὺς θύννους ἐλλοχώσας. καὶ δελφίνων δὲ γένη δύο φασὶν εἶναι, τὸ μὲν ἄγριον καὶ κάρχαρον καὶ ἀφειδέστατον ἐς τοὺς ἁλιέας καὶ σφόδρα ἄνοικτον,[2] τὸ δὲ πρᾶόν τε καὶ τιθασὸν φύσει. περισκιρτᾷ γοῦν καὶ περινήχεται, καὶ ἔοικε κυνιδίῳ αἰκάλλοντι, καὶ ψηλαφήσεις,[3] ὁ δὲ ὑπομένει·[4] κἂν τροφὴν ἐμβάλῃς, ἀσμένως λήψεται.

19. Λαγὼς θαλάττιος (τῆς μέντοι Μεγάλης· τὸν γὰρ ἕτερον εἶπον τὸν ἐκ τῆς ἑτέρας) ἀλλ' οὗτός γε ἔοικε τῷ χερσαίῳ πάντα πάντῃ πλὴν τῶν τριχῶν. τοῦ μὲν γὰρ ἠπειρώτου ἡ λάχνη ἔοικεν ἁπαλή τε εἶναι καὶ ἐπαφωμένῳ[5] μὴ ἀντίτυπος· ἔχει δὲ οὗτος ἀκανθώδεις τὰς τρίχας καὶ ὀρθάς, καὶ εἴ τις προσάψαιτο,[6] ἀμύσσεται. φασὶ δὲ αὐτὸν ἐπ' ἄκρᾳ τῇ φρίκῃ τῆς θαλάττης νήχεσθαι καὶ μὴ καταδύνειν ἐς βάθος, ὤκιστον δὲ εἶναι τὴν νῆξιν. ζῶν δὲ οὐκ ἂν ἁλῴη ῥᾳδίως. τὸ δὲ αἴτιον, οὐκ ἐμπίπτει ποτὲ ἐς δίκτυον, οὐ μὴν οὐδὲ καλάμου πρόσεισιν ὁρμιᾷ καὶ δελέατι. ὅταν δὲ ἄρα νοσήσας ὅδε ὁ λαγὼς εἶτα ἥκιστος ὢν νήχεσθαι ἐκβρασθῇ, πᾶς ὅστις ἂν αὐτοῦ προσάψηται τῇ χειρὶ ἀπόλλυται ἀμεληθείς. ἀλλὰ καὶ τῇ βακτηρίᾳ ἐὰν θίγῃ τοῦ λαγὼ τοῦδε, καὶ δι' αὐτῆς πάσχει τὸ αὐτό, ὥσπερ οὖν καὶ οἱ τοῦ βασιλίσκου προσαψάμενοι. ῥίζαν

[1] ὑπολάμπειν αὐτό.

[2] *Schn*: τῶν μὲν ἀγρίων . . . καρχάρων . . . ἀφειδεστάτων . . . ἀνοίκτων.

[3] κἂν ψηλαφήσῃς *Cobet*, *H* (1876).

[4] ὑπομενεῖ *H*.

[5] *Reiske*: ἐπαφωμένῃ.

begins to glow. They say that there are also numerous Whales which lie in wait for the tunnies; they do not however come up on to the land. They also say that there are two kinds of Dolphin, the one savage, sharp-toothed, and absolutely merciless and without pity towards fishermen, the other naturally gentle and tame. At any rate it gambols and swims around, and resembles a fawning puppy, and if you handle it, it will allow you to do so, and if you throw food to it, it will receive it gladly.

Whales

Dolphins

19. The Sea-hare [a] (I mean that which is found in the Great Sea [b]; the other kind in the other sea I have mentioned above) resembles the land animal in every respect except in its fur. For the fur of the land-hare seems smooth and is not hard to the touch. Whereas the Sea-hare's fur is prickly and erect and if one touches it one is stabbed. They say that it swims on the surface ripples of the sea and does not dive into the depths, and that it swims very fast. It is not easily caught alive, the reason being that it never falls into a net, nor yet will it approach the line and bait of a fishing-rod. When however this Hare through sickness and inability to swim is cast up on shore, anyone who touches it with his hand dies if he is not treated. Moreover even if he touches this Hare with a stick, he suffers the same fate thereby, just like those who touch a

The Sea-hare of the Indian Ocean

[a] Not the 'Sea-hare' of 2. 45 and 9. 51; this seems to be 'one of the spiny Globe-fishes (*Diodon*)' (Thompson, *Gk. fishes*).
[b] See above, ch. 17.

[6] προσάψεται.

δὲ ἐν τῇ νήσῳ τῇ κατὰ τὴν Μεγάλην θάλατταν φύεσθαί φασι καὶ εἶναι πᾶσιν εὔγνωστον, ἥπερ οὖν τῇ λιποθυμίᾳ ἀντίπαλός ἐστιν. προσενεχθεῖσα γοῦν τῇ τοῦ λιποψυχοῦντος ῥινὶ ἀναβιώσκεται τὸν ἄνθρωπον. ἐὰν δὲ ἀμεληθῇ, καὶ μέχρι θανάτου πρόεισι τῷ ἀνθρώπῳ τὸ πάθος· τοσαύτην ἄρα ἐς τὸ κακὸν ὅδε ὁ λαγὼς ἔχει τὴν ἰσχύν.

20. Ἐν τοῖς χωρίοις τοῖς ἐν Ἰνδίᾳ (λέγω δὲ [1] τοῖς ἐνδοτάτω) ὄρη φασὶν εἶναι δύσβατά τε καὶ ἔνθηρα, καὶ ἔχειν ζῷα ὅσα καὶ ἡ καθ᾽ ἡμᾶς τρέφει γῆ, ἄγρια δέ· καὶ γάρ τοι καὶ τὰς οἶς τὰς ἐκεῖ φασιν εἶναι καὶ ταύτας θηρία, καὶ κύνας καὶ αἶγας καὶ βοῦς, αὐτόνομά τε ἀλᾶσθαι καὶ ἐλεύθερα, ἀφειμένα νομευτικῆς ἀρχῆς. πλήθη δὲ αὐτῶν καὶ ἀριθμοῦ πλείω φασὶν [2] οἱ τῶν Ἰνδῶν λόγιοι. ἐν δὲ [3] τοῖς καὶ τοὺς Βραχμᾶνας ἀριθμεῖν ἄξιον· καὶ γάρ τοι καὶ ἐκεῖνοι ὑπὲρ τῶνδε ὁμολογοῦσι τὰ αὐτά. λέγεται δὲ καὶ ζῷον ἐν τούτοις εἶναι μονόκερων, καὶ ὑπ᾽ αὐτῶν ὀνομάζεσθαι καρτάζωνον. καὶ μέγεθος μὲν ἔχειν ἵππου τοῦ τελείου καὶ λόφον, καὶ λάχνην ἔχειν ξανθήν, ποδῶν δὲ ἄριστα εἰληχέναι.[4] καὶ τοὺς μὲν πόδας ἀδιαρθρώτους τε καὶ ἐμφερεῖς ἐλέφαντι πεφυκέναι,[5] τὴν δὲ οὐρὰν ⟨ἔχειν⟩ [6] συός· μέσον δὲ τῶν ὀφρύων ἔχειν

[1] δή.
[2] φασὶν οἱ τούτων συγγραφεῖς καί.
[3] *Jac*: δή.
[4] εἰληχέναι καὶ εἶναι ὤκιστον.
[5] συμπεφυκέναι.
[6] ⟨ἔχειν⟩ *add. H.*

---

[a] 'Cartazonus' may be presumed to be a corruption of some Indian word. In Sanskrit 'the one-horned animal' is the Rhinoceros; *Khaḍga* and *Khaḍgin* = rhinoceros. A

basilisk. But they say that there is a root which grows in the island by the Great Sea and that it is well-known to everybody, and is an antidote to fainting. At any rate if it is applied to the nose of the fainting man it revives him. But if he is not treated, his malady grows worse until the man dies. Such power, you see, has this Hare to work destruction.

20. In certain regions of India (I mean in the very heart of the country) they say that there are impassable mountains full of wild life, and that they contain just as many animals as our own country produces, only wild. For they say that even the sheep there are wild, the dogs too and the goats and the cattle, and that they roam at their own sweet will in freedom and uncontrolled by any herdsman. Indian historians assert that their numbers are past counting, and among the historians we must reckon the Brahmins, for they also agree in telling the same story.

Wild animals of India

And in these same regions there is said to exist a one-horned beast which they call *Cartazonus*.[a] It is the size of a full-grown horse, has the mane of a horse, reddish hair, and is very swift of foot. Its feet are, like those of the elephant, not articulated and it has the tail of a pig. Between its eyebrows it has a horn growing out; it is not smooth but has

The 'Cartazonus'

fuller form was *Khaḍgadanta*, whence came the Persian *Kargadan*. The Greek καρτάζωνος may have replaced some such Indian-Prakrit word. See H. W. Bailey, *Zoroastrian problems*, 110, and *Bull. of School of Or. & Afr. studies* 10 (1940–42) 899; F. Edgerton, *Buddhist hybrid Sanskrit dict.* 202; E. Sachau, *Alberuni's India*, l. 204, and *Indo-europ. Studien* (Abh. Berl. Ak. Wiss. 1888), p. 18; O. Shepard, *Lore of the Unicorn*, 36.

ἐκπεφυκὸς κέρας οὐ λεῖον ἀλλὰ ἑλιγμοὺς ἔχον τινὰς καὶ μάλα αὐτοφυεῖς, καὶ εἶναι μέλαν τὴν χρόαν· λέγεται δὲ καὶ ὀξύτατον εἶναι τὸ κέρας ἐκεῖνο. φωνὴν δὲ ἔχειν τὸ θηρίον ἀκούω τοῦτο πάντων ἀπηχεστάτην τε καὶ γεγωνοτάτην. καὶ τῶν μὲν ἄλλων αὐτῷ ζῴων προσιόντων φέρειν καὶ πρᾶον εἶναι, λέγουσι δὲ ἄρα πρὸς τὸ ὁμόφυλον δύσεριν εἶναί πως. καὶ οὐ μόνον φασὶ τοῖς ἄρρεσιν εἶναί τινα συμφυῆ κύριξίν τε πρὸς ἀλλήλους καὶ μάχην, ἀλλὰ καὶ πρὸς τὰς θηλείας ἔχουσι θυμὸν τὸν αὐτόν, καὶ προάγοντες τὴν φιλονικίαν καὶ μέχρι θανάτου ⟨τοῦ⟩ [1] ἡττηθέντος ἐξάγουσιν. ἔστι μὲν οὖν καὶ διὰ παντὸς τοῦ σώματος ῥωμαλέον, ἀλκὴ δέ οἱ τοῦ κέρατος ἄμαχός ἐστι. νομὰς δὲ ἐρήμους ἀσπάζεται, καὶ πλανᾶται μόνον· ὥρᾳ δὲ ἀφροδίτης τῆς σφετέρας συνδυασθεὶς πρὸς τὴν θήλειαν πεπράυνται, καὶ μέντοι καὶ συννόμω ἐστόν. εἶτα ταύτης παραδραμούσης καὶ τῆς θηλείας κυούσης, ἐκθηριοῦται αὖθις, καὶ μονίας ἐστὶν ὅδε ὁ Ἰνδὸς καρτάζωνος. τούτων οὖν πώλους πάνυ νεαροὺς κομίζεσθαί φασι τῷ τῶν Πρασίων βασιλεῖ, καὶ τὴν ἀλκὴν ἐν ἀλλήλοις ἐπιδείκνυσθαι κατὰ τὰς θέας τὰς πανηγυρικάς. τέλειον δὲ ἁλῶναί ποτε οὐδεὶς μέμνηται.

21. Ὑπερελθόντι τὰ ὄρη τὰ γειτνιῶντα τοῖς Ἰνδοῖς κατὰ τὴν ἐνδοτάτω πλευρὰν φανοῦνταί φασιν αὐλῶνες δασύτατοι, καὶ καλεῖταί γε ὑπ' Ἰνδῶν ὁ χῶρος Κόλουνδα. ἀλᾶται [2] δὲ ἄρα φασὶν ἐν τοῖσδε τοῖς αὐλῶσι ζῷα Σατύροις ἐμφερῆ τὰς μορφάς, τὸ πᾶν σῶμα λάσια, καὶ ἔχει κατὰ τῆς

spirals of quite natural growth, and is black in colour. This horn is also said to be exceedingly sharp. And I am told that the creature has the most discordant and powerful voice of all animals. When other animals approach, it does not object but is gentle; with its own kind however it is inclined to be quarrelsome. And they say that not only do the males instinctively butt and fight one another, but that they display the same temper towards the females, and carry their contentiousness to such a length that it ends only in the death of their defeated rival. The fact is that strength resides in every part of the animal's body, and the power of its horn is invincible. It likes lonely grazing-grounds where it roams in solitude, but at the mating season, when it associates with the female, it becomes gentle and the two even graze side by side. Later when the season has passed and the female is pregnant, the male Cartazonus of India reverts to its savage and solitary state. They say that the foals when quite young are taken to the King of the Prasii and exhibit their strength one against another in the public shows, but nobody remembers a full-grown animal having been captured.

Satyr-like creatures in India

21. When one has passed the mountains that border upon India there will come into view densely wooded glens on the inner side of the mountains, and the Indians call the region Colunda. And in these glens, they say, creatures resembling Satyrs roam at large; their whole body is shaggy and they

[1] ⟨τοῦ⟩ *add. H.* [2] ἁλῶνται.

ἰξύος ἵππουριν. καὶ καθ' ἑαυτὰ μὲν μὴ ἐνοχλούμενα διατρίβει ἐν τοῖς δρυμοῖς ὑλοτραγοῦντα· ὅταν δὲ αἴσθωνται κυνηγετῶν κτύπου, καὶ ἀκούσωσι κυνῶν ὑλακῆς, ἀναθέουσιν ἐς τὰς ἀκρωρείας αὐτὰς ἀμάχῳ [1] τῷ τάχει· καὶ γάρ εἰσι ταῖς ὀρειβασίαις ἐντριβεῖς. καὶ ἀπομάχονται πέτρας τινὰς κυλινδοῦντες κατὰ τῶν ἐπιόντων, καὶ καταλαμβανόμενοί γε πολλοὶ διαφθείρονται. καὶ ἐκ τούτων εἰσὶν ἐκεῖνοι δυσάλωτοι, καὶ μόλις ποτὲ καὶ διὰ μακροῦ τινὰς αὐτῶν ἐς Πρασίους κομίζεσθαι λέγουσι. καὶ τούτων μέντοι ἢ τὰ νοσοῦντα ἐκομίσθη ἢ θήλεά τινα κύοντα· [2] καὶ συνέβη γε θηραθῆναι τοῖς μὲν διὰ τὴν νωθείαν, ταῖς δὲ διὰ τὸν τῆς γαστρὸς ὄγκον.

22. Ἔστι δὲ καὶ Σκιρᾶται πέραν Ἰνδῶν ἔθνος καὶ τοῦτο, καὶ εἰσὶ σιμοὶ τὰς ῥῖνας, εἴτε οὕτως ἐκ βρεφῶν ἁπαλῶν ἐνθλάσει τῇ τῆς ῥινὸς διαμείναντες, εἴτε καὶ τοῦτον τὸν τρόπον τίκτονται. γίνονται δὲ ὄφεις παρ' αὐτοῖς μεγέθει μέγιστοι, ὧν οἱ μὲν ἁρπάζουσι τὰς ποίμνας καὶ σιτοῦνται, οἱ δὲ ἐκθηλάζουσι τὸ αἷμα, ὥσπερ οὖν παρὰ τοῖς Ἕλλησιν οἱ αἰγοθῆλαι, ὧνπερ οὖν καὶ ἀνωτέρω οἶδα ποιησάμενος μνήμην εὐκαιροτάτην.

23. Ἵππου δὲ ἄρα καὶ τὸ εὐμαθὲς ἴδιον ἦν, καὶ τούτου μαρτύριον ἐκεῖνο. Συβαρίτας τοὺς ἐν Ἰταλίᾳ τρυφῆς ἀκούω ποιήσασθαι φροντίδα ὑπερ-

[1] ἀμηχάνῳ. [2] καὶ κύοντα.

[a] A primitive race of Pygmies, long-haired and with a light-coloured skin, living in the N and NE of India.

have a horse's tail at their waist. And if left to themselves and not troubled, they live among the thickets and subsist off the trees, but whenever they hear the sound of huntsmen or the baying of dogs they run up to the mountain ridges with a speed that none can overtake, for they are inured to roaming the mountains. And from there they fight by rolling down rocks upon their assailants, and many are they that are caught and destroyed. These are the reasons why they are hard to capture, so they say that few indeed, and these at long intervals, are despatched to the Prasii, and of these few it was either sick animals or pregnant females that were despatched: the accident of their capture was due in the case of the males to their tardiness, in the case of the females to their being big-bellied.

22. The Sciratae [a] also are a people on the other side of India, and they are snub-nosed, and are permanently so either from having their noses dinted in tender infancy or because they are born like that. And in their country there occur Snakes of enormous size, some of which seize and devour the flocks, while others suck out their blood, just as the goatsuckers do in Greece [b]: the latter I know I have mentioned earlier on [c] at the most appropriate place.

The Sciratae

Snakes of their country

23. Docility, it seems, is another characteristic of the Horse; witness the following account. I have heard that the inhabitants of Sybaris in Italy

The people of Sybaris and their Horses

[b] This is a complete fiction; see Thompson, *Gk. birds*, s.v. αἰγοθήλας.

[c] See 3. 39.

βάλλουσαν, καὶ τῶν μὲν ἄλλων ἔργων τε καὶ σπουδασμάτων ἀμαθῶς ἔχειν, πάντα δὲ τὸν ἑαυτῶν βίον διάγειν ῥᾳστωνεύοντας ἐν ἀργίᾳ καὶ πολυτελείᾳ. περιηγεῖσθαι μὲν οὖν ἕκαστα τῶν ἐν Συβάρει μακρὸν ἂν εἴη νῦν, ἐκεῖνο δ' οὖν ὁμολογεῖ τρυφὴν ἄμαχον. δεδιδαγμένοι ἦσαν αὐτοῖς οἱ ἵπποι παρὰ τὸν τῆς εὐωχίας καιρὸν ὀρχεῖσθαι πρὸς αὐλόν ἐν ῥυθμῷ.[1] τοῦτο οὖν εἰδότες οἱ Κροτωνιᾶται (ἐπολέμουν δὲ αὐτοῖς) σάλπιγγα μὲν καὶ ἦχον σύντονον καὶ παρακλητικὸν ἐς ὅπλα κατεσίγασαν, αὐλοὺς δὲ καὶ αὐλητὰς παραλαβόντες, ἐπεὶ ὁμοῦ ἦσαν καὶ τόξευμα ἐξικνεῖτο ἤδη, ἐνέδοσαν ἐκεῖνοι τὸ μέλος τὸ ὀρχηστικόν, ὅπερ οὖν ἀκούσαντες οἱ τῶν Συβαριτῶν ἵπποι,[2] ὡς ἐν μέσοις ὄντες τοῖς συμποσίοις, ἀπεσείσαντο μὲν τοὺς ἀναβάτας, ἐσκίρτων δὲ καὶ ἐχόρευον. καὶ τῇ τε ἄλλῃ [3] τὴν τάξιν συνέχεαν καὶ τὸν πόλεμον ἐξωρχήσαντο.

24. Ὑπὲρ τῶν ἵππων τῶν καλουμένων λυκοσπάδων εἶπον καὶ ἀνωτέρω, καὶ νῦν δὲ εἰρήσεται ὅσα προσακήκοα ἴδια. τὴν ὄψιν ἔχουσι συνεστραμμένην καὶ βραχεῖαν, ἔτι δὲ σιμήν. λέγουσι δὲ αὐτὰς εἶναι καὶ φιλέλληνας, καὶ ἔχειν τοῦ γένους τοῦδε σύνεσίν τινα ἀπόρρητον, καὶ συμφυῆ πρὸς αὐτοὺς ἀποσώζειν φιλίαν, καὶ προσιόντων τε καὶ

[1] *Schn*: τῷ ῥυθμῷ MSS, *del. H.*
[2] ἵπποι κατὰ τὴν οἴκοι μνήμην.
[3] τῇ τε ἄλλῃ] ταύτῃ *H*, τῇ τε ἄλῃ *Bernard*.

devoted an excessive amount of thought to delicate living; of other matters and pursuits they knew nothing, but spent their entire time in easy-going sloth and extravagance. To explain in detail all that went on in Sybaris would make a long story now; the following tale however attests their unsurpassed luxuriousness. Their horses had been trained to dance in time to the music of the pipe at their hour for banqueting. Accordingly the inhabitants of Croton knowing this (they were at war with Sybaris), had their trumpet with its piercing note that summons to arms silenced; they collected pipes and pipe-players, and when they were at close quarters and within a bowshot, the players struck up the dance-music. At the sound the horses of the people of Sybaris, imagining that they were in the midst of a wine-party, shook off their riders and began to leap about and dance. And they not only threw the ranks into confusion but also 'danced away' the war.[a]

24. I have spoken earlier on of the horses which are called *lycospades*,[b] and I will now describe some further characteristics of which I have heard. Their face is compact, short, and snub-nosed. They are said to be fond of the Greek people, to understand them by some mysterious means, and to maintain a natural friendship for them, so that if Greeks approach them, touch them, and pat them

The 'lycospad' horse

[a] Sybaris was annihilated by the people of Croton, 510 B.C. Efforts to re-found it were unsuccessful.

[b] A breed of horses from the S of Italy. Ael. has not mentioned them before, though they share some of the characteristics mentioned in 11. 36.

ἐπαφωμένων καὶ κοίλῃ τῇ χειρί πως ἐπικροτούντων μὴ ἄχθεσθαι μηδὲ ἀποσκιρτᾶν, καὶ συνδιημερεύειν μὲν αὐτοῖς [1] ὥσπερ οὖν δεδεμένους, καθευδόντων δὲ καθεύδειν πλησίον. ἐὰν δὲ προσέλθῃ βάρβαρος, ὥσπερ οὖν αἱ ῥινηλατοῦσαι κύνες ἐκ τῶν ἰχνῶν συνιᾶσι τὰ θηρία, οὕτω τοι καὶ αἱ ἵπποι ἐκεῖναι [2] γινώσκουσι τὸ γένος, καὶ χρεμετίζουσι καὶ φεύγουσιν οἷα δήπου δεδοικυῖαι θηρίον. τοῖς μὲν οὖν συνήθεσι καὶ χιλὸν ἐμβάλλουσι καὶ τὴν ἄλλην κομιδὴν προσφέρουσι πάνυ ἥδονται, καὶ βούλονται ὡραῖαι φαίνεσθαι, καὶ ἔτι μᾶλλον τοῖς ἑαυτῶν ἡνιόχοις. καὶ τὸ μαρτύριον, ὅταν νήχωνται, ἐνδοτέρω προχωροῦσιν ἢ τῆς λίμνης ἢ τῆς θαλάττης ἢ τῆς κρήνης, τὸ πρόσωπον φαιδρύνειν βουλόμεναι, ἵνα μή τι ἄμορφον ἢ ἀκαλλὲς ἐκ τῆς φάτνης ἢ ἐκ τῆς ὁδοῦ προσπεσὸν εἶτα ἐπιθολώσῃ τὸ κάλλος. λυκοσπάδι δὲ ἄρα ἵππῳ καὶ ἀλοιφαὶ εὐώδεις ὡς νύμφῃ φίλαι καὶ μύρων ὀσμή. καὶ Ὅμηρος δὲ τὸ φιληδεῖν ἵππους χρίσμασι φύσει πάντας ὁμολογεῖ λέγων

τοίου γὰρ σθένος ἐσθλὸν ἀπώλεσαν ἡνιόχοιο
ἠπίου, ὅς σφῶιν μάλα πολλάκις ὑγρὸν ἔλαιον
χαιτάων κατέχευε, λοέσσας ὕδατι λευκῷ.

καὶ Σημωνίδης [3] δὲ ἐκ παντοδαπῶν θηρίων λέγων τὰς γυναῖκας γενέσθαι τε καὶ διαπλασθῆναί φησιν ἐνίαις ἐκ τῶν ἵππων τό τε φιλόκοσμον καὶ φιλόμυρον συντεχθῆναι καὶ ἐκείναις [4] φύσει.[5] ἃ δὲ λέγει, ταῦτά ἐστιν·

[1] *Reiske* : αὐτούς.
[2] οὕτω . . . ἐκεῖναι] *Jac* : ἐκεῖνα, οὕτω . . . ἵπποι γ.
[3] Σιμ- MSS, *H*.

with the hollow of their hand, they do not resent it or shy away, but pass their days at their side as though they were tethered, and when the Greeks lie down to sleep they will sleep at their side. If however some foreigner approaches, then, just as hounds on the scent recognise animals by their tracks, so do these mares know the man's origin, and neigh and flee away as though they were afraid of some wild beast. But their delight is in familiar friends who give them fodder and generally tend them, and they are anxious to appear beautiful, especially in the eyes of their drivers. The proof of this is that when they go swimming they advance far into the lake or sea or spring in their eagerness to sleek their faces, so that nothing disfiguring or unlovely from the manger or from their journey may befoul their beauty. Fragrant unguents and the scent of perfumes are as dear to a lycospad horse as they are to a bride. And Homer testifies to the natural love which all horses have for unguents when he says [*Il.* 23. 280]

> 'For so mighty a charioteer and so gentle have they lost, who right often would pour upon their manes smooth oil when he had washed them in clear water.'

And Semonides describing how women are born and moulded after animals of all kinds, says that the horse's love of ornament and of perfumes is innate in some women also. These are his words [*fr.* 7. 57 Diehl]:

---

[4] κατ' ἐκείνους.

[5] *Ges*: φύσαις, φῦναι, *or* φυούσαις.

τὴν δ᾿ ἵππος ἁβρὴ χαιτέεσσ᾿ [1] ἐγείνατο,
ἣ δούλι᾿ ἔργα καὶ δύην [2] περιτρέπει·
κοὔτ᾿ ἂν μύλης ψαύσειεν, οὔτε κόσκινον
ἄρειεν, οὔτε κόπρον ἐξ οἴκου βάλοι,
οὔτε πρὸς ἰπνὸν ἀσβόλην ἀλευμένη
ἵζοιτ᾿. ἀνάγκῃ δ᾿ ἄνδρα ποιεῖται φίλον.
λοῦται δὲ πάσης ἡμέρας ἄπο ῥύπον
δίς, ἄλλοτε τρίς, καὶ μύροις ἀλείφεται.
αἰεὶ [3] δὲ χαίτην ἐκτενισμένην φορεῖ
βαθεῖαν, ἀνθέμοισιν ἐσκιασμένην.
καλὸν μὲν οὖν θέημα τοιαύτη γυνὴ
ἄλλοισι, τῷ δ᾿ ἔχοντι γίγνεται κακόν,
ἢν μή τις ἢ τύραννος ἢ σκηπτοῦχος ᾖ,
ὅστις [4] τοιούτοις θυμὸν ἀγλαΐζεται.

25. Ἴδια δὲ ἵππων καὶ ἐκεῖνα δήπου. οἱ Πέρσαι, ἵνα μὴ ὦσιν αὐτοῖς οἱ ἵπποι καταπλῆγες, ψόφοις αὐτοὺς καὶ ἤχοις χαλκοῖς [5] προσεθίζουσι, καὶ κωδωνίζουσιν,[6] ὡς μή ποτε ἐν τῷ πολέμῳ δείσωσι τοὺς τῶν πανοπλιῶν ἀραγμοὺς καὶ τὸν τῶν ξιφῶν πρὸς τὰς ἀσπίδας δοῦπον. εἴδωλά τε νεκρῶν δὴ σεσαγμένα ἀχύροις ὑποβάλλουσιν αὐτοῖς, ἵνα προσεθισθῶσι νεκροὺς ἐν τῷ πολέμῳ πατεῖν, καὶ μὴ δεδιότες ὥς τι ἐκπληκτικὸν εἶτα μέντοι [7] ἐν τοῖς ἔργοις τοῖς ὁπλιτικοῖς ἀχρεῖοι ὦσιν. οὐκ ἐλελήθει δὲ Ὅμηρον οὐδὲ τοῦτο, ὡς αὐτὸς [8] δείκνυσιν. ὅτι γοῦν ἀνῃρέθη μὲν ὁ Θρᾷξ Ῥῆσος, σὺν αὐτῷ δὲ καὶ οἱ ἑταῖροι, ἀκούομεν [9] ἐν Ἰλιάδι ἐκ παίδων· ἃ δὲ ἀκούομεν, ταῦτά ἐστιν. ἀποσφάττει μὲν ὁ τοῦ Τυδέως τοὺς Θρᾷκας, ὁ δὲ τοῦ

[1] *Mein*: ἁβρὰ χαιτάεσσ᾿. [2] *Stobaeus*: ἄτην.
[3] *Bergk*: ἀεί. [4] ὅς τοῖς *H*.

'But another is born of a dainty, long-maned mare: she turns away from servile tasks and drudgery; she will never touch a mill or pick up a sieve or cast muck out of the house, nor, since she would escape the soot, will she sit by the oven. Only by constraint does she take a man to her bosom. And every day she washes off the dirt twice, sometimes thrice, and anoints herself with perfumes. And always she wears her deep tresses combed and shaded with flowers. Such a woman is fair to look upon—for others, but to her husband, a plague, unless he be a despot or sceptred lord who delights his heart with such gauds.'

The Horse trained for battle

25. Here, I think, are further characteristics of Horses. In order that their Horses may not panic, the Persians accustom them to noises and the clang of bronze, and sound them so that in war they may never be afraid of the rattle of full armour and the clash of swords upon shields. And they throw dummy corpses stuffed with straw beneath their feet in order that they may get used to trampling on corpses in war and may not through terror at some unnerving occurrence be useless in encountering men-at-arms. Nor did this escape the notice of Homer, as he himself shows. At any rate we learn in our childhood from the *Iliad* [10. 486] how the Thracian Rhesus and his companions with him were slain. This is the story we learn. The son of Tydeus [a] slaughters the Thracians, while the son

[a] Diomedes.

[5] χαλκοῦ *Reiske*.
[6] *Schn*: κωδωνοῦσιν.
[7] μέντοι δυσωπούμενοι.
[8] *Ges*: αὐτά.
[9] *Jac*: οὓς ἀκούομεν.

Λαέρτου τοὺς ἀνῃρημένους ὑπάγει τῶν ποδῶν, ἵνα μή ποτε ἄρα νεήλυδες ὄντες οἱ Θρᾷκες ἵπποι εἶτα μέντοι ἐκπλήττωνται [1] τοῖς νεκροῖς ἐμπαλασσόμενοι,[2] καὶ ἀήθως κατ' αὐτῶν ὡς τινων φοβερῶν βαίνοντες ἀποσκιρτῶσιν. οἱ δὲ ἅπαξ μαθόντες, οὐκ ἂν αὐτοὺς λάβοι τοῦ μαθήματος λήθη· οὕτως εἰσὶν ἀγαθοὶ μαθεῖν ὁτιοῦν τῶν λυσιτελῶν οἱ ἵπποι. φιλεῖν δὲ οἷοί ποτέ εἰσι καὶ ἐς ὅσον, ἐμοὶ μὲν εἴρηται ἐν λόγοις τοῖς ἄνω.

26. Ἐν τοῖς κρυμώδεσι τόποις τὰ πρόβατα τῆς χιόνος ἐπιρρεούσης καὶ τοῦ κρύους ἐνακμάζοντος ἄχολά ἐστι (καθειργμένα δὲ ἄρα καὶ τοῦ χιλοῦ τοῦ νέου μὴ μεταλαμβάνοντα εἶτα μέντοι τοιαῦτα εὑρίσκεται), ὑπαρχομένου δὲ τοῦ ἦρος προϊόντα [3] ἐπὶ τὰς νομὰς τῆς χολῆς ὑποπίμπλαται. τοῦτο δὲ ἄρα ἔτι καὶ μᾶλλον φιλεῖ παρακολουθεῖν τοῖς Σκυθικοῖς προβάτοις φασίν.

27. Ἀγαθαρχίδης φησὶν εἶναι γένος ἐν τῇ Λιβύῃ τινῶν ἀνθρώπων, καὶ μέντοι καὶ καλεῖσθαι αὐτοὺς Ψύλλους. καὶ ὅσα μὲν κατὰ τὸν ἄλλον βίον τῶν λοιπῶν ἀνθρώπων διαφέρειν [4] οὐδὲ ἕν, τὸ δὲ σῶμα ἔχειν ξένον τε καὶ παράδοξον ὡς πρὸς τοὺς ἑτεροφύλους ἀντικρινόμενον· τὰ γάρ τοι ζῷα τὰ δακετὰ καὶ τὰ ἐγχρίμπτοντα [5] πάμπολλα ὄντα μηδὲν αὐτοὺς μόνους ἀδικεῖν. οὔτε γοῦν ὄφεως δακόντος ἐπαΐουσιν οὔτε φαλαγγίου νύξαντος τοὺς ἄλλους [6] ἐς θάνατον οὔτε μὴν σκορπίου ⟨τὸ⟩ [7] κέντρον ἀπερείσαντος. ἐπὰν δὲ ἄρα τούτων προσ-

[1] *Ges*: ἐκπλήττονται. [2] ἐμπλαττ- *and* ἐμπαλαττ-.
[3] *Jac*: καὶ προϊόντα. [4] *Ges*: διαφέρει.

of Laertes [a] draws the slain men away by the feet for fear lest the Thracian horses, being newcomers, get entangled among the dead bodies and panic, and through being unused to them may leap aside as though they were treading upon some terrifying objects. But once Horses have learnt a thing, they will not forget what they have learnt, so clever are they at learning whatever is of any advantage. I have spoken earlier on [b] of their capacity for affection and how far they will feel it.

Sheep in cold weather

26. In frosty regions when the snow falls and the cold is at its worst the Sheep have no gall (they are found to be in this condition when penned up and unable to get fresh fodder), but at the beginning of spring they go out to the pastures and become filled with gall. And this, they say, is a constant occurrence especially in the Sheep of Scythia.

The Psylli

27. Agatharcides asserts that there is in Libya a certain race of men who are called Psylli. So far as their general way of life is concerned they differ not a whit from other men, except that, compared with men of other nations, their bodies have an unusual and marvellous quality: they alone are uninjured by the numerous creatures that bite or strike. At any rate they do not feel either the bite of a snake or the prick of a spider which is fatal to others, or even the sting planted by a scorpion, and whenever one of these creatures comes near and

[a] Odysseus. [b] See 6. 44.

[5] ἐγχρίπτοντα. [6] ὡς τοὺς ἄ.
[7] ⟨τό⟩ *add.* H.

πελάσῃ τι καὶ παραψαύσῃ τοῦ σώματος καὶ ἅμα καὶ τῆς ὀσμῆς τῆς ἐκείνων σπάσῃ,[1] ὥσπερ οὖν φαρμάκου γευσάμενον[2] κάρωσίν τινα ἑλκτικὴν ἐς ἀναισθησίαν ἐμποιοῦντος, ἐξασθενεῖ καὶ παρεῖται, ἔστ᾽ ἂν παραδράμῃ ὁ ἄνθρωπος. ὅπως δὲ ἐλέγχουσι τὰ ἑαυτῶν βρέφη εἴτε ἐστὶ γνήσια εἴτε καὶ νόθα, ἐν τοῖς ἑρπετοῖς βασανίζοντες ὡς ἐν τῷ πυρὶ τὸν χρυσὸν οἱ βάναυσοι,[3] ἀνωτέρω εἶπον.

28. Καλλίας ἐν τῷ δεκάτῳ τῶν περὶ τὸν Συρακόσιον Ἀγαθοκλέα λόγων φησὶ τοὺς κεράστας ὄφεις δεινοὺς εἶναι τὸ δῆγμα· ἀναιρεῖν γὰρ καὶ ζῷα ἄλογα καὶ ἀνθρώπους, εἰ μὴ παρείη Λίβυς ἀνήρ, Ψύλλος ὢν τὸ γένος. οὗτος γοῦν ἐάν τε κλητὸς ἀφίκηται ἐάν τε καὶ παρῇ κατὰ τύχην καὶ θεάσηται πράως ἔτι ἀλγοῦντα, τῇ πληγῇ[4] μόνον προσπτύσας εἶτα μέντοι τὴν ὀδύνην ἐπράυνε, καὶ κατεγοήτευσε τὸ δεινὸν τῷ σιάλῳ. ἐὰν δὲ εὕρῃ δυσανασχετοῦντα καὶ ἀτλήτως φέροντα, ὕδωρ ἀθρόον σπάσας ἔσω τῶν ὀδόντων καὶ χρησάμενος αὐτῷ τοῦ στόματος κλύσματι, εἶτα τοῦτο ἐς κύλικα ἐμβαλὼν δίδωσι ῥοφῆσαι τῷ τρωθέντι. ἐὰν δὲ περαιτέρω καὶ τοῦδε τοῦ φαρμάκου κατισχύῃ τὸ κακόν, ὁ δὲ τῷ νοσοῦντι παρακλίνεται γυμνῷ γυμνός, καὶ τοῦ χρωτὸς οἱ τοῦ ἰδίου προσανατρίψας τὴν ἰσχὺν τὴν συμφυῆ, εἶτα μέντοι τοῦ κακοῦ πεποίηκε τὸν ἄνθρωπον ἐξάντη.[5] ἀποχρῶν δὲ ἄρα ὑπὲρ τούτου εἴη ἂν μάρτυς καὶ Νίκανδρος ὁ Κολοφώνιος λέγων

[1] *Jac*: ψαύσῃ ἢ σπάσῃ.
[2] γευσάμενον ὑπνοποιοῦ.
[3] βάναυσοι χρυσουργοί.

touches a Psyllian and inhales the odour from him, it is as though it had tasted some drug that brings on a drowsiness inducing insensibility, for it becomes quite weak and relaxed until the man has passed by. And their manner of proving that their children are either their own or bastards by testing them among reptiles, just as artizans test gold in the fire, I have described earlier on.[a]

The Psylli and snake-bites

28. Callias in the tenth book of his *History of Agathocles of Syracuse* says that the Cerastes inflicts a terrible bite, for it kills dumb animals and human beings unless a Libyan belonging to the race of Psylli happens to be at hand. At any rate if a Psyllian comes in answer to a summons or is present by chance and sees that the victim is still only in slight pain, by simply spitting on the wound he alleviates the pain and conjures away the poison by his spittle. If however he finds the man in a sore plight and in intolerable suffering, he takes a large mouthful of water behind his teeth, and this same water with which he has rinsed his mouth he pours into a cup and gives to the stricken man to swallow. But if the poison is too strong even for this medicine, the Psyllian lies down naked beside the sick man also naked, and applying to him by friction the innate power of his own skin, renders the man free of the poison. And Nicander of Colophon should be sufficient witness to this when he says [*fr.* 32]

[a] See 1. 57.

---

[4] τὴν πληγὴν ἢ τὸ δῆγμα.
[5] ἐξάντην.

ἔκλυον ὡς Λιβύης [1] Ψύλλων γένος οὔτε τι θηρῶν
αὐτοὶ κάμνουσιν μυδαλέῃσι [2] τυπαῖς,
οὓς Σύρτις βοσκεῖ θινοτρόφος,[3] εὖ δὲ καὶ ἄλλοις
ἀνδράσιν ἤμυναν τύμμασιν ἀχθομένοις,
οὐ ῥίζαις ἔρδοντες,[4] ἑῶν δ' ἀπὸ σύγχροα γυίων

καὶ τὰ ἐπὶ τούτοις.

29. Ἐμπεδοκλῆς ὁ φυσικός φησι, περὶ ζῴων ἰδιότητος λέγων καὶ ἐκεῖνος δήπου, γίνεσθαί τινα συμφυῆ καὶ κράσει μορφῆς μὲν διάφορα, ἑνώσει δὲ σώματος συμπλακέντα. ἃ δὲ λέγει, ταῦτά ἐστι·

πολλὰ μὲν ἀμφιπρόσωπα καὶ ἀμφίστερνα φύεσθαι,
βουγενῆ ἀνδρόπρῳρα, τὰ δ' ἔμπαλιν ἐξανατέλλειν
ἀνδροφυῆ βούκρανα μεμιγμένα τῇ μὲν ἀπ' ἀνδρῶν,
τῇ δὲ γυναικοφυῆ σκιεροῖς [5] ἠσκημένα γυίοις.

30. Ἐν Λυκίᾳ λέγει Καλλισθένης ὁ Ὀλύνθιος κείρεσθαι καὶ τὰς αἶγας ὥσπερ οὖν πανταχῇ τὰ πρόβατα· γίνεσθαι γὰρ δασυτάτας καὶ εὔτριχας δεινῶς,[6] ὡς εἰπεῖν βοστρύχους ἤ τινας ἕλικας κόμης ἐξηρτῆσθαι αὐτῶν, καὶ μέντοι καὶ τοὺς τεχνίτας τῆς τῶν νεῶν ἐργασίας καταχρῆσθαι αὐταῖς τὰς σειρὰς συμπλέκοντας.[7]

[1] *Bergk*: Λίβυες.
[2] *OSchn*: μυδαλέαισι.
[3] θηροτρόφος A, *H* (1858).
[4] ῥίζας ἄρδοντες *H*.
[5] θιμβροῖς *H*, στείροις or σκιροῖς *Diels*.
[6] δεινῶς τὰς αἶγας.
[7] συμπλέκοντας ὡς καὶ τοὺς κάλως ἄλλοι.

'I have heard how the race of Psylli in Libya suffer not at all from the festering wounds inflicted by the creatures that are nurtured by Syrtis, mother of sands, and well-skilled are they to succour others also when afflicted by their blows; not working with simples, but from their own limbs, skin touching skin—'

and so on.

29. Empedocles, the natural philosopher, who of course also speaks about the characteristics of animals, says that there are some creatures of composite nature, differing in so far as they are two forms combined, but conjoined in a single body. These are his words:[a]

Different natures combined

'Many creatures are begotten with two faces and two breasts: some born of a cow have the fore-parts of a man; others on the contrary spring up begotten of a man but with the head of a cow; others again mingle the limbs of a man with those of a woman, being endowed with parts veiled in shadow.'[b]

30. Callisthenes of Olynthus asserts that in Lycia the Goats are shorn just as sheep are everywhere else, for they have such wonderfully thick, fine fleeces that one might say that their hair hung down in curls or ringlets. Moreover those who make tackle for ships use them for weaving ropes.

The Goats of Lycia

[a] The lines are from his poem Περὶ Φύσεως, *fr.* 61, Diels *Vorsok.*[6] 1. 334.

[b] σκιεροῖς: both text and interpretation are uncertain.

31. Λέγει δὲ ἄρα Κτησίας ἐν λόγοις Ἰνδικοῖς τοὺς καλουμένους Κυναμολγοὺς τρέφειν κύνας πολλοὺς κατὰ τοὺς Ὑρκανοὺς τὸ μέγεθος, καὶ εἶναί γε ἰσχυρῶς κυνοτρόφους. καὶ τὰς αἰτίας ὁ Κνίδιος ἐκείνας λέγει. ἀπὸ τῶν θερινῶν τροπῶν ἐς μεσοῦντα χειμῶνα ἐπιφοιτῶσιν αὐτοῖς ἀγέλαι βοῶν, οἱονεὶ μελιττῶν σμῆνος ἢ σφηκιὰ κεκινημένη, πλῆθος ἀριθμοῦ κρείττους οἱ βόες· εἰσὶ δὲ ἄγριοι καὶ ὑβρισταί, καὶ ἐς κέρας θυμοῦνται δεινῶς. οὔκουν ἔχοντες αὐτοὺς ἀναστέλλειν ἑτέρως οἵδε τοὺς τροφίμους αὑτῶν κύνας ἐπ᾿ αὐτοὺς ἀφιᾶσιν ἐς τοῦτο ἀεὶ τρεφομένους, οἵπερ οὖν καταγωνίζονταί τε καὶ διαφθείρουσιν αὐτοὺς ῥᾷστα. εἶτα τῶν κρεῶν τὰ μὲν δοκοῦντά σφισιν ἐς ἐδωδὴν ἐπιτήδεια ἐξαιροῦσι, τὰ δὲ λοιπὰ τοῖς κυσὶν ἀποκρίνουσι, καὶ μάλα γε ἀσμένως κοινωνοῦντες αὐτοῖς, ὥσπερ οὖν εὐεργέταις ἀπαρχόμενοι. κατὰ τὴν ὥραν δὲ καθ᾿ ἣν οὐκέτι φοιτῶσιν οἱ βόες, συνθήρους αὐτοὺς ἐπὶ τοῖς ζῴοις τοῖς ἄλλοις ἔχουσι. καὶ τῶν θηλειῶν ἀμέλγουσι τὸ γάλα, καὶ ἐκ τούτου κέκληνται· πίνουσι γὰρ αὐτό, ὥσπερ οὖν ἡμεῖς τὸ τῶν οἰῶν τε καὶ τῶν αἰγῶν.

32. Αἰσχυλίδης ἐν τοῖς περὶ γεωργίας κατὰ τὴν Κείων [1] γῆν πρόβατα γίνεσθαι ὀλίγα ἑκάστῳ τῶν γεωργῶν φησι. τὸ δὲ αἴτιον, λεπτόγεών τε εἶναι τὴν Κέω [2] ἰσχυρῶς καὶ νομὰς οὐκ ἔχειν· κύτισον δὲ καὶ θρία ἐμβάλλειν, καὶ τῆς ἐλαίας τὰ ῥεύσαντα φύλλα, καὶ μέντοι καὶ ὀσπρίων [3] ἄχυρα ποικίλων,

[1] Κίων.
[2] Κίον.
[3] τῶν ὀσπρίων.

31. Ctesias in his account of India asserts that the people called *Cynamolgi* [a] (dog-milkers) keep a great number of hounds as large as those of Hyrcania, and, in particular, that they are keen dog-breeders. The Cnidian writer gives the reasons as follows. From the summer solstice up to mid-winter herds of cattle come roaming; like a swarm of bees or a wasps' nest that has been disturbed these cattle are past numbering. And they are wild and aggressive and vent their fury with their horns in a terrible fashion. Being unable to check them by any other means the Cynamolgi let loose their hounds, which they always breed for this purpose, upon them, and the hounds overcome and destroy them without any difficulty. Thereupon the men select such portions of the flesh as they consider suitable for eating, the residue they set aside for the hounds and are glad indeed to give them a share, an offering as it were to benefactors. And during the season when these cattle are no longer on the move the Cynamolgi have the hounds to help them in their pursuit of other beasts. The bitches they milk; hence their name, for they drink hounds' milk just as we drink that of sheep and goats.

The Cynamolgi and their Dogs

32. In his work on agriculture Aeschylides [b] says that in Ceos each of the farmers owns but few sheep, the reason being that the soil of Ceos is exceedingly poor and has no pasture-land. So they throw tree-medick and fig-leaves and the fallen leaves of the olive to the flocks, also the husks of various kinds of

The Sheep of Ceos

[a] A tribe in Ethiopia.

[b] All that is known of him is that his work was in at least three books; see Athen. 14. 650 D.

παρασπείρειν δὲ καὶ ἀκάνθας, καὶ ἐκείνοις ἀγαθὸν εἶναι ταῦτα δεῖπνον.[1] γίνεσθαι δὲ ἐξ αὐτῶν γάλα, καὶ τοῦτο τρεφόμενον τυρὸν ἐργάζεσθαι κάλλιστον· καλεῖσθαι δὲ αὐτὸν Κύθνιον[2] ὁ αὐτὸς λέγει, καὶ μέντοι καὶ τὸ τάλαντον αὐτοῦ πιπράσκεσθαι δραχμῶν καὶ ἐνενήκοντα. γίνεσθαι δὲ καὶ ἄρνας τὴν ὥραν διαπρεπεῖς, καὶ πιπράσκεσθαι οὐ κατὰ τοὺς ἑτέρους, ἀλλὰ καὶ σοβαρωτέρᾳ τῇ τιμῇ.

33. Φοίνικες λέγουσι λόγοι τὰς βοῦς τὰς ἐπιχωρίους τοσαύτας εἶναι τὸ μέγεθος, ὡς ἑστάναι τοὺς ἀμέλγοντας ὄντας μεγίστους ἢ δεῖσθαι θρανίου, ἵνα ἀναβάντες ἐφίκωνται τῶν μαζῶν. Λιβύων δὲ ἄρα τῶν γειτνιώντων Ἰνδοῖς ὀπισθονόμων βοῶν ἀγέλας εἶναί τινας ἀκούω. τὸ δὲ αἴτιον, ἡ φύσις ὡς ἐξήμαρτε τὸ πρῶτον ἢ ὠλιγώρησεν, ἐπεὶ πρὸ τῶν ὀφθαλμῶν αὐτοῖς ἐστι τὰ κέρατα, καὶ ὁρᾶν οὐκ ἐᾷ τὰ πρὸ ποδῶν, ἡ δὲ ἐς οὐρὰν ἐπανάγει τὴν βάδισιν αὐτοῖς, εἶτα ἐπικύπτοντες κείρουσι τὴν πόαν. Ἀριστοτέλης γε μήν φησι τῶν βοῶν τῶν ἐν Νευροῖς[3] ἐκπεφυκέναι τὰ κέρατα καὶ τὰ ὦτα ἔκφυσιν τὴν αὐτήν, καὶ εἶναι συνυφασμένα. ὁ δὲ αὐτὸς ἐν χώρῳ τινὶ Λιβύων τὰς αἶγας τοῦ στήθους φησὶ τοὺς μαζοὺς ἔχειν ἀπηρτημένους. εἴη δ' ἂν τοῦ παιδὸς τοῦ Νικομάχου καὶ ταῦτα· ἐν τοῖς

[1] ποικίλων . . . δεῖπνον] ποικίλα τε καὶ ποικίλον ἐκείνοις ἀγαθὰ εἶναι τ. δ. παρασπείρειν δὲ καὶ ἀκάνθας.
[2] *Bochart*: Κύνθιον.
[3] *Rose*: Λεύκτροις.

---

[a] That is, £3 7*s*. 6*d*. for 74 lb. avoirdupois, the drachma being reckoned at 9*d*.

pulse, and they even sow thistles among their crops, all of which afford excellent feeding for the sheep. And from them they obtain milk which when curdled produces the finest cheese. And the same writer says that it is called *Cythnian* and that it is sold at the rate of ninety drachmas a talent.[a] And lambs also are produced that are of remarkable beauty and are sold not at the price of ordinary lambs but for a far more impressive figure.

The Cattle and Sheep of various countries

33. Phoenician histories state that the Cows of that country are so tall that the milkers for all their great height have to stand or else need a stool to enable them to reach the teats. And among the Libyans who border upon India I learn that there are herds of cattle that graze moving backwards, the reason being that Nature made an initial blunder or failed to pay attention, because their horns grow in front of their eyes and prevent them from seeing what lies immediately ahead, and so she obliges them to move backwards, and they lower their heads and crop the grass. Again, Aristotle says[b] that among the Neuri[c] the horns and ears of the cattle spring from the same source and are knit together. And the same writer says that in a certain place in Libya the goats have their teats attached to the chest. Let me add the following statement also from the son of Nicomachus: he says that among the Budini who live on the banks of

[b] Not in any extant work; the nearest approach to these two statements is to be found (for the cattle) in *HA* 517 a 28 and (for the goats) 500 a 15. See *frr.* 313, 314 (Rose p. 331).

[c] A Slav people who in the 6th cent. B.C. settled in the region about modern Kiev.

περὶ τὸν Καρίσκον Βουδίνοις [1] οἰκοῦσιν οὐ γίνεσθαί φησι πρόβατον λευκόν, ἀλλὰ μέλανα πάντα.

34. Νυμφόδωρος λέγει τὴν Σαρδὼ εἶναι θρεμμάτων μητέρα ἀγαθήν. θαυμάσαι δὲ ἄξιον τίκτει ζῷον τὰς αἶγας αὕτη. τὰς γάρ τοι δορὰς τοὺς ἐπιχωρίους [2] ἠσθῆσθαι, καὶ εἶναι ταῦτα σκέπην· καὶ διὰ χειμῶνος μὲν ἀλεαίνειν, ψύχειν δὲ ἐν τῷ θέρει φύσει τινὶ ἀπορρήτῳ· συμπεφυκέναι δὲ ἄρα ταῖς δοραῖς καὶ πήχεως τὴν τρίχα. τοῦ φορήματος δὲ [3] τούτου ἔοικε χρῆναι διὰ μὲν τοῦ κρυμοῦ τὰς τρίχας ἐς τὸν χρῶτα ἐπιστρέφειν [4] διὰ δὲ τοῦ θέρους ἔξω, εἰ μέλλοι ὁ ἠσθημένος διὰ μὲν τοῦ κρυμοῦ θάλπεσθαι, διὰ δὲ τοῦ θέρους μὴ ἀποπνίγεσθαι.

35. Τί δαί; Ὀρθαγόραν μνήμης ἄμοιρον ἐάσομεν; ὅσπερ οὖν ἐν Ἰνδοῖς λόγοις φησὶ Κώυθα μὲν οὕτως εἶναι κώμην τὸ ὄνομα λαβοῦσαν,[5] ταῖς δὲ αἰξὶ ταῖς ἐπιχωρίοις ἔνδον ἐν τοῖς σηκοῖς παραβάλλειν τοὺς νομέας ἰχθῦς ξηροὺς χιλόν.

36. Ὅτι δέδοικεν ὗν ἐλέφας ἀνωτέρω εἶπον· τὸ δὲ ἐν Μεγάροις γενόμενον Μεγαρέων ὑπ' Ἀντιγόνου [6] πολιορκουμένων ἐθέλω εἰπεῖν, καὶ μέντοι ⟨καὶ⟩ [7] τὸ εἰρησόμενον τοῦτό ἐστι. τῶν Μακεδόνων βιαίως ἐγκειμένων, ὗς πίττῃ χρίσαντες ὑγρᾷ [8] καὶ ὑποπρήσαντες αὐτὰς ἀφῆκαν ἐς τοὺς

[1] *LDindorf*: Βουδιανοῖς.
[2] *Schn*: τοῖς ἐπιχωρίοις.
[3] δὲ ἄρα.
[4] ἀποστρέφειν.
[5] τόδε ὄνομα λαχοῦσαν.
[6] *Ed.*: Ἀντιπάτρου MSS, *edd.*
[7] ⟨καί⟩ *add. H.*
[8] *Ges*: ψυχρᾷ.

the Cariscus[a] a white sheep does not occur, they are all black.

The Goats of Sardinia

34. Nymphodorus says that Sardinia is an excellent mother of flocks. The Goats which she nourishes are animals deserving admiration, for the natives clothe themselves in their skins and these afford them protection; and in the winter the skins keep them warm, and in the summer by some mysterious natural property keep them cool. The hair on the hides actually grows to the length of a cubit. And it seems that during frosty weather the wearer must turn the hairs of this garment inwards to the skin, but in summer outwards, if he wants to keep warm during the frost and not to be suffocated in the summer.

Fish as food for Goats

35. What? Are we to leave the name of Orthagoras without a mention? He says in his account of India that there is a village which has been given the name of Coÿtha, and that the herdsmen give dried fish as fodder to the goats of that country when in their pens.

Elephants routed by Pigs

36. I have stated earlier on[b] that the Elephant dreads a pig; I now wish to tell what happened at Megara when the Megarians were besieged by Antigonus,[c] and the story I have to tell is as follows. When the Macedonians were pressing them hard, they smeared some pigs with liquid pitch, set a light

[a] Unidentified. [b] See 1. 38; 8. 28.

[c] It was Antigonus (not Antipater) who besieged and took Megara; see 11. 14. The story of the pigs is given by Polyaenus, 4. 5. 3.

πολεμίους. ἐμπεσοῦσαι δὲ ἄρα ἐκεῖναι οἰστρημέναι ταῖς τῶν ἐλεφάντων ἴλαις καὶ βοῶσαι, ἅτε ἐμπιπράμεναι, ἐξέμαινον τοὺς θῆρας καὶ ἐτάραττον δεινῶς. οὔτε οὖν [1] ἔμενον ἐν τάξει, οὔτε ἦσαν ἔτι πρᾶοι, καίτοι καὶ ἐκ νηπίων πεπωλευμένοι, εἴτε φύσει τινὶ οἱ ἐλέφαντες ἰδίᾳ μισοῦντες τὰς ὗς καὶ μυσαττόμενοι, εἴτε καὶ τῆς φωνῆς αὐτῶν τὸ ὀξὺ καὶ ἀπηχὲς πεφρικότες ἐκεῖνοι. συνιδόντες [2] οὖν ἐκ τούτου [3] οἱ πωλοτρόφοι τῶν ἐλεφάντων ὗς παρατρέφουσιν αὐτοῖς, ὥς φασιν, ἵνα γε ἐκ τῆς συνηθείας ἧττον ὀρρωδῶσιν αὐτάς.

37. Παρά γε τοῖς Ψύλλοις καλουμένοις τῶν Ἰνδῶν (εἰσὶ γὰρ καὶ Λιβύων ἕτεροι) [4] ⟨οἱ⟩ [5] ἵπποι γίνονται τῶν κριῶν οὐ μείζους, καὶ τὰ πρόβατα ἰδεῖν μικρὰ κατὰ τοὺς ἄρνας, καὶ οἱ ὄνοι δὲ τοσοῦτοι γίνονται τὸ μέγεθος καὶ οἱ ἡμίονοι καὶ οἱ βοῦς καὶ πᾶν κτῆνος ἕτερον ὁτιοῦν. ὗν δὲ ἐν Ἰνδοῖς οὔ φασι γίνεσθαι οὔτε ἥμερον οὔτε ἄγριον· μυσάττονται δὲ καὶ ἐσθίειν τοῦδε τοῦ ζῴου Ἰνδοί, καὶ οὐκ ἂν γεύσαιντό ποτε ὑείων, ὥσπερ οὖν οὐδὲ ἀνθρωπείων οἱ αὐτοί.

38. Ἐν Μητροπόλει ἀκούω τῇ Ἐφεσίᾳ λίμνην εἶναι καὶ πρὸς αὐτῇ σπήλαιον· ἔχει δὲ ἄρα τὸ ἄντρον ὄφεων πλῆθος ἄμαχον, καὶ εἶναι τούτους μεγίστους φασὶ καὶ δεινοὺς τὸ δῆγμα. προϊέναι μὲν οὖν αὐτοὺς τοῦ ἄντρου λόγος ἔχει, ὅσον ἐς τὴν λίμνην ἐξερπύσαι τὴν παρακειμένην καὶ

[1] γοῦν.
[2] συνειδότες.
[3] τούτων.
[4] (εἰσὶ . . . ἕτεροι) *gloss*, *H.*

to them, and let them loose against the enemy. Goaded with pain and shrieking because of their burns, the pigs fell upon the troops of Elephants, driving them mad and throwing them into terrible confusion. So the Elephants broke ranks and were no longer tractable in spite of having been trained since they were small, either because Elephants by some instinct hate and loathe pigs, or because they dread the shrill and discordant sound of their voices. In consequence those who train young Elephants, being aware of this, keep pigs along with them, so it is said, in order that through herding together the Elephants may get to fear them less.

The Psylli of India and their horses, etc.

37. Among the people called Psylli in India (there are other Psylli in Libya also) the Horses are no bigger than rams, the Sheep look as small as lambs, while the Asses, Mules, Cattle, and domestic animals of every kind are proportionately small. They say that neither the domestic nor the wild Pig exists in India, and the Indians revolt at the idea of eating this animal: they would no more eat pork than they would human flesh.

Snakes and Crabs at Metropolis

38. I have heard that in Metropolis [a] near Ephesus there is a lake and near to it a cavern. Now this cave contains a host of Snakes past numbering, and they say that their size is enormous and their bite terrible. The story goes that they emerge from the cave, crawl out as far as the lake near by, and

[a] Metropolis lay on the road between Ephesus and Smyrna somewhat nearer to the former.

---

[5] ⟨οἱ⟩ *add. H.*

νήχεσθαι, πειρωμένους δὲ περαιτέρω τοῦ ὕδατος προελθεῖν οὐ δύνασθαι· μέλλοντας γὰρ ἐπιβαίνειν τῆς γῆς ἐλλοχᾶν καρκίνους μεγάλους, οἵπερ οὖν ἀνατείναντες τὰς χηλὰς συλλαμβάνουσιν ἐς πνῖγμα αὐτοὺς καὶ ἀναιροῦσι. δεδοικότες οὖν οἱ ὄφεις τοὺς ἐχθροὺς ἡσυχάζουσι, καὶ ἐς τὴν γῆν αὐτοῖς ἄβατά ἐστι· πεφρίκασι γὰρ τὴν ἐκ τῶν καρκίνων φρουρὰν καὶ κόλασιν. καὶ [1] πάντως ἂν ἀπολώλεσαν [2] οἱ περὶ τὸν χῶρον ἐκ πολλοῦ, εἰ μὴ φύσει τινὶ ἀπορρήτῳ περιειληφότες οἱ προειρημένοι καρκίνοι τῆς λίμνης τὰ χείλη καὶ ἀπείργοντες [3] εἶτα εἰρηναῖα ἀπέφαινον τὰ ἐν τῷ τόπῳ πάντα.

39. Ὀνησίκριτος ὁ Ἀστυπαλαιεὺς λέγει ἐν Ἰνδοῖς κατὰ τὴν Ἀλεξάνδρου τοῦ παιδὸς [4] Φιλίππου ⟨ἀνάβασιν⟩ [5] γενέσθαι δράκοντας δύο, οὓς Ἀβισάρης [6] ὁ Ἰνδὸς ἔτρεφεν, ὧν ὁ μὲν ἦν πήχεων τετταράκοντα καὶ ἑκατόν, ὁ δὲ ὀγδοήκοντα· καί φησι ἐπιθυμῆσαι δεινῶς Ἀλέξανδρον θεάσασθαι αὐτούς. λέγουσι δὲ Αἰγύπτιοι [7] λόγοι καὶ ἐπὶ τοῦ Φιλαδέλφου ἐξ Αἰθιοπίας ἐς τὴν Ἀλεξάνδρου πόλιν κομισθῆναι δράκοντας δύο ζῶντας, καὶ τὸν μὲν αὐτῶν ⟨εἶναι⟩ [8] πήχεων δεκατεττάρων [9] τὸν δὲ δεκατριῶν· ἐπί γε μὴν τοῦ Εὐεργέτου τρεῖς κομισθῆναι, καὶ τὸν μὲν εἶναι πήχεων ἐννέα, τὸν δὲ ἑπτά, τὸν δὲ τρίτον ἑνὶ ἀπολείπεσθαι· καὶ τρέφεσθαί γε ἐν Ἀσκληπιοῦ σὺν πολλῇ τῇ κομιδῇ αὐτοὺς Αἰγύπτιοί φασι. καὶ ἀσπίδας δὲ τετραπή-

[1] ἤ. [2] ἀπολώλασιν.
[3] ἀνείργοντες. [4] παιδός *gloss, H.*
[5] ⟨ἀνάβασιν⟩ *add. Ges.*
[6] *Reinesius* : Ἀπεισάρης, Ἀποσεισάρης.

swim about, but if they try to go further afield than the water they cannot, for while they are about to pass on to the land huge Crabs lie in wait for them, and these raise their claws, seize, throttle, and kill the Snakes. And so through fear of their enemy the Snakes remain where they are, and the land for them is inaccessible, for they dread the vigilance of the Crabs and the punishment which they inflict. And the people round about would long ago have been utterly destroyed, had not the aforesaid Crabs by some mysterious instinct encircled the margin of the lake and by keeping off the Snakes had ensured that all was peace thereabouts.

39. Onesicritus of Astypalaea says that at the time of the expedition of Alexander, the son of Philip, there were in India two Snakes kept by Abisares the Indian, and that one of them measured a hundred and forty cubits, the other eighty. He says also that Alexander had a great desire to see them.

Monstrous Snakes in India,

Egyptian histories relate that in the time of Ptolemy Philadelphus[a] there were brought from Ethiopia to Alexandria two live Snakes and that one of them was fourteen cubits long, the other thirteen; and in the time of Ptolemy Euergetes three were brought, one was nine cubits long, the second seven, and the third snake one cubit less. And the Egyptians assert that they were tended with great care in the temple of Asclepius. And the same people maintain that Asps four cubits in length frequently

from Ethiopia

[a] Ptolemy Philadelphus, 308–246 B.C.; P. Euergetes, 182–116 B.C.

[7] Αἰγυπτίων. [8] ⟨εἶναι⟩ *add. H.*
[9] -τεσσάρων.

χεις γίνεσθαι πολλάκις οἱ αὐτοὶ λέγουσι. μνήμην δὲ ἄρα τούτων ἐποιησάμην, τὸ ἴδιον τῶν ζῴων ἐπεξελθὼν καὶ ἐς ὅσον πέφυκε μῆκος προϊέναι δεῖξαι βουληθείς. λέγουσιν οὖν [1] καὶ οἱ τὰς ὑπὲρ τῆς Χίου συγγράψαντες ἱστορίας γενέσθαι ἐν τῇ νήσῳ παρὰ τὸ ὄρος τὸ καλούμενον Πελινναῖον ἐν αὐλῶνί τινι δασεῖ καὶ δένδρων ὑψηλῶν πεπληρωμένῳ δράκοντα μεγέθει μέγιστον, οὗπερ οὖν καὶ τὸν συριγμὸν ἐπεφρίκεσαν οἱ τὴν Χίον κατοικοῦντες. οὔκουν οὐδὲ ἐτόλμων τινὲς ἢ τῶν γεωργούντων ἢ τῶν νεμόντων πλησίον γενόμενοι καταγνῶναι τὸ μέγεθος, ἀλλὰ ἐκ μόνης τῆς σύριγγος πελώριόν τε [2] καὶ ἐκπληκτικὸν τὸ θηρίον εἶναι ἐπίστευον· ἐγνώσθη δ' οὖν ὅσος ποτὲ τὸ μέγεθος ἦν δαιμονίᾳ τινὶ καὶ θαυμαστῇ μηχανῇ ναὶ μὰ Δία. σφοδροῦ γὰρ ἀνέμου καὶ σκληροῦ προσπεσόντος ἐρρίφη [3] πρὸς ἄλληλα τὰ ἐν τῷ αὐλῶνι δένδρα, καὶ οἱ κλάδοι βιαίως ἀραττόμενοι τίκτουσι πῦρ, καὶ αἴρεται μεγίστη φλόξ, καὶ περιλαμβάνει τὸν πάντα χῶρον καὶ κυκλοῦται τὸν θῆρα· ὁ δὲ ἀποληφθεὶς καὶ ἐξερπύσαι μὴ δυνάμενος καταπίμπραται. οὐκοῦν γυμνωθέντος τοῦ τόπου γίνεται κάτοπτα πάντα· καὶ οἱ Χῖοι ἐλευθερωθέντες τοῦ δέους ἧκον ἐπὶ θέαν, καὶ καταλαμβάνουσι τὰ ὀστᾶ μεγέθει μέγιστα καὶ ⟨τὴν⟩ [4] κεφαλὴν ἐκπληκτικήν, ὡς ἐκ τούτων ἔχειν συμβαλεῖν ὅσος τε ἐκεῖνος ἦν καὶ ὁποῖος ἔτι περιών.

40. Ὄφεως ὄνομα σήψ, καὶ ἔχει θαυμάσαι ἄξιον ἐκεῖνο. τὴν χρόαν ἐκτρέπει τὴν ἑαυτοῦ, καὶ ἔοικε τοῖς τόποις καθ' οὓς ἔρχεται. ὀδόντας δὲ ἄρα

[1] λέγουσι γοῦν. [2] τι.

occur. And so I have mentioned these facts in the course of discussing animal characteristics from a wish to demonstrate the length to which by nature they attain.

A Snake in Chios

Now historians of Chios also assert that in that island near the mountain named Pelinnaeus in a wooded glen filled with tall trees there was a snake of gigantic size whose very hiss made the inhabitants of Chios shudder. None of the farmers, none of the herdsmen dared to approach the spot and observe its size, but they were confident simply from its hiss that the beast was a monster to strike one with consternation. Now the discovery how large in fact it was, was due to a miraculous and truly wonderful contrivance. A furious and violent wind assailed the trees in the glen and they were hurled one against the other, and the boughs crashed together with such force that they generated flames, and a huge fire was kindled which embraced the entire region and encircled the monster. It was cut off, and being unable to creep out was burned to death. And so when the place was stripped, all lay bare to view. And the inhabitants of Chios, freed now from their dread, came to see, and discovered the bones to be of gigantic size and the head terrifying. From these they were able to guess how large and how awful the brute was while still alive.

The 'Sêps'

40. There is a snake called the *Sêps* and it has this remarkable quality: it changes the colour of its body so as to match the places through which it

---

[3] ἐρρίφθη.
[4] ⟨τήν⟩ add. Schn.

τέτταρας τοὺς κάτω φέρει κοίλους, ἐφ' ὧν ὑμενώδεις ἐπίκεινται χιτῶνες, καλύπτοντες τὰ κοιλώματα. ἐκ τούτων οὖν πατάξαν τὸ θηρίον εἶτα ἀφίησι τὸν ἰόν· σήπει δὲ παραχρῆμα καὶ ἀναιρεῖ τάχιστα.

41. Μεγασθένης φησὶ κατὰ τὴν Ἰνδικὴν σκορπίους γίνεσθαι πτερωτοὺς μεγέθει μεγίστους, τὸ κέντρον δὲ ἐγχρίμπτειν [1] τοῖς Εὐρωπαίοις παραπλησίως. γίνεσθαι δὲ καὶ ὄφεις αὐτόθι καὶ τούτους πτηνούς· ἐπιφοιτᾶν δὲ οὐ μεθ' ἡμέραν ἀλλὰ νύκτωρ, καὶ ἀφιέναι ἐξ αὑτῶν οὖρον, ὅπερ οὖν ἐὰν κατά τινος ἐπιστάξῃ σώματος, σῆψιν ἐργάζεται παραχρῆμα. καὶ τὰ μὲν τοῦ Μεγασθένους ταῦτα. Πολύκλειτός γε μήν φησιν ἐν τῇ αὐτῇ γῇ σαύρους γίνεσθαι μεγίστους καὶ πολυχρόους, καὶ βαφαῖς τισιν εὐανθέσι τὰς δορὰς πεποικίλθαι δεινῶς, εἶναι δὲ καὶ ἅψασθαι ταύτας ἁπαλωτάτας. σαύρους δὲ Ἀριστοτέλης ἐν τῇ τῶν Ἀράβων γῇ τίκτεσθαί φησι, καὶ δύο πήχεις ἔχειν αὐτούς.

42. Παμμένης ἐν τῷ περὶ θηρίων σκορπίους λέγει γίνεσθαι πτερωτοὺς καὶ δικέντρους ἐν Αἰγύπτῳ (καὶ οὔ φησιν ἀκοὴν λέγειν, ἀλλὰ ἑαυτοῦ τήνδε τὴν ἱστορίαν ὁμολογεῖ) καὶ ὄφεις δικεφάλους, καὶ ἔχειν δύο πόδας κατὰ τὸ οὐραῖον τούτους. Κτησίας γε μὴν ὁ Κνίδιός φησι περὶ τὴν Περσικὴν Σιττάκην ποταμὸν εἶναι Ἀργάδην

[1] ἐγχρίπτειν.

[a] The fangs of the Asp are similarly described in 9. 4; cp. Nic. *Th.* 182–5. See W. Morel in *Philol.* 83. 361.

passes. The four fangs of its lower jaw are hollow, and membrane-like veils cover them and conceal the hollows. Directly the creature has struck, it projects its poison through these ducts,[a] which at once makes a festering wound and very soon causes death.

41. Megasthenes states that in India there are winged Scorpions of immense size and that they give a sting somewhat like the Scorpions of Europe. He also says that there are Snakes there with wings, and that their visitations occur not during the daytime but by night, and that they emit urine which at once produces a festering wound on any body on which it may happen to drop. This is what Megasthenes says. Further, Polycleitus says that in the same country there are Lizards of very great size and of many colours, and that their skins are wonderfully dappled with bright hues, and that they are extremely soft to the touch. And Aristotle says [*HA* 606 b 5] that there are Lizards in Arabia two cubits long.

Winged Scorpions, Snakes, and Lizards of India

42. Pammenes in his work *Concerning wild animals* says that in Egypt there are Scorpions with wings and a double sting (this, he says, is not mere hearsay, but professes that it is his personal observation): there are also two-headed Snakes which have two feet in the region of the tail. Further, Ctesias of Cnidus says that in the neighbourhood of Sittace[b] in Persia there is a river called the Argades, and that

Winged Scorpions, two-headed Snakes

river Snakes in Persia

[b] Sittace, town on the Tigris, at the N end of the province of Babylonia. The Argades has not been identified.

ὄνομα. ὄφεις δὲ ἄρα ἐν αὐτῷ γίνεσθαι πολλούς, μέλανας τὸ σῶμα πλήν γε τῆς κεφαλῆς· εἶναι δὲ αὐτοῖς λευκὴν ταύτην. προϊέναι δὲ ἐς ὀργυιὰν τὸ μῆκος τοὺς ὄφεις τούσδε.[1] καὶ μεθ' ἡμέραν μὲν μὴ ὁρᾶσθαι, ὑφύδρους δὲ νήχεσθαι, νύκτωρ δὲ ἢ τοὺς ὑδρευομένους ἢ τοὺς τὴν ἐσθῆτα φαιδρύνοντας διαφθείρειν. πολλοὺς δὲ ἄρα πάσχειν τοῦτο ἢ χρείᾳ ὕδατος ἐπιλείποντος ἢ μεθ' ἡμέραν ἀσχοληθέντας ἀποπλῦναι[2] τὴν ἐσθῆτα μὴ δεδυνημένους.

[1] τούσδε. τοὺς οὖν ὑπὸ τούτων δηχθέντας ἀποθνήσκειν.
[2] καὶ ἀποπλῦναι.

it contains a great number of Snakes whose bodies are entirely black except for the head, and this is white. There Snakes attain to as much as six feet in length. By day they are not visible, for they swim under water, but at night they kill those who come either to draw water or to wash their clothes. And the victims are numerous, either because they need water when their supply fails, or because they were busy during the day-time and unable to wash their clothes then.

BOOK XVII

ΙΖ

1. Ἀλέξανδρος ἐν τῷ περίπλῳ τῆς Ἐρυθρᾶς θαλάττης λέγει[1] ὄφεις ἑορακέναι[2] τετταράκοντα πήχεων τὸ μῆκος,[3] καὶ γένος καρκίνων, οἷς τὸ μὲν ὄστρακον τὴν περιφέρειαν εἶχε πανταχόθεν πόδα, χηλαὶ[4] δὲ ἠρτημέναι μέγισται προεῖχον, ἐπιβουλεύεσθαι δὲ ὑπ' οὐδενὸς αὐτούς. τὸ δὲ αἴτιον, ἱεροὶ λέγονται Ποσειδῶνος. καὶ ἀφιέρωνται τῷ θεῷ, οἷον ἀναθήματα εἶναι ἐκείνου ἀσινῆ τε καὶ ἀνεπιβούλευτα οἱ καρκίνοι.

2. Κλείταρχος ἐν τῇ . . .[5] περὶ τὴν Ἰνδικήν φησι γίνεσθαι ὄφεις πήχεων ἑκκαίδεκα. γίνεσθαι δὲ καὶ ἄλλο τι γένος ὄφεων ὑμνεῖ, οὐ κατὰ τοὺς ἑτέρους τὸ εἶδος· βραχυτέρους μὲν γὰρ εἶναι[6] πολλῷ, ποικίλους γε μὴν τὴν χρόαν ὁρᾶσθαι, ὥσπερ οὖν φαρμάκοις καταγραφέντας· τοὺς μὲν γὰρ χαλκοειδεῖς ταινίας ἔχειν ἀπὸ ⟨τῆς⟩[7] κεφαλῆς ἐς τὴν οὐρὰν καθερπούσας, τοὺς δὲ ἀργύρῳ[8] προσεικασμένας, πεφοινιγμένας ἄλλους, καὶ μέντοι καὶ χρυσοφαεῖς τινας. δακεῖν δὲ ἄρα καὶ ἀποκτεῖναι ὤκιστα δεινοὺς αὐτοὺς λέγει οὗτος.

[1] λέγει οὕτως·.
[2] ἑωρακέναι.
[3] μῆκος, πλάτος δὲ καὶ πάχος κατὰ τὸ μῆκος δηλονότι καὶ γ. κ.
[4] καὶ χηλαί.

BOOK XVII

1. Alexander[a] in his *Voyage round the Red Sea* says that he has seen Snakes forty cubits long, and a species of Crab whose shell measured one foot across in all directions, with claws attached and projecting to an enormous length. But nobody has designs upon them, the reason being that they are said to be sacred to Poseidon. And they are consecrated to the god, so that, as offerings to him, they are free from harm and immune from attack.

Monstrous Snakes and Crabs

2. Cleitarchus in his work on India says that there are Snakes sixteen cubits long. He also relates that there is another species of Snake different in appearance from the rest, for it is a great deal shorter and its colour looks mottled as though it had been painted with pigments: some have stripes of bronze descending from the head to the tail, others look like silver, others again are stained red, and there are even some with a golden sheen. The same writer asserts that they give a terrible bite which kills very speedily.

Snakes of India

[a] This 'Alexander' has not certainly been identified with Alexander of Myndus, although Wellmann (*Hermes* 26. 565) shows reasons for regarding them as one and the same.

---

[5] *Lacuna*; *or read* ἐν τοῖς *H.*
[6] εἶναι τῷ μεγέθει.
[7] ⟨τῆς⟩ *add. Ges.*
[8] ἀργυρίῳ.

3. Ἐν τῷ ἐννάτῳ τῶν περὶ Πτολεμαίων [1] λόγων λέγει Νύμφις ἐν ⟨τῇ⟩ [2] γῇ τῇ Τρωγλοδύτιδι γίνεσθαι ἔχεις ἄμαχόν τι μέγεθος, εἰ πρὸς τοὺς ἄλλους ἔχεις ἀντικρίνοιντο· εἶναι γὰρ πήχεων καὶ πεντεκαίδεκα· τάς γε μὴν χελώνας εἶναι τοσαύτας τὸ χελώνιον, ὡς χωρεῖν μεδίμνους Ἀττικοὺς καὶ ἓξ αὐτό.

4. Ἔστι δὲ καὶ πρηστὴρ ὄφεων γένος, ὅσπερ οὖν εἰ δάκοι, τὰ μὲν πρῶτα νωθεῖς ἀπεργάζεται καὶ ἥκιστα κινητικούς, εἶτα μέντοι κατ᾽ ὀλίγον ἀρρώστους [3] καὶ ἀναπνεῖν ἀδυνάτους· καὶ μέντοι καὶ λήθην καταχεῖ τῆς γνώμης [4] τὸ δῆγμα, καὶ τὴν κύστιν ἐπέχει, καὶ λιπότριχας [5] ἀποφαίνει, εἶτα ἕπεται πνιγμός, καὶ σπᾶσθαι ποιεῖ, καὶ τὸ τέλος τοῦ βίου ἀλγεινότατον.

5. Φύλαρχος ἐν τῇ δωδεκάτῃ ὑπὲρ τῶν Αἰγυπτίων ἀσπίδων ᾄδει τοιαῦτα. τιμᾶσθαί φησιν αὐτὰς ἰσχυρῶς, καὶ ἐκ ταύτης γε τῆς τιμῆς ἡμερωτάτας τε καὶ χειροήθεις γίνεσθαι.[6] τοῖς παιδίοις οὖν συντρεφομένας μηδὲν ἀδικεῖν, καλουμένας [7] δὲ ἐξέρπειν τῶν φωλεῶν καὶ ἀφικνεῖσθαι. κλῆσις δὲ αὐταῖς ὁ τῶν δακτύλων ἐστὶ κρότος. προτείνουσι δὲ ἄρα οἱ Αἰγύπτιοι καὶ ξένια αὐταῖς. ἐπὰν γὰρ ἀπὸ δείπνου γένωνται, ἄλφιτα οἴνῳ καὶ μέλιτι ἀναδεύσαντες κατὰ τῆς τραπέζης τιθέασιν, ἐφ᾽ ἧς ἔτυχον δεδειπνηκότες· εἶτα μέντοι κροτήσαντες

[1] Πτολεμαῖον A, *H*. [2] ⟨τῇ⟩ *add. H*.
[3] ἀγνώστους. [4] τῇ γνώμῃ.
[5] *Ges*: λειπότριχας.
[6] γίνεσθαι ἐκ τῆς τροφῆς πεπωλευμένας.

3. Nymphis in the ninth book of his *History of the Ptolemies* says that in the country of the Troglodytes [a] there are Vipers of surpassing size if compared with other vipers, for they measure as much as fifteen cubits. Moreover the Tortoises have shells large enough to contain six Attic *medimni*.[b]

Monstrous Vipers and Tortoises

4. The *Prester* [c] also is a species of snake and if it bites, to begin with it makes men lethargic and quite incapable of bestirring themselves, and in the next place they gradually weaken and are unable to breathe. Further, the bite induces loss of memory, stops the flow from the bladder, and causes the hair to fall; then there ensues a choking which causes convulsions, and life ends in agonies.

The 'Prester'

5. Phylarchus in his twelfth book gives the following account of the Asps of Egypt. He says that they are treated with great respect, and as a result of this respect they become extremely gentle and tame. And so, being fed along with the children, they do no harm, but creep out of their lairs when called and come to the spot. And the way to call them is to snap one's fingers. Then the Egyptians give them presents in the way of friendship, for when they have finished their meal they soak barley in wine and honey and place it on the table off which they happen to have dined. Then they snap their fingers

The Asps of Egypt

[a] The Red Sea coasts of Egypt and of Arabia.
[b] See n. at 16. 14.
[c] In 6. 51 identified with the *Dipsas*; what its modern equivalent may be, is impossible to say.

[7] καὶ καλουμένας.

οἱονεὶ δαιτυμόνας καλοῦσι. καὶ ἐκεῖναι ὥσπερ οὖν ὑπὸ συνθήματι παραγίνονται, καὶ ἄλλη ἀλλαχόθεν ἐξέρπει, καὶ περιστᾶσαι τὴν τράπεζαν τὴν μὲν λοιπὴν σπεῖραν ἐῶσι κατὰ τοῦ δαπέδου, ἄρασαι δὲ τὴν κεφαλὴν περιλιχμῶνται, καὶ ἡσυχῇ καὶ κατ' ὀλίγον ἐμπίπλανται τῶν ἀλφίτων, καὶ καταναλίσκουσιν αὐτά. νύκτωρ δὲ ἐὰν ἐπείγῃ τι τοὺς Αἰγυπτίους, κροτοῦσι πάλιν· ὑποσημαίνει δὲ ἄρα αὐταῖς ὅδε ὁ ψόφος ἐξίστασθαί τε[1] καὶ ἀναχωρεῖν. οὐκοῦν συνιᾶσιν ἐκεῖναι[2] τὴν τοῦ κτύπου διαφορὰν καὶ ἐφ' ὅτῳ τοῦτο δρᾶται, καὶ παραχρῆμα ἀναστέλλονται καὶ ἀφανίζονται, ἐς τοὺς χηραμούς τε καὶ φωλεοὺς ἕρπουσαι. ὁ οὖν[3] ἀνιστάμενος οὔτε ἐμβαίνει τινὶ αὐτῶν οὔτε περιπίπτει.

6. Ὁ κροκόδιλος γίνεται μήκιστος πολλάκις. ἐπὶ γοῦν Ψαμμιτίχου τοῦ Αἰγυπτίων βασιλέως πέντε καὶ εἴκοσι πήχεων κροκόδιλον φανῆναί φασιν, ἐπὶ δὲ Ἀμάσιδος παλαιστῶν τεττάρων καὶ πήχεων ἓξ καὶ εἴκοσι. κήτη δὲ περὶ τὴν Λάκαιναν θάλατταν ἀκούω γίνεσθαι μεγέθει μέγιστα, καί τινές γε τῶν κριτικῶν Ὅμηρον[4] ἐντεῦθέν φασιν εἰπεῖν Λακεδαίμονα κητώεσσαν. περὶ τὰ Κύθηρα δὲ ἔτι καὶ μείζω τὰ κήτη ὑμνοῦσι γίνεσθαι. ἔοικε δὲ αὐτῶν καὶ τὰ νεῦρα λυσιτελῆ εἶναι ἐς τὰς τῶν ψαλτηρίων καὶ τῶν ἄλλων ὀργάνων χορδοστροφίας καὶ μέντοι καὶ ἐς τὰ πολεμικὰ ὄργανα.[5] ἐν δὲ τῇ Ἐρυθρᾷ θαλάττῃ πρὸς τοῖς

[1] αὐταῖς . . . ἐξίστασθαί τε] ὅδε ὁ ψ. ἐξ. τε αὐτάς.
[2] καὶ ἐκεῖναι. [3] γοῦν. [4] τὸν Ὅ.
[5] ὄργανα αἱ τούτων νεῦραι δοκοῦσι λυσιτελέσταται.

and summon 'the guests,' so to call them. And the Asps as at a signal assemble, creeping out from different quarters, and as they encircle the table, while the rest of their coils remain on the floor, they rear their heads up and lick the food; gently and by degrees they take their fill of the barley and eat it up. And if some need causes the Egyptians to rise during the night, they again snap their fingers: this is a signal for the Asps to make way for them and to withdraw. So the snakes realise the difference between this sound and the other and the reason for it, and promptly retire and disappear, creeping into their holes and lairs. Accordingly the man who has got out of bed neither treads upon nor encounters any of them.

6. The Crocodile often attains to an immense length. At any rate they say that in the reign of Psammitichus, King of Egypt,[a] there appeared a Crocodile twenty-five cubits long, and in the reign of Amasis[b] there appeared one of twenty-six cubits and four palms.[c] And I have heard that in the Gulf of Laconia there are sea-monsters of immense size; that is why according to some grammarians Homer speaks of 'Lacedaemon with its sea-monsters'[d] [*Il.* 2. 581, *Od.* 4. 1]. And round about Cythera there are said to be sea-monsters still larger. And it appears that their sinews are useful for the stringing of harps and other instruments, and even for engines of war. And in addition to

The Crocodile

Sea-monsters

[a] 7th cent. B.C. [b] 6th cent. B.C.

[c] A palm = about 3 in.

[d] So Ael. understood κητώεσσαν, now generally taken to mean 'full of ravines.'

ἤδη προειρημένοις γίνονται καὶ σκορπίοι καὶ κωβιοὶ δύο πήχεων καὶ μέντοι καὶ τριῶν. Ἀμώμητος δέ φησιν ἐν τῇ Λιβύῃ πόλιν εἶναί τινα, ἐν ᾗ τοὺς ἱερέας ἔκ τινος λίμνης ἐπαοιδαῖς καταγοητεύοντας εὖ μάλα ἑλκτικαῖς ἐξάγειν κροκοδίλους πήχεων ἑκκαίδεκα. Θεοκλῆς δὲ ἐν τῇ τετάρτῃ περὶ τὴν Σύρτιν λέγει γίνεσθαι κήτη τριήρων μείζονα. περὶ δὲ τὴν Γεδρωσίων χώραν (ἔστι δὲ μοῖρα τῆς γῆς τῆς Ἰνδικῆς οὐκ ἄδοξος) Ὀνησίκριτος λέγει καὶ Ὀρθαγόρας γίνεσθαι κήτη ἥμισυ ἔχοντα σταδίου τὸ μῆκος.[1] τοσαύτην δέ φασιν ἔχειν δύναμιν αὐτά, ὡς πολλάκις, ὅταν ἀναφυσήσῃ τοῖς μυκτῆρσιν, ἐς τοσοῦτον ἀναρρίπτειν τῆς θαλάττης τὸ κλυδώνιον, ὡς δοκεῖν τοῖς ἀμαθέσι καὶ ἀπείροις πρηστῆρας εἶναι ταῦτα.

7. Ἀριστοτέλης ἐν τῷ ὀγδόῳ περὶ ζῴων φησὶ τοὺς ἐλέφαντας ἐσθίειν κριθῶν μεδίμνους Μακεδονικοὺς ἐννέα, ἀλφίτων δὲ ἐπὶ τούτοις ἕξ, εἰ δὲ δοίης, καὶ ἑπτά·[2] πίνειν δὲ αὐτοὺς ὁ αὐτὸς λέγει μετρητὰς Μακεδονικοὺς τετταρεσκαίδεκα, καὶ πάλιν τῆς δείλης ἐπιπίνειν ὀκτώ. βιοῦν δὲ ἐλέφαντας ἔτη διακόσιά φησι, προϊέναι δὲ ἔστιν οὓς καὶ ἐς τρεῖς ἑκατοντάδας.

Διειδὲς δὲ ὕδωρ καὶ ἀκραιφνὲς καμήλῳ πιεῖν ἔχθιστόν ἐστι, τεθολωμένον δὲ καὶ ῥυπαρὸν ἥδιστον πωμάτων ἡγεῖται. καὶ μέντοι καὶ ἐς ποταμὸν ἐὰν ἀφίκηται ἢ λίμνην, οὐ πρότερον ἐπικύπτει πιεῖν, πρὶν ἢ τοῖς ποσὶν ἀναταράξῃ[3] τὴν ἰλὺν καὶ

[1] μῆκος, πλάτος δὲ κατὰ λόγον τοῦ μήκους καὶ τοῦτο δηλονότι.
[2] ἑπτά, καὶ χιλὸν φύλλα καὶ κλάδους ἁπαλούς.
[3] ἀναταράξει V, ἐπιταράξῃ *other* MSS.

those that I have mentioned before there occur in the Red Sea Scorpion-fish [a] and Gobies two and even three cubits long. And Amometus says that in Libya there is a certain city where the priests by their powerful spells draw Crocodiles sixteen cubits long from a certain lake. And Theocles in his fourth book says that round about Syrtis there are Sea-monsters larger than a trireme. And Onesicritus and Orthagoras say that round the coast of Gedrosia [b] (this is no inconsiderable part of India) there are Sea-monsters half a *stade* [c] in length, and so powerful are they that, when they blow with their nostrils, they often hurl up a wave from the sea to such a height that ignorant and inexperienced people take it for a waterspout.

7. Aristotle says in the eighth book of his *History of Animals* [*HA* 596 a 3] that Elephants eat nine Macedonian *medimni* [d] of barley, and in addition six of barley-groats, or even seven if you give it them. And he also says that they drink fourteen Macedonian *metretae* [e] of water, and again eight more in the afternoon. Elephants, he says, live for two hundred years, and there are some that even attain to three hundred.

The Elephant

The Camel [*Id. HA* 595 b 31] greatly dislikes clear, pure water for drinking, and regards muddy, dirty water as the pleasantest. Indeed if it comes to a stream or a lake, it does not bend down to drink until it has stirred up the slime with its feet and destroyed

The Came

[a] Not to be identified with the Bullhead or Sculpin (*Scorpaena* sp.), Thompson, *Gk. fishes*, p. 246.
[b] See 15. 25 n.
[c] *Stade* = 600 ft.
[d] *Medimnus*, see 16. 14 n.
[e] *Metretes* = about 8½ gallons.

ἀφανίσῃ τὸ κάλλος τοῦ ὕδατος. ἄποτος δὲ ἐὰν μείνῃ, διακαρτερεῖ καὶ ὀκτὼ ἡμέρας.

8. Πυθαγόρας ἐν τοῖς περὶ τῆς Ἐρυθρᾶς θαλάττης λέγει ζῷόν τι γίνεσθαι χερσαῖον περὶ τὸ πέλαγος ἐκεῖνο, τὸν [1] καλούμενον κῆπον. φερώνυμον δὲ εἶναι· [2] ἔχειν γὰρ χρόας πολλάς. καὶ μέγεθος μὲν εἰληχέναι τὸν τέλειον κατὰ τοὺς κύνας τοὺς Ἐρετρικούς. περιελθεῖν δὲ αὐτοῦ τὸ ποικίλον ἐθέλω καὶ δεῖξαι τῷ λόγῳ, ὡς ἐκεῖνος γράφει. τὰ μὲν δὴ περὶ τὴν κεφαλὴν αὐτῷ καὶ τὸ [3] νῶτον καὶ τὴν ῥάχιν ἐς τὴν οὐρὰν τελευτῶντα ἀκράτως πυρρά [4] ἐστι, θεάσαιο δ' ἂν καὶ τρίχας χρυσοειδεῖς τινας διεσπαρμένας· λευκὸν δὲ τὸ πρόσωπόν οἱ μέχρι τῶν παρειῶν,[5] ἐντεῦθέν γε μὴν ταινίαι χρυσοειδεῖς κατίασιν ἐς τὴν δέρην. ταύτης δὲ τὰ κάτω μέχρι τῶν στέρνων καὶ οἱ πόδες δὲ οἱ πρόσθιοι λευκανθίζει πάντα. μαζοὶ δὲ χειροπληθεῖς δύο κυανοῖ, γαστὴρ δὲ λευκὴ πᾶσα, πόδες δὲ οἱ κατόπιν μέλανές εἰσι. προσώπου δὲ μορφή,[6] κυνοκεφάλῳ παραβαλὼν αὐτὴν ἀληθεύσεις, εὖ ἴσθι.

9. Ὀνοκενταύραν καλοῦσι ζῷόν τι, καὶ ταύτην ὅστις εἶδεν, οὐκ ἂν ἠπίστησεν [7] ὅτι καὶ Κενταύρων φῦλα ἦν, καὶ οὐ κατεψεύσαντο οἱ χειρουργοὶ [8] τῆς φύσεως, ἀλλὰ καὶ ἐκείνους ἤνεγκεν ὁ χρόνος κράσει σωμάτων οὐχ ὁμοίων ἑνωθέντας. καταλείπωμεν δὲ [9] αὐτούς, εἴτε ἐγένοντο ὄντως ἐπιδημίᾳ

[1] τό. [2] εἶναι, καὶ εἰκότως.
[3] τὸν νῶτον. [4] πυρσά.
[5] παρειῶν αὐτῶν. [6] μορφὴν ἐκείνου.

the beauty of the water. And if it goes unwatered, it can endure for as much as eight days.[a]

The 'Kepos

8. In his writings about the Red Sea Pythagoras says that there is an animal that lives on the shores and is called *Kêpos*.[b] And it is well-named (*kêpos*, garden), for it is of many colours. When full-grown it is the size of an Eretrian hound. But I wish to return to the subject of its varied colouring and to describe it as he writes. Its head, its back, and its spine down as far as the tail are a pure red, though you may observe a sprinkling of golden hairs. But its face including the cheeks is white, and from there golden stripes descend as far as the neck. The lower portions down to its chest and its fore-feet are all white; its two breasts, which would fill your hand, are dark, but its belly is entirely white; its hind feet are black. As to the shape of its face, be sure you will not go wrong if you liken it to that of a baboon.

The 'Onocentaura'

9. There is a certain creature which they call an *Onocentaura*,[c] and anybody who has seen one would never have doubted that the race of Centaurs once existed, and that artificers did not falsify Nature, but that time produced even these creatures by blending dissimilar bodies into one. But whether in fact they came into being and visited us at one

[a] Ael. has doubled Aristotle's number.
[b] Or *Kêbos*; the spelling varies. It is a long-tailed monkey.
[c] A tailless ape, identified by Gossen (§ 241) with the Gorilla; more probably the Chimpanzee.

---

[7] ἠπίστησεν, ὡς λόγος MSS, ἠ. ὡς λόγος, *Jac.*
[8] χειρουργοὶ περὶ πλαστικήν τε καὶ γραφικήν.
[9] δή.

μιᾷ [1] καὶ τῇ αὐτῇ, εἴτε ἡ φήμη κηροῦ παντὸς οὖσα εὐπλαστοτέρα τε καὶ εὐπειθεστέρα διέπλασεν αὐτούς, καὶ ἀνέμιξεν ἵππου καὶ ἀνθρώπου δαιμονίᾳ τινὶ συναφῇ ἡμίτομα, καὶ ἔδωκε μίαν ψυχήν. αὕτη δὲ ὑπὲρ ἧς ὥρμηται λέγειν ὅδε ὁ λόγος, ἐς ἀκοὴν τὴν ἐμὴν τοιάδε ἀφίκετο. ἀνθρώπῳ τὸ πρόσωπον εἴκασται, περιέρχονται δὲ αὐτὸ [2] βαθεῖαι τρίχες. τράχηλός τε ὑπὸ τῷ προσώπῳ καὶ στέρνα, καὶ ταῦτα ἀνθρωπικά· μαζοὶ δὲ ἠρμένοι καὶ κατὰ τοῦ στήθους ἐφεστῶτες, ὦμοι δὲ καὶ βραχίονες καὶ πήχεις, ἔτι δὲ χεῖρες καὶ . . .[3] στέρνα ἐς ἰξύν, καὶ ταῦτα ἀνθρωπικά· [4] ῥάχις δὲ καὶ πλευραὶ καὶ γαστὴρ καὶ πόδες οἱ κατόπιν ὄνῳ καὶ μάλα ἐμφερῆ, καὶ τεφρώδης κατ' ἐκεῖνον [5] ἡ χρόα, τὰ δὲ ὑπὸ τὰς λαπάρας [6] ἡσυχῇ λευκανθίζει. αἱ χεῖρες δὲ τῷδε τῷ ζῴῳ διπλῆν παρέχουσι χρείαν· ἔνθα μὲν γὰρ τάχους δεῖ, προθέουσι τῶν ὀπίσω σκελῶν, καὶ τῶν λοιπῶν τετραπόδων οὐχ ἡττᾶται τὸν δρόμον· δεῖ δὲ πάλιν [7] ἢ ἀφελεῖν τι ἢ καταθέσθαι ἢ συλλαβεῖν καὶ σφίγξαι, καὶ οἱ πόδες οἱ τέως χεῖρες ἐγένοντο, καὶ οὐ βαδίζει, κάθηται δέ. βαρύθυμον δὲ ἰσχυρῶς τὸ ζῷόν ἐστιν. ἐὰν γοῦν ἁλῷ, δουλείαν μὴ φέρον καὶ τῆς τέως ἐλευθερίας γλιχόμενον τροφὴν ἀπέστραπται πᾶσαν, καὶ ἀποθνήσκει λιμῷ. Πυθαγόρας λέγει καὶ ταῦτα, ὥσπερ οὖν τεκμηριοῖ Κράτης ὁ ἐκ τοῦ Μυσίου Περγάμου.

[1] πάντες μιᾷ.

[2] *Schn*: αὐτοῦ.

[3] *Lacuna.*

[4] *The words* στέρνα *and* καὶ ταῦτα ἀνθρωπικά, *repeated from*

and the same period,[a] or whether rumour, more ductile than any wax and too credulous, fashioned them and by some miraculous combination fused the halves of a horse and a man while endowing them with a single soul—let us pass them by. But this creature of which my discourse set out to speak, I have heard described as follows. Its face is like that of a man and is surrounded by thick hair. Its neck below its face, and its chest are also those of a man, but its teats are swelling and stand out on the breast; its shoulders, arms, and forearms, its hands too . . . chest down to the waist are also those of a man. But its spine, ribs, belly, and hind legs closely resemble those of an ass; likewise its colour is ashen, although beneath the flanks it inclines to white. The hands of this creature serve a double purpose, for when speed is necessary they run in front of the hind legs, and it can move quite as fast as other quadrupeds. Again, if it needs to pluck something, or to put it down, or to seize and hold it tight, what were feet become hands; it no longer walks but sits down. The creature has a violent temper. At any rate if captured it will not endure servitude and in its yearning for freedom declines all food and dies of starvation.

This also is the account given by Pythagoras and attested by Crates of Pergamum in Mysia.

[a] That is, they were a temporary phenomenon, did not propagate their kind, and soon became extinct.

---

three lines above, can hardly proceed from Ael., and have been condemned by edd.

[5] ἐκείνους.

[6] ταῖς λαπάραις.

[7] πάλιν τροφῆς.

10. Ἡ Βοιωτῶν γῆ ἀσπαλάκων ἀφεῖται, καὶ αὐτὴν οὐ διορύττει τὸ ζῷον τοῦτο κατὰ Λεβάδειαν· ἐὰν δέ πως καὶ ἀλλαχόθεν ἐσκομισθῶσιν, ἀποθνήσκουσι. [περὶ μὲν οὖν τὴν Ὀρχομενίων γίνονται καὶ πολλοί.][1] ἐν δὲ Λιβύῃ συῶν ἀγρίων ἀπορία ἐστὶ καὶ ἐλάφων. ἐν δὲ τῷ Πόντῳ οὔτε μαλάκια οὔτε ὀστρακόδερμα γίνεται, εἰ μὴ σπανίως καὶ ὀλίγα. λέγει δὲ Δείνων ἐν Αἰθιοπίᾳ γίνεσθαι τοὺς ὄρνιθας τοὺς μονόκερως καὶ ὗς τετράκερως καὶ πρόβατα ἐρίων μὲν ψιλά, τρίχας δὲ καμήλων ἔχοντα.

11. Ἐν Ζακύνθῳ λέγουσιν οἱ δεινοὶ τὰ τοιαῦτα βασανίσαι τε καὶ ἀνιχνεῦσαι τοῖς ὑπὸ τῶν φαλαγγίων δακνομένοις μὴ μόνον τοσαῦτα[2] ἀπαντᾶν, ὅσα καὶ τοῖς ἀλλαχόθι δηχθεῖσιν, ἀλλὰ ἐκείνων[3] πλείω. ὅλα γὰρ αὐτοῖς τὰ σώματα γίνεται νάρκης ἀνάπλεω καί πως ὑπότρομα καὶ ψυχρὰ ἰσχυρῶς, καὶ ἔμετοι . . .[4] σπασμὸν ἀναφύοντες, καὶ ὀρθοῦται τὸ σκεῦος αὐτοῖς· ἀλγοῦσι δὲ καὶ τὰ ὦτα ἰσχυρῶς, καὶ τοῦ ποδὸς ἑκατέρου τὸ θέναρ καὶ τοῦτο ὀδυνῶνται. ἐνδείκνυνται[5] δὲ ἄρα αὐτὰ[6] ὅσα εἶπον ἕκαστα[7] ⟨καὶ οἱ⟩[8] τὰς χεῖρας ἐπιβάλλοντες αὐτοῖς.[9] ὃ δέ ἐστι καὶ ἀκοῦσαι ἐκπληκτικὸν καὶ μέντοι καὶ θαυμασιώτερον[10] ἰδεῖν, ὅταν τινὲς τῶν ἀδήκτων ἢ ἐμβῶσι τοῖς ἀπολούτροις[11] τῶν

[1] περὶ μὲν . . . πολλοί] *interpolation*, *H*.
[2] ταῦτα.
[3] ἐκεῖνα.
[4] *Lacuna* : ⟨ἕπονται⟩ *or* ⟨παρακολουθοῦσι⟩ *ex. gr. H*.
[5] *Ges* : ἐνδείκνυται.
[6] ταῦτα καὶ ἀλγοῦντες.
[7] ἕκαστος.
[8] ⟨καὶ οἱ⟩ *add. H*.
[9] αὐτῶν.
[10] θαυμασιώτατον ? *H*.
[11] ἀπολου τρ' V, -λουτρίοις *other* MSS.

10. Boeotia is free of Moles, and this animal does not burrow through at Lebadea, and if by some chance Moles are introduced from elsewhere they die. [But in the neighbourhood of Orchomenus [a] they abound.]

The Mole, in Boeotia

In Libya there is an absence of wild swine and of stags. In the Euxine there are neither cephalopod mollusca nor testacea, except on rare occasions and in small numbers. And Dinon says that in Ethiopia there occur the one-horned birds,[b] swine with four horns,[c] and sheep destitute of wool but with the hair of camels.

Peculiarities of Libya, the Euxine, and Ethiopia

11. Those who are skilled at testing and investigating such matters assert that in Zacynthus [d] people who are bitten by Malmignattes [e] are not only assailed by all the symptoms that assail other victims elsewhere but by even more, for their entire body is infected with a torpor and a kind of trembling and a violent chill, and ⟨there follow⟩ vomitings which produce convulsions, and their member stands up. They have violent earache too, and the sole of either foot is painful. Moreover even those who touch them with their hands exhibit all the symptoms which I have enumerated. But it is startling to learn, and even more amazing to see, how when some persons unbitten tread in the water in which the

The Malmignatte

[a] Orchomenus was in Boeotia, about 5 mi. NE of Lebadea.

[b] The Hornbill.

[c] Perhaps the Warthog is intended, its four prominent tusks being mistaken for horns.

[d] Island off W coast of Peloponnese.

[e] A kind of spider, small, black, and spotted with red; its bite is poisonous and may even be fatal.

δηχθέντων ἢ καὶ νὴ Δία ἀπονίψωνται τοὺς πόδας (οἷα δήπου γίνεσθαι φιλεῖ πολλάκις· ἤδη δὲ ἄρα ἀπαντᾷ τὰ τοιαῦτα καὶ κατά τινας ἐπιβουλὰς ἐχθρῶν), πάντα καὶ ἐκείνοις γίνεται τὰ ἀλγήματα, ὅσα καὶ τοῖς δηχθεῖσι δήπου.

12. Γένος τι φρύνης ἀκούω καὶ πιεῖν δεινὸν καὶ πικρὸν ἰδεῖν. πιεῖν μέν, εἴ τις αὐτὴν συντρίψας εἶτα μέντοι τὸ αἷμα δοίη τῳ πιεῖν, κατ᾿ ἐπιβουλὴν ἐμβαλὼν εἴτε ἐς οἶνον εἴτε ἐς ἄλλα πώματα,[1] ὧνπερ οἱ τούτων[2] κατάρατοι σοφισταὶ[3] ἐπιτήδειον ἥγηνται τὴν πρὸς ἐκεῖνο τὸ αἷμα κρᾶσιν. καὶ ποθὲν ἀπέκτεινεν οὐκ ἐς ἀναβολὰς ἀλλὰ παραχρῆμα. ἰδεῖν δὲ ἡ φρύνη κακόν ἐστι τοιοῦτον. ἐάν τις θεάσηται τὴν θῆρα,[4] εἶτα αὐτῇ ἀντίος ὁρῶν προσβλέψῃ δριμύ, καὶ ἐκείνη κατὰ τὴν ἑαυτῆς φύσιν ἰταμὸν ἀντιβλέψῃ, καί τι καὶ φύσημα ἐμπνεύσῃ ἑαυτῇ μὲν συμφυές, χρωτὶ δὲ ἐχθρὸν ἀνθρωπίνῳ, ὠχρὸν ἐργάζεται, ὡς εἰπεῖν τὸν οὐκ ἰδόντα[5] ἀλλὰ ἐντυχόντα πρῶτον ὅτι νοσήσαντα εἶδεν ἄνθρωπον. μένει τε ἡ ὠχρότης ἡμερῶν οὐ πολλῶν, εἶτα ἀφανίζεται.

13. Χαραδριοῦ δὲ ἦν ἄρα δῶρον τοῦτο, ὃ οὐ μὰ Δία ἀτιμάζειν ἄξιον. εἰ[6] γοῦν ὑπαναπλησθεὶς τὸ σῶμα ἰκτέρου τις εἶτά οἱ δριμὺ ἐνορῴη,[7] ὁ δὲ ἀντιβλέπει καὶ μάλα γε ἀτρέπτως, ὥσπερ οὖν ἀντιφιλοτιμούμενος,[8] καὶ ἡ τοιάδε ἀντίβλεψις[9] ἰᾶται τὸ προειρημένον πάθος τῷ ἀνθρώπῳ.

[1] ἄλλο πόμα τι.
[2] τῶν τοιούτων? H.
[3] σοφισταὶ τὴν πονηρὰν ἀκριβοῦντες σοφίαν.

victims have washed, or simply bathe their feet in it (as of course frequently happens; indeed this has been brought about before now through the evil designs of enemies), they too suffer all the pains incurred by the victims of the bite.

12. I learn that there is a species of Toad which it is fatal to drink and dangerous to look at. It is fatal to drink if a man crushes a Toad and then offers the blood to another to drink after he has with malicious intent poured it into wine or such other beverages as accursed practitioners of these arts deem suitable for mixing with it. The draught brings not a lingering but an instant death. To gaze at a Toad is harmful in this way. If a man sees the beast and then looks intently at it, face to face, while it, following its nature, retaliates with a bold gaze and also breathes forth the breath which though natural to it has an adverse effect on the human skin, it turns the man pale, so that anyone who had not seen him but met him for the first time would say that he had seen a sick man. And the pallor lasts for a few days only and then disappears.

A poisonous Toad

13. The Stone-curlew, it seems, has this gift, which assuredly is by no means to be despised. At any rate if a man who has become infected with jaundice gazes intently at it and it returns the gaze without flinching, as though it were moved by jealousy against the man, this retaliatory gaze heals the man of the aforesaid complaint.

The Stone-curlew

[4] *Ges*: θήραν.
[5] εἰδότα.
[6] ἦν.
[7] *Apostolius*, *Ges*: ἐνορῶν.
[8] ἀντιθυμούμενος.
[9] *Gron*: ἀνάβλεψις.

14. Ἐγὼ μὲν οὐ πεπίστευκα, εἰ δὲ ἕτερος Εὐδόξῳ πείθεται, πιστευέτω ὅ φησιν Εὔδοξος, ὑπερβαλὼν τὰς Ἡρακλείους στήλας ἐν λίμναις ἑορακέναι [1] ὄρνιθάς τινας καὶ μείζους βοῶν. καὶ ὅτι μὲν οὐ πείθει με ὁ λέγων, ἤδη εἶπον· ἃ δ' οὖν ἤκουσα, οὐκ ἐσίγησα.

15. Ἀριστοτέλης λέγει πέρδικα θῆλυν, ὅταν κατὰ ἄνεμον [2] γένηται τοῦ ἄρρενος, ἐγκύμονα γίνεσθαι φύσει τινὶ ἀπορρήτῳ.[3] διαπλέκει δὲ ἄρα ὁ ὄρνις οὗτος ἐν ἡμέραις τὴν νεοττιὰν ἑπτά, καὶ ἐν ἑπτὰ μέντοι τίκτει, ἐν δὲ ταῖς τοσαύταις καὶ ἐκτρέφει τὰ νεόττια.

Τίμαιος δὲ καὶ Ἡρακλείδης καὶ Διοκλῆς [4] ὁ ἰατρὸς λέγουσι τοὺς φρύνους δύο ἥπατα ἔχειν, καὶ τὸ μὲν ἀποκτείνειν, τὸ δὲ ἐκείνου πεφυκέναι ἀντίπαλον· σώζειν γάρ.

16. Θεόπομπος λέγει τοὺς περὶ τὸν Ἀδρίαν οἰκοῦντας Ἐνετούς, ὅταν τοῦ τρίτου ἀρότου καὶ σπόρου ᾖ ὥρα,[5] τοῖς κολοιοῖς ἀποστέλλειν δῶρα· εἴη δ' ἂν τὰ δῶρα ψαιστὰ ἄττα καὶ μεμαγμέναι μᾶζαι καλῶς τε καὶ εὖ. βούλεται δὲ ἄρα ἡ τῶνδε τῶν δώρων πρόθεσις μειλίγματα τοῖς κολοιοῖς εἶναι καὶ σπονδῶν ὁμολογίαι, ὡς ἐκείνους τὸν καρπὸν τὸν Δημήτρειον [6] μὴ ἀνορύττειν καταβληθέντα ἐς τὴν γῆν μηδὲ παρεκλέγειν. Λύκος

[1] ἑωρακέναι.
[2] *Schn* : νώτου.
[3] ἀρρήτῳ.
[4] *Wellmann* : Νεοκλῆς MSS, *H*.
[5] *Jac* : ὅταν περὶ τὸν ἄροτον τρίτον καὶ σπόρον ἡ ὥρα ᾖ most MSS, ᾖ ὥρα V, τοῦ τρίτου *del*. *H*.
[6] Δημήτριον.

14. For my part I do not believe Eudoxus, but if others are persuaded by him, then they may believe Eudoxus when he says that after passing the Pillars of Heracles [a] he saw upon some meres certain birds larger than oxen. That his statement fails to convince me I have already remarked. But what I have heard I do not suppress.

Gigantic birds

15. Aristotle says [*HA* 541 a 27] that when the female Partridge gets to leeward of the male bird, by some mysterious process of nature she becomes impregnated. This bird builds its nest in seven days, and in seven days lays its eggs, and in the same number of days rears its chicks.

The hen Partridge

Timaeus, Heraclides, and Diocles the physician state that Toads have two livers, and that one of them is deadly, while the other is its natural rival, for it brings health.

The Toad's two livers

16. Theopompus says that at the season of the third ploughing and sowing [b] the Veneti who live on the shores of the Adriatic despatch presents to the Jackdaws, and these presents would be cakes of ground barley with honey and oil well and truly kneaded. The purpose of these presents is to placate the Jackdaws and to declare a truce, so that they shall refrain from digging up and collecting here and there the fruits of Demeter sown in the

The Veneti and Jackdaws

[a] Straits of Gibraltar.

[b] The '*third* ploughing' began early in Sept.; the *fourth* shortly before the equinox when the soil was ribbed for the reception of the seed. Sowing began at the autumnal equinox (Sept. 22), or more usually after the setting of the Pleiades (Oct. 23); see Smith, *Dict. Antiqu.* 1. 60, 62, art. 'Agricultura.'

δὲ ἄρα καὶ ταῦτα μὲν ὁμολογεῖ, καὶ ἐκεῖνα δὲ ἐπὶ τούτοις προστίθησι . . .[1] καὶ φοινικοῦς ἱμάντας τὴν χρόαν, καὶ τοὺς μὲν προθέντας ταῦτα εἶτα ἀναχωρεῖν. καὶ τὰ μὲν τῶν κολοιῶν νέφη τῶν ὅρων ἔξω καταμένειν, δύο δὲ ἄρα ἢ τρεῖς προῃρημένους κατὰ τοὺς πρέσβεις τοὺς ἐκ τῶν πόλεων πέμπεσθαι κατασκεψομένους τῶν ξενίων τὸ πλῆθος· οἵπερ οὖν ἐπανίασι θεασάμενοι, καὶ καλοῦσιν αὐτούς,[2] ᾗ πεφύκασιν οἱ μὲν καλεῖν, οἱ δὲ ὑπακούειν. ἔρχονται μὲν ⟨οὖν⟩[3] κατὰ νέφη· ἐὰν δὲ γεύσωνται τῶν προειρημένων, ἴσασιν οἱ Ἐνετοὶ ὅτι ἄρα αὐτοῖς πρὸς τοὺς ὄρνιθας τοὺς προειρημένους ἔνσπονδά ἐστιν· ἐὰν δὲ ὑπερίδωσι καὶ ἀτιμάσαντες ὡς εὐτελῆ μὴ γεύσωνται, πεπιστεύκασιν οἱ ἐπιχώριοι ὅτι τῆς ἐκείνων ὑπεροψίας ἐστὶν αὐτοῖς λιμὸς τὸ τίμημα. ἄγευστοι γὰρ μένοντες[4] οἱ προειρημένοι καὶ ἀδέκαστοί γε[5] ὡς εἰπεῖν ἐπιπέτονταί τε ταῖς ἀρούραις καὶ τό γε πλεῖστον τῶν κατεσπαρμένων συλῶσι πικρότατά γε ἐκεῖνοι, σὺν τῷ θυμῷ καὶ ἀνορύττοντες καὶ ἀνιχνεύοντες.

17. Ἀμύντας ἐν τοῖς ἐπιγραφομένοις οὕτως ὑπ' αὐτοῦ Σταθμοῖς κατὰ τὴν γῆν τὴν Κασπίαν καὶ βοῶν ἀγέλας λέγει πολλὰς καὶ ἵππων,[6] καὶ κρείττονας ἀριθμοῦ εἶναι. ἐπιλέγει δὲ ἄρα καὶ ἐκεῖνο,[7] ἐν ὡρῶν τισι περιτροπαῖς μυῶν ἐπιδημίας γίνεσθαι πλῆθος ἄμαχον, καὶ τὸ μαρτύριον ἐπάγει

[1] *Lacuna.*
[2] τοὺς ἄλλους ? *H.*
[3] ⟨οὖν⟩ *add. Jac.*
[4] ὄντες.
[5] και ἀ. γε] ἀδεκατεύτοις *H.*
[6] καὶ ἵππων *after* εἶναι *in* MSS.

soil. And Lycus confirms this adding further the following details . . .[a] scarlet thongs, and after setting them out they withdraw. And the clouds of Jackdaws remain outside the boundaries, while two or three birds, selected like ambassadors from cities, are sent to take a good look and see how many presents there are. After their inspection they return and summon the birds, giving the call which is natural for them to utter and for the others to respond to. And the birds come in clouds, and if they eat the aforesaid presents, the Veneti know that there is a truce between them and the aforesaid birds. If however they ignore and scorn them as skimpy and refuse to eat them, the inhabitants are confident that a famine will be the price they have to pay for this rejection. For if the aforesaid birds remain unfed and, so to say, unbribed, they swoop upon the ploughlands and pillage in the most distressing way the greater part of what has been sown, digging up and tracking out the seeds in their anger.

The Caspii their land invaded by Rats

17. Amyntas in the work which he entitles *Stages* says that in Caspian territory[b] there are numerous herds of cattle and of horses and that they are past counting. And he adds the following statement: at certain changes of the seasons Rats visit the land in countless hordes, and he adduces as evidence the

[a] The sense of the missing words was perhaps: 'They mark the boundaries of their fields with scarlet thongs.'

[b] The region lying below the S end of the Caucasus through which the river Cyrus flows and is joined not far from its mouth by the Araxes; it corresponds to the modern Transcaucasian province of Azerbaijan.

[7] ἐκεῖνα.

λέγων, τῶν ποταμῶν τῶν ἀενάων [1] σὺν πολλῷ τῷ ῥοίζῳ φερομένων, τοὺς δὲ καὶ μάλα ἀτρέπτως ἐπινήχεσθαί τε αὐτοῖς καὶ τὰς οὐρὰς ἀλλήλων ἐνδακόντας ἕρμα τοῦτο ἴσχειν, καὶ τοῦ διαβάλλειν τὸν πόρον σύνδεσμόν σφισιν [2] ἰσχυρότατον ἀποφαίνειν τόνδε. ἐς τὰς ἀρούρας δὲ ἀπονηξάμενοί φησι καὶ τὰ λήια ὑποκείρουσι, καὶ διὰ τῶν δένδρων ἀνέρπουσι, καὶ τὰ ὡραῖα δεῖπνον ἔχουσι, καὶ τοὺς κλάδους δὲ διακόπτουσιν, οὐδὲ ἐκείνους κατατραγεῖν ἀδυνατοῦντες. οὐκοῦν ἀμυνούμενοι [3] οἱ Κάσπιοι τὴν ἐκ τῶν μυῶν ἐπιδρομήν τε ἅμα καὶ λύμην φείδονται τῶν γαμψωνύχων, οἵπερ οὖν καὶ αὐτοὶ κατὰ νέφη πετόμενοι εἶτα αὐτοὺς ἀνασπῶσιν,[4] καὶ ἰδίᾳ τινὶ φύσει τοῖς Κασπίοις ἀναστέλλουσι τὸν λιμόν.

Ἀλώπεκες δὲ αἱ Κάσπιαι, τὸ πλῆθος αὐτῶν τοσοῦτόν ἐστιν ὡς καὶ ἐπιφοιτᾶν οὐ μόνον τοῖς αὐλίοις τοῖς κατὰ τοὺς ἀγρούς, ἤδη γε μὴν καὶ ἐς τὰς πόλεις παριέναι. καὶ ἐν οἰκίᾳ ἀλώπηξ φανεῖται οὐ μὰ Δία ἐπὶ λύμῃ οὐδὲ ἁρπαγῇ, ἀλλὰ οἷα τιθασός· καὶ ὑποσαίνουσί τε [5] καὶ ὑπαικάλλουσι . . .[6] τῶν παρ᾽ ἡμῖν κυνιδίων. οἱ δὲ μύες οἱ τοῖς Κασπίοις ἐπίδημον [7] ὄντες κακόν, μέγεθος αὐτῶν ὅσον κατά γε τοὺς Αἰγυπτίων ἰχνεύμονας ὁρᾶσθαι· ἄγριοι δὲ καὶ δεινοὶ καὶ καρτεροὶ τοὺς ὀδόντας, καὶ διακόψαι τε καὶ διατραγεῖν οἷοί τε εἰσὶ καὶ σίδηρον. τοιοῦτοι δὲ ἄρα καὶ οἱ μύες οἱ ἐν τῇ Τερηδόνι τῆς Βαβυλωνίας εἰσίν, ὧνπερ οὖν καὶ τὰς δορὰς οἱ τούτων κάπηλοι ἐς Πέρσας ἄγουσι φόρτον. εἰσὶ δὲ

[1] ἀεννάων.
[2] *Wytt*: φησιν.
[3] ἀμυνόμεναι.
[4] *Corrupt*: *perh.* διασπῶσιν *or* ἀναρπάζουσιν *H.*

fact that when the perennial rivers come roaring down, the Rats have no hesitation in swimming them, and by fixing their teeth in one another's tails acquire support and make an unbreakable chain for the crossing of the strait. And when they have swum across to the ploughlands they cut the crops at the foot, creep up all over the trees, make a meal off the fruits, and cut through the branches, for they are capable of eating up even these. And so the Caspii to protect themselves against these raids and the ruin caused by the Rats, refrain from killing birds of prey, which in their turn come flying in clouds and snatch up the Rats and by some natural instinct of their own avert famine from the Caspii.

The Fox in Caspia

The Foxes in Caspian territory are so numerous that they not only constantly visit the sheepfolds in the country but actually come up into the towns. And a Fox will appear in a house not, you may be sure, with any mischievous or thievish intent but as though it were tame. And they fawn and wag their tails ⟨just like⟩ lapdogs in our country. And the Rats, which are a chronic plague to the Caspii, are as large as the ichneumons of Egypt. And they are savage, destructive, and have strong teeth, and are even able to cut and eat through iron. And the Rats of Teredon [a] in Babylonia are just the same, and the traders there bring their skins to the Persians, for they are soft and when sewn together make tunics that keep men warm. And these garments

The Rats of Teredon

[a] Coastal town at the NW end of the Persian Gulf.

---

[5] γε οἱ Κάσπιοι.
[6] *Lacuna*: ⟨δίκην⟩ *Bernard*, ⟨τρόπον⟩ *Jac.*
[7] *Jac*: ἐπίδημοι.

ἁπαλαί, καὶ συνερραμμέναι χιτῶνές τε ἅμα γίνονται καὶ ἀλεαίνουσιν αὐτούς. καλοῦνται δὲ ἄρα οὗτοι κανδύτανες,[1] ὡς ἐκείνοις φίλον. θαυμάσαι δὲ τῶν μυῶν τῶνδε ἄξιον ἄρα καὶ τοῦτο. ἐὰν ἁλῷ μῦς κύουσα, κᾆτα ἐξαιρεθῇ τὸ ἔμβρυον, αὐτῆς δὲ διατμηθείσης ἐκείνης εἶτα μέντοι καὶ αὐτὸ διανοιχθῇ, καὶ ἐκεῖνο ἔχει βρέφος.

18. Τῆς θαλαττίας τρυγόνος ἴδιον καὶ τοῦτο προσακήκοα. ἐπὶ τῆς ἁλιάδος [2] ὀρχεῖταί τις, ὅταν αὐτὴν ὑπονέουσαν θεάσηται, καὶ μέντοι καὶ ἀπέσκωψέ τι κέρτομον, καὶ πρὸς ἐπὶ τούτοις, ἐάνπερ αὐλητικὸς ᾖ, καὶ τὸν αὐλὸν ὡς δέλεαρ φέρει καὶ ὑπαυλεῖ· ἡ δὲ ὑπερήδεται (καὶ γάρ τοι καὶ ὦτα ἔχει μουσικῆς ἐπαΐοντα, ὥς φασι, καὶ ὄμματα συνιέντα ὀρχηστικῆς) εἶτα κηλουμένη ἡσυχῇ πως ἀναπλεῖ. καὶ ὁ μὲν τὰς ἴυγγας τὰς προειρημένας ἐνεργότατα ⟨προσείει⟩,[3] ἔθηκε δὲ τὸ φέρνιόν τις ἕτερος, καὶ τὸν ἰχθὺν ἀνάγει· καὶ (τοῦτο δήπου τὸ καινότατον) κηλουμένη εἶτα ἑαυτὴν διαλέληθεν ᾑρημένη.

19. Γαλάτας Εὔδοξος τοὺς ἑῴους λέγει δρᾶν τοιαῦτα, καὶ εἰ φανεῖταί τῳ πιστά, πιστευέτω, εἰ δὲ ἧττον τοιαῦτα, μὴ προσεχέτω. ὅταν αὐτῶν τῇ γῇ νέφη παρνόπων ἐπιφοιτήσαντα εἶτα λυπήσῃ τοὺς καρπούς, οἵδε [4] εὐχάς τινας εὔχονται, καὶ ἱερουργίας καταθύουσιν ὀρνίθων κατακηλητικάς· οἱ δὲ ὑπακούουσι, καὶ ἔρχονται στόλῳ κοινῷ, καὶ

[1] *W Dindorf*: καναυτᾶνες. [2] *Reiske*: ἁλιάδος νεώς.
[3] ἐνεργότατός ἐστιν MSS, ἐ. ἐ. ⟨προσείων⟩ *Schn.*
[4] οἵδε οἱ Γαλάται.

they call *candytanes* or ' clothes-presses ' according to custom. And here is another amazing phenomenon about these Rats. If a pregnant Rat is caught and the foetus is removed, and after the dissection of the female the foetus in turn is opened, it too is found to contain a young Rat.

18. Here is another characteristic of the Sting-ray which I have learnt. When a man sees it swimming below the surface, if he begins to dance in his fishing-boat and utters taunts and jibes, and moreover, should he chance to be a pipe-player, if he has his pipe as an attraction and will play a tune, the Sting-ray is delighted (you know it has ears that are sensitive to music, so they say, and eyes that can appreciate dancing) and in answer to the spell floats gently to the surface. Meantime the fisherman continues to put forth all his enchantments as described, while some other hand manages the creel and draws up the fish. And what is, I think, the most extraordinary feature is that the fish is so beguiled that it is unaware that it has been caught.

The Sting-ray and music

19. Eudoxus says that the eastern Galatians [a] act as follows, and if anyone regards his account as credible, he may believe it; if not, let him pay no attention to it. When Locusts invade their country in clouds and damage the crops, they put up certain prayers and offer sacrifices warranted to charm birds.[b] And the birds lend an ear and come in a

The Locust in Galatia

[a] Galatia, province in the centre of Asia Minor.

[b] The birds in question are σελευκίδες, *Rose-coloured Pastors* cp. Plin. *HN* 10. 75.

τοὺς πάρνοπας ἀφανίζουσιν. ἐὰν δὲ τούτων τινὰ [1] θηράσηται Γαλάτης, τίμημά οἱ ἐκ τῶν νόμων τῶν ἐπιχωρίων θάνατός ἐστιν. ἐὰν δὲ συγγνώμης τύχῃ [2] καὶ ἀφεθῇ, ἐς μῆνιν ἐμβάλλει τοὺς ὄρνιθας, καὶ τιμωροῦντες τῷ ἑαλωκότι οὐκ ἀξιοῦσιν ὑπακοῦσαι, ἐάν γε καλῶνται αὖθις.

20. Ἀριστοτέλης λέγει γίνεσθαι ἐν Σάμῳ λευκὴν χελιδόνα· ταύτης γε μὴν ἐάν τις ἐκκεντήσῃ [3] τοὺς ὀφθαλμούς, γίνεσθαι μὲν αὐτὴν παραχρῆμα τυφλήν, μετὰ ταῦτα δὲ ἐξωμμάτωται καὶ λελάμπρυνται κόρας [4] καὶ ἐξ ὑπαρχῆς ὁρᾷ, ὡς ἐκεῖνός φησι.

21. Τὸν κιννάμωμον ὄρνιν ἀκούω εἶναι, καὶ μέντοι καὶ κομίζειν κάρφη φυτοῦ τοῦ ὁμωνύμου ἐκ τῶν τῆς γῆς τερμάτων, καὶ καλιὰς ὑποπλέκειν ἔνθα Ἡρόδοτοί τε ᾄδουσι καὶ ἄλλοι, φιλοῦσι δέ πως οἵδε οἱ ὄρνιθες τὰς ἑαυτῶν εὐνάς τε καὶ καταγωγὰς ὑφαίνειν.[5] οὐκοῦν οἷσπερ μέλει τῶνδε τῶν καρφῶν, οἰστοὺς βαρεῖς ῥοίζῳ βιαιοτάτῳ καὶ νευρᾶς ἐντάσει σφοδρᾷ [6] κατὰ τῶν καλιῶν [7] ἀφιᾶσιν· αἱ δὲ ῥήγνυνται, καὶ κατολισθάνει [8] τὰ κάρφη, καὶ μέντοι καὶ τὸ ᾀδόμενον δήπου κιννάμωμον ταῦτά ἐστιν.

22. Καὶ Κλειτάρχῳ χῶρον δῶμεν. λέγει δὲ Κλείταρχος ἐν Ἰνδοῖς γίνεσθαι ὄρνιν, καὶ εἶναι

[1] *Reiske*: τις. [2] τύχῃ τινός.
[3] *Valck*: κεντήσῃ. [4] ἐξομματοῦται καὶ τὰς κ. λ.
[5] *Some words are missing in the sentence*: καταγωγὰς ⟨ἐν πάγοις ἀποτόμοις⟩ *H*, φιλοῦσι δ' ⟨ἐπ' ὄρεσιν⟩ *or* ⟨ἐν σκοπέλοις⟩ *Jac*, *cp*. Hdt. 3. 111 πρὸς ἀποκρήμνοισι οὔρεσι.
[6] σφοδρᾷ ἰσχυρᾶς. [7] *Reiske*: κλάδων.

united host and destroy the Locusts. If however some Galatian should capture one of the birds, his punishment as laid down by the laws of the land is death. But if he is pardoned and let off, this throws the birds into a passion, and to avenge the captured bird they do not deign to respond if they do happen to be invoked again.

20. Aristotle says [*HA* 519 a 6; *Col.* 798 a 27] that a white Swallow occurs in Samos,[a] and that if one puts out its eyes, it immediately becomes blind, but that later on 'sight is restored and the eyes are enlightened' [Soph. *fr.* 701 P], and once again it can see, according to his account.

A white Swallow

21. I have heard that the *Cinnamomus* is a bird; also that it fetches twigs of the tree that bears its name from the ends of the earth and builds nests in places which our historians, Herodotus [3. 111] and others, describe. And these birds seem to like constructing their couches and lodgings ⟨among sheer crags⟩. Accordingly those who are anxious to obtain these twigs shoot heavy arrows that go with a tremendous whizz from a bowstring strained to the utmost, at the nests. And the nests are shattered and the twigs come tumbling down, and they are the celebrated *Cinnamon*.

The Cinnamon-bird

22. Let us make room for Cleitarchus also. He says that in India there occurs a bird with strongly

The 'Orion'

[a] Ar. mentions white swallows, but Samos is not named in either passage, nor is anything said about the blinding and restoration of its sight. See *fr.* 524 (Rose, p. 520).

---

[8] κατολισθαίνει.

σφόδρα ἐρωτικόν, καὶ τὸ ὄνομα αὐτοῦ λέγει ὠρίωνα εἶναι. φέρε δὲ καὶ διαγράψωμεν [1] αὐτὸν τῷ λόγῳ, ὡς ἐκεῖνος διδάσκει. τοῖς μὲν καλουμένοις ἐρῳδιοῖς [2] ὅμοιος τὸ μέγεθος ὅδε ⟨ὁ⟩ [3] ὠρίων ἐστίν, ἔστι δὲ καὶ τὰ σκέλη ὡς ἐκεῖνοι φοίνιξ, ὀφθαλμοὺς δὲ κυανοῦς ἔχει (τοῦτο μὲν οὐχ ὡς ἐκεῖνοι), μέλος δὲ μουσουργεῖν ὑπὸ τῆς φύσεως πεπαίδευται, οἷα δήπου μέλη ὑμεναιοῦται γλυκέα καὶ [4] προσείοντα σειρῆνας.

23. Κατρέα ⟨τὸ⟩ ὄνομα, Ἰνδὸν ⟨τὸ⟩ [5] γένος, τῇ φύσει ὄρνιν λέγει Κλείταρχος εἶναι τὸ [6] κάλλος ὑπερήφανον· τὸ μέγεθος γὰρ εἴη ἂν κατὰ [7] τὸν ταὼν, τὰ δὲ ἄκρα τῶν πτερῶν ἔοικε σμαράγδῳ. καὶ ὁρῶντος μὲν ἄλλοσε [8] οὐκ οἶσθα [9] οἵους ὀφθαλμοὺς ἔχει· εἰ δὲ ἐς σὲ ἀπίδοι, ἐρεῖς κινναβάρινον [10] εἶναι τὸ ὄμμα πλὴν τῆς κόρης· ἐκείνη δὲ μηλιάδι [11] τὴν χρόαν προσείκασται [12] καὶ βλέπει ὀξύ. τό γε μὴν τοῖς ἁπάντων ὀφθαλμοῖς λευκόν, ἀλλὰ τοῖς τοῦ κατρέως τοῦδε ὠχρόν ἐστι. τὰ ⟨δὲ⟩ [13] τῆς κεφαλῆς πτίλα γλαυκωπά, καὶ ἔχει ῥανίδας οἱονεὶ κρόκῳ προσεικασμένας [14] εἶτα ἄλλην ἄλλῃ διεσπαρμένας. πόδες δὲ αὐτῷ σανδαράκινοι. ἔχει δὲ καὶ φώνημα εὔμουσον καὶ κατὰ τὴν ἀηδόνα τορόν. Ἰνδοὶ δὲ ἄρα [15] τὴν ἐξ ὀρνίθων τροφὴν . . . [16] εἶχον, ἵνα καὶ οἱ ὁρῶντες ἑστιᾶν τὴν ὄψιν δύνωνται. ἰδεῖν γοῦν αὐτοῖς πάρεστι καὶ

[1] γράψωμεν. [2] *Ges*: καλοῦσιν ἐρωδιόν.
[3] ⟨ὁ⟩ *add.* Bernhardy.
[4] ὑμεναιοῦται γλυκέα καί] ὑμνεῖται ταῦτα γλυκέα καὶ πρὸς τὸν ᾀδόμενον ὑμέναιον βλέπει θέλγοντα γονὴν ὕπνῳ [ὕμνῳ *Toup*] τινὶ γαμικῷ.
[5] ⟨τό⟩ . . . ⟨τό⟩ *add.* H.

amorous propensities and that it is called the *Orion*.[a] Well now, let us depict it as he has described it. This ' Orion ' is the same size as the birds they call herons and its legs are red like theirs; its eyes are dark (in this repect it is unlike them), and Nature has taught it to make melody sweet as any bridal song with its alluring charms.

23. Cleitarchus says that the *Catreus*,[b] as it is called, is a native of India, and is a bird of magnificent beauty. It might be about the same size as a peacock; the tips of its feathers are the colour of an emerald, and when it looks in another direction you cannot tell what its eyes are like. If however it looks you in the face, you will pronounce them to be vermilion all except the pupil, and this has a grey hue and a keen glance. And what is white in the eyes of all other birds is pale brown[c] in the Catreus. And its head feathers are a blue-grey with saffron-coloured speckles sprinkled here and there. Its legs are an orange colour, and its note is as melodious and clear as the nightingale. Now the use of these birds for food is ⟨prohibited⟩ by the Indians, in order that spectators may feast their eyes upon them. At any rate there are to be seen in India

The 'Catreus

[a] Otherwise unknown, and fabulous.
[b] Probably the ' Manâl pheasant.'
[c] See W. Beebe, *Monog. of the Phasianidae*, 1. 113ff.

[6] *Ges* : τι.
[7] πρός.
[8] ὁρῶν μὲν ἄλλους.
[9] οἶδας.
[10] κιννάβαριν.
[11] μήλῳ.
[12] παρείκασται.
[13] ⟨δέ⟩ *add. H.*
[14] παρεικασμένας.
[15] ἄρα καί.
[16] *Lacuna* : ⟨ἀπόρρητον⟩ *conj. H.*

ὅλους πορφυροῦς καὶ τῇ καθαρωτάτῃ φλογὶ προσεοικότας· καὶ τούτων αἱ πτήσεις κατὰ πλῆθός εἰσιν, ὡς νομίζειν νέφη· ἄλλοι γε μὴν ποικίλοι καὶ οὐ πάνυ τι τὸ εἶδος εὔρητοι,[1] μελῳδίαν δὲ καὶ εὐστομίαν καὶ εὐγλωττίαν ἄμαχοι,[2] ὡς εἶναι, ⟨εἰ⟩ [3] μή πη καὶ τραχύτερόν [4] ἐστιν εἰπεῖν, Σειρῆνάς τινας.[5] κατάπτεροι γὰρ ὡς ἦσαν αἱ [6] τοῦ μύθου κόραι,[7] ποιηταί τε ᾄδουσιν καὶ ζωγράφοι δεικνύουσιν.

24. Κύκνου δὲ ἤθη καὶ διατριβαὶ λίμναι τε καὶ ἕλη καὶ τενάγη καὶ ἀέναοι [8] ποταμοὶ πράως καὶ ἡσυχῇ ῥέοντες. εἰρηναῖοι δέ εἰσι καὶ ἐς γῆρας προϊασιν ἑαυτοῖς κοῦφον. εἰσὶ δὲ καὶ ἐς [9] ῥώμην ἄλκιμοι, καὶ θαρροῦσιν αὐτῇ, οὐ μὴν ὥστε ἄρχειν ἀδίκων ἀλλ' ἀμύνεσθαι τὸν ἄρξαντα. ῥᾳδίως οὖν καὶ τῶν ἀετῶν περιγίνονται, ὅταν ἐκεῖνοι τολμήσωσιν ἐπιθέσθαι αὐτοῖς. καὶ εἶπον ἀνωτέρω τῆς μάχης τὸν τρόπον.

25. Λέγει δὲ Κλείταρχος πιθήκων ἐν Ἰνδοῖς εἶναι γένη ποικίλα τὴν χρόαν, μεγέθει δὲ μέγιστα. ἐν δὲ τοῖς χωρίοις τοῖς ὀρείοις τοσοῦτον αὐτῶν τὸ πλῆθος [10] εἶναι, ὡς Ἀλέξανδρόν φησι τὸν Φιλίππου καὶ πάνυ καταπλαγῆναι σὺν καὶ τῇ οἰκείᾳ δυνάμει, οἰόμενον ἀθρόους ἰδόντα στρατιὰν ὁρᾶν συνειλεγ-

[1] ἄλλοι . . . εὔρητοι] καὶ ἄλλα μὴν ποικίλα καὶ οὐ πάντη . . . εὔρητα.
[2] ἄμαχα.
[3] ⟨εἰ⟩ *add. Schn.*
[4] παχύτερον.
[5] τινας ἢ καὶ τοῦτό γε ἐγγύθεν.
[6] καί MSS, καὶ ⟨αἱ⟩ *Abresch.*

birds entirely scarlet, the colour of the purest flame, and they fly in such multitudes that one would take them for clouds. Others however are mottled and it is not very easy to say what they look like, but for beauty and clarity of tone their singing is unsurpassed; they might be, if the expression is not too strong, Sirens, for these fabled maidens as celebrated by poets and portrayed by artists had wings.

24. The Swan's customary haunts are lakes, marshes, pools, and rivers with a ceaseless, gentle, tranquil flow. They are creatures of peace and attain to an old age that has no burdens for them. Their strength is redoubtable and that gives them confidence, but not to the extent that they are the aggressors in an injury; against an aggressor they will defend themselves. And so they have no difficulty in getting the better of eagles when the latter venture to attack them. I have described earlier on [a] how they do battle.

The Swan

25. Cleitarchus says that in India there are Monkeys of a mottled hue and immense size. And in mountainous districts they are so numerous that, says Cleitarchus, Alexander, the son of Philip, and the army under his command also were quite terrified at the sight of their massed numbers, imagining that they saw an army marshalled and waiting in

An Indian Monkey

[a] See 5. 34.

---

[7] *κόραι καὶ τοὺς πόδας ὄρνιθες ἐδόκουν.*
[8] *ἀένναοι.*
[9] *πρός.*
[10] *Jac* : *μέγεθος.*

μένην καὶ ἐλλοχῶσαν αὐτόν. ὀρθοὶ δὲ ἄρα ἦσαν οἱ πίθηκοι κατὰ τύχην ἡνίκα ἐφάνησαν. θηρῶνται δὲ οὗτοι οὔτε δικτύοις οὔτε κυνῶν ῥινηλατούντων σοφίᾳ καὶ μάλα ἀγρευτικῇ. ἔστι δὲ τὸ ζῷον ὀρχηστικόν, εἰ θεῷτο ὀρχούμενον· καὶ θέλει γε αὐλεῖν, εἰ καταπνεῖν μάθοι.[1] πρὸς τούτοις εἰ θεάσαιτό τινα ὑποδήματα τοῖς ποσὶ περιτιθέντα, μιμεῖται τὴν ὑπόδεσιν· καὶ ὑπογράφοντος[2] τὼ ὀφθαλμὼ μέλανι,[3] καὶ τοῦτο δρᾶσαι θέλει. οὐκοῦν ὑπὲρ τῶν εἰρημένων μολίβου πεποιημένα κοῖλα καὶ βαρέα ὑποδήματα προτιθέασι, βρόχους[4] αὐτοῖς ὑποβαλόντες, ὡς ἐσβαλεῖν μὲν τὼ πόδε, ἔχεσθαι δὲ τῇ πάγῃ καὶ μάλα ἀφύκτῳ· δέλεαρ δὲ αὐτοῖς ὀφθαλμῶν πρόκειται ὑπὲρ τοῦ μέλανος[5] ἰξός. κατόπτρῳ δὲ χρησάμενος ὁ Ἰνδὸς ὁρώντων ἐκείνων . . .[6] οὐκ εἰσὶ δ' ἔτι τὰ κάτοπτρα, ἀλλὰ ἕτερα προτιθέντες·[7] εἶτα καὶ τούτοις ἕρματα ἰσχυρὰ ὑποπλέκουσι· καὶ μὴν τὰ σκεύη[8] τοιαῦτά ἐστιν. οἱ μὲν ⟨οὖν⟩[9] ἔρχονται, καὶ ἀτενῶς[10] ὁρῶσι[11] κατὰ μίμησιν[12] ὧν[13] εἶδον· ἐκπηδᾷ δὲ ἰσχύς τις κολλητικὴ βλεφάρων ἐκ τῆς πρὸς τὴν αὐγὴν ἀντιτυπίας,[14] ὅταν ἴδωσιν ἀτενές· εἶτα οὐχ ὁρῶντες αἱροῦνται ῥᾷστα· φυγεῖν γὰρ ἔτι

[1] μάθοι εἰδέναι.
[2] ὑπογράφοντα MSS, ⟨εἰ⟩ ὑπογρ. *Schn.*
[3] μέλος τι.
[4] καὶ βρόχους.
[5] μέλλοντος.
[6] *Lacuna.*
[7] *Schn* : προστιθέντες MSS, *H.*
[8] *Jac* : καὶ μέντοι καί MSS, *H.*
[9] ⟨οὖν⟩ *add. Jac.*
[10] *Gron* : ἀγεννῶς MSS, γενναίως *H.*
[11] δρῶσι *Jac, H.*
[12] τὴν μίμησιν.
[13] *Jac* : ἥν.
[14] ἐκ . . . ἀντιτυπίας *corrupt Jac.*

ambush for them. You see, the Monkeys happened to be standing upright when they appeared. These creatures are not to be caught with nets or by means of hounds following a scent, however great their skill in hunting. But this Monkey is ready to dance if it sees a man dancing; it is even willing to play the pipe if it could learn how to blow. Further, if it catches sight of someone putting on his shoes, it imitates the action; and if a man underlines his eyes with lamp-black,[a] it is anxious to do this too. Accordingly in place of the aforesaid objects men put out hollow, heavy shoes made of lead, to which they attach a noose underneath, so that when the Monkeys slip their feet into them they are caught in the snare and cannot escape. And as a bait for their eyes men put out bird-lime in place of lamp-black. And an Indian after using a mirror in sight of the Monkeys . . .[b] displaying not genuine mirrors but ones of a different kind, on to which they lace strong nooses. Such then is the apparatus which they employ. And so the Monkeys come and gaze steadily, imitating what they have seen. And from the reflecting surface opposite their sight there is a surge of strongly gluey substance that gums up their eyelids, when they gaze intently into it. Then being unable to see, they are caught without any difficulty, for they are no longer able to escape.

its capacity for imitation

how caught

[a] Cp. Alexis *fr.* 98. 16. The *kohl* of modern India is a mixture of lamp-black and castor oil.

[b] The text is defective; to fill the gap one might conjecture something on these lines: '[withdraws, leaving behind him an object resembling it. By such means the Indians attract the creatures,] though what they display are not genuine, *etc.*'

εἰσίν ἥκιστοι. εἴρηται μὲν ὑπὲρ πιθήκων καὶ ἄλλα, Ἰνδῶν τε καὶ οὐκ Ἰνδῶν· καὶ ταῦτα δὲ ἔχει τινὰ τῷ συνιέντι οὐκ ἀσπούδαστα, οὐ μὰ Δία.

26. Λέοντας ἐν Ἰνδοῖς γίνεσθαι μεγίστους οὐ διαπορῶ· τὸ δὲ αἴτιον, τῶν ζῴων τῶν ἑτέρων ἥδε ἡ γῆ μήτηρ ἐστὶν ἀγαθή.[1] εἰσὶ δὲ ἀγριώτατοι καὶ θηριωδέστατοι. δέρη ⟨δὲ⟩[2] ἐκείνων ⟨τῶν⟩[3] λεόντων μέλαινά τε ἰδεῖν, καὶ φρίξασα ὀρθή τε ἀνίσταται καὶ συνεκπέμπει δέος οἷον ἐκπληκτικόν. εἰ δὲ ἁλῶναι δυνηθεῖεν, πραΰνονται,[4] ἀλλ' οὐχ οἱ μέγιστοι· καὶ ἡμεροῦνταί τε καὶ γίνονται[5] ῥᾷστα τιθασοί, ὡς ἄγειν γε[6] ἀπὸ ῥυτῆρος[7] ἐπὶ θήραν κεμάδων καὶ ἐλάφων καὶ συῶν καὶ ταύρων καὶ ἀγρίων ὄνων. εἰσὶ γὰρ καὶ ῥινηλατῆσαι ὡς ἀκούω δεινοί.

27. Ἐν τῇ Λιβύων χώρᾳ ἔθνος ἦν φασι τὸ καλούμενον Νόμαιον. καὶ τὰ μὲν ἄλλα διευτυχοῦντες εὐνόμου μάλα καὶ εὐδαίμονος ναὶ μὰ Δία λήξεως εἶτα ἠφανίσθησαν[8] τελέως, λεόντων αὐτοῖς ἐπελθόντων πλήθει τε παμπόλλων καὶ μεγέθει μεγίστων καὶ τὴν τόλμαν ἀμάχων, ὑφ' ὧν πανδημεὶ τε καὶ παγγενεὶ διαφθαρέντες, εἶτα ἐς τὸ παντελὲς ἀπώλοντο.[9] λεόντων γὰρ ἀθρόων ἐπιδημία χρῆμα ἀπρόσμαχον.

[1] ἀγαθὴ ὅσα γε ἐντυχεῖν κατὰ πρόσωπον MSS, ἀγαθή. εἰσὶ δέ, ὅσα γε ἐ. κ. π., *Jac.*

[2] ⟨δέ⟩ *add. H.*

[3] ⟨τῶν⟩ *add. Reiske.*

[4] πραΰνονταί γε.

[5] γε καὶ γ. γε.

[6] τε.

Now touching Monkeys both Indian and non-Indian I have written an account elsewhere,[a] but the foregoing chapter contains facts that must assuredly interest any man of intelligence.

26. I have no doubt that in India the Lions are of the very largest, the reason being that this country is an excellent mother of other animals. And they are exceedingly wild and savage. The mane of these Lions is black in appearance, and when it bristles and stands upright it inspires such fear as to unnerve a man. But if once they can be captured, they can be tamed, though not the largest of them. And they become gentle and are easily domesticated, so that they can be led by a rein to hunt prickets, deer, swine, bulls, and wild asses, for they are (so I have heard) clever at tracking by scent.

The Indian Lion

27. It is said that in Libya there used to exist a race of men called the Nomaei. They continued generally prosperous in a territory where the pastures were good and the land unquestionably rich, until finally they were wiped out when a vast horde of Lions of the very largest size and of irresistible boldness attacked them. The whole race to a man was destroyed by the Lions and perished utterly. A visitation by Lions in a mass is something that no creature can withstand.

The Nomaei and Lions

[a] See 5. 26; 7. 21; 6.10; 17. 39.

---

[7] ῥυτῆρος καὶ κατὰ κυναγωγούς.
[8] ἠφανίσθη.
[9] ἀπώλοντο τὸ ἔθνος.

28. Εὐφορίων δὲ ἐν τοῖς Ὑπομνήμασι λέγει τὴν Σάμον ἐν τοῖς παλαιτάτοις χρόνοις ἐρήμην γενέσθαι· φανῆναι γὰρ ἐν αὐτῇ θηρία μεγέθει μὲν μέγιστα, ἄγρια δέ, καὶ προσπελάσαι τῳ δεινά, καλεῖσθαί γε [1] μὴν νηάδας. ἅπερ οὖν καὶ μόνῃ τῇ βοῇ ῥηγνύναι τὴν γῆν. παροιμίαν οὖν ἐν τῇ Σάμῳ διαρρεῖν τὴν λέγουσαν 'μεῖζον βοᾷ τῶν νηάδων.' ὀστᾶ δὲ ἔτι καὶ νῦν αὐτῶν δείκνυσθαι μεγάλα ὁ αὐτός φησι.

29. Τοῦ Ἰνδῶν βασιλέως ἐλαύνοντος ἐπὶ τοὺς πολεμίους δέκα μυριάδες ἐλεφάντων προηγοῦνται μαχίμων. ἑτέρους δὲ ἀκούω τρισχιλίους τοὺς μεγίστους τε καὶ ἰσχυροτάτους ἕπεσθαι, οἵπερ οὖν εἰσι πεπαιδευμένοι τὰ τείχη τῶν πολεμίων ἀνατρέπειν, ἐμπεσόντες ὅταν κελεύσῃ ⟨ὁ⟩ [2] βασιλεύς· ἀνατρέπουσι δὲ τοῖς στήθεσι. καὶ λέγει μὲν ταῦτα Κτησίας,[3] ἀκοῦσαι γράφων. ἰδεῖν δὲ ἐν Βαβυλῶνι ὁ αὐτὸς λέγει τοὺς φοίνικας αὐτορρίζους ἀνατρεπομένους ὑπὸ τῶν ἐλεφάντων τὸν αὐτὸν τρόπον, ἐμπιπτόντων τῶν θηρίων αὐτοῖς βιαιότατα· δρῶσι δὲ ἄρα, ἂν [4] ὁ Ἰνδὸς ὁ πωλεύων αὐτοὺς κελεύσῃ δρᾶσαι τοῦτο αὐτοῖς.

30. Ζηνόθεμις λέγει Παιονίδα λίμνην τινὰς φέρειν ἰχθῦς, οὕσπερ οὖν εἰ παραβάλοι τις ἀσπαίροντας τοῖς βουσίν, οἱ δε ἐμφοροῦνται αὐτῶν μάλα ἀσμένως, ὡς οἱ λοιποὶ τοῦ χόρτου. νεκρῶν δὲ τῶν ἰχθύων οὐκ ἂν πάσαιντο ἔτι οἱ βόες, ἐκεῖνος λέγει.

[1] *Mein* : δέ. [2] ⟨ὁ⟩ *add. H.*
[3] καὶ K.
[4] *Jac* : ἂν ἄρα.

28. Euphorion says in his *Commentaries* that in primaeval times Samos was uninhabited, for there appeared in the island animals of gigantic size, which were savage and dangerous for a man to approach, and they were called *Neades*. Now these animals with their mere roar split the ground. So there is a proverbial saying current in Samos, 'He roars louder than the Neades.' And the same writer asserts that their huge bones are displayed even to this day.

The Neades of Samos

29. When the Indian King goes to battle against his enemies a hundred thousand Elephants of war form the vanguard. And I learn that another three thousand of the largest and strongest bring up the rear, and these have been trained to overturn the enemies' walls by attacking them when the King gives the order; and they overturn them by the weight of their chest. Such is the account given by Ctesias, who writes that this is hearsay. But the same writer says that in Babylon he has seen date-palms completely uprooted by Elephants in the same way, the animals falling upon them with all their force. This they do if their Indian trainer orders them to do so.

Indian Elephants of war

30. Zenothemis says that a lake in Paeonia [a] produces certain Fish, and if these are given, while still gasping, to cattle, the cattle are glad to take their fill of them, as others do of fodder. But if the Fish are dead the cattle refuse to touch them, so he says.

Fish as cattle-fodder

[a] Mountainous district to the N of Macedonia. The lake is unknown.

31. Παρὰ Ἀρμενίοις ἀκούω πέτραν εἶναι ὑψηλήν, εἶτα ταύτην ὕδωρ ἐκβάλλειν πάμπολυ. ὑποκεῖσθαι δὲ τῇ πέτρᾳ πυνθάνομαι κρήνην τετράγωνον πάντῃ, καὶ ἑκάστην πλευρὰν σταδίου ἥμισυ ἔχειν, βάθος δὲ τριῶν ὀργυιῶν εἶναι. συνεκπίπτειν δὲ τῷ ὕδατι προσακούω τῷ προειρημένῳ καὶ ἰχθύας πολλάκις ἔχοντας τὸ μῆκος καὶ πήχεως καὶ ἔτι μείζους καὶ μέντοι καὶ ἐλάττονας, ἀλλ' οὐ κατὰ πολύ. καὶ τοὺς μὲν αὐτῶν κατολισθαίνειν[1] ἡμιθνῆτας, τοὺς δὲ ἀσπαίροντας καὶ μάλα γε ἰσχυρῶς ἀποθνήσκειν. εἶναι δὲ αὐτοὺς ἡ φήμη λέγει πάνυ σφόδρα μέλανας καὶ ἰδεῖν ἀειδεῖς. ἂν δὲ τούτων[2] γεύσηται ἢ ἄνθρωπος ἢ θηρίον, παραχρῆμα ἀπόλλυται. τοὺς μὲν οὖν Ἀρμενίους διὰ τὸ ἔνθηρον αὐτοῖς εἶναι καὶ πολύθηρον τὴν γῆν ἀθροίζειν αὐτοὺς καὶ αὐαίνειν ὑπὸ τῇ εἵλῃ τοῦ ἡλίου, εἶτα κόπτειν ἐπιδήσαντάς[3] τι[4] ταῖς ῥισὶ καὶ τῷ στόματι, ἵνα μὴ τῷ ἄσθματι σπάσαντες[5] τὸν ἐκ τῶν πτισσομένων[6] ἀέρα[7] εἶτα ἀποθάνωσι. ποιήσαντες οὖν ἄλευρα τοὺς ἰχθῦς ἐν τοῖς μάλιστα θηριωδεστάτοις[8] χωρίοις κατασπείρουσιν αὐτά,[9] σῦκά γε μὴν[10] παραμιγνύναι τοῖς ἀλεύροις ἔθος ἔχουσιν. οὕτω μὲν οὖν διαφθείρονται οἵ τε σῦς οἱ ἄγριοι καὶ αἱ δορκάδες καὶ οἱ ἔλαφοι καὶ οἱ ἄρκτοι καὶ οἱ ὄνοι οἱ ἄγριοι καὶ ⟨οἱ⟩[11] αἶγες, ἄγριοι μέντοι καὶ οὗτοι· συκοτράγα γὰρ ταῦτα καὶ ἀλφιτοφάγα τὰ ζῷά ἐστιν.[12] λέοντας δὲ καὶ παρδάλεις καὶ λύκους σαρκοφάγα ὄντα ἑτέρως ἀναιροῦσι. τῶν γὰρ οἰῶν τῶν ἡμέρων καὶ τῶν

[1] κατολισθάνειν *H*.
[2] τι τούτων.
[3] ὑποδήσαντας.
[4] τι κάτω.
[5] *Jac*: ἐκσπάσαντες.
[6] ἐκ τῶν π.] *Jac*: ἐκπτισσομένων.

31. I have heard that in Armenia there is a lofty rock which discharges a copious stream of water. And I am told that at the foot of the rock there is a square fountain, each side measuring half a *stade*, and the depth is three fathoms. I learn further that along with the aforesaid water there descend Fish often a cubit long and even more, but sometimes less, though not much less. Some of them collapse half dead, others fall gasping and die a violent death. And report states that they are a deep black and unsightly to look at. And if man or beast eats of them, death follows immediately. Accordingly the Armenians, since their country is infested with numerous wild animals, collect these Fish and dry them by the heat of the sun; they then mince them, after bandaging nose and mouth in order to prevent themselves from inhaling the odours given off by the Fish in the process of being brayed, and so catching their death. Then after making the Fish into meal they sprinkle it about in the districts that are most infested with wild beasts; they even have a custom of mixing figs with the meal. And this is the way in which they destroy wild swine, gazelles, deer, bears, wild asses, and goats, and these too are wild. For these animals eat figs and meal. But they adopt a different device for killing lions, leopards, and wolves, which are carnivorous. They make a

A poisonous Fish in Armenia

[7] ἀέρα, ἢ τὴν ἐγειρομένην ἐκ λεπτῶν τινῶν κόνιν ἀλφίτων.

[8] *Ges*: θηριωτάτοις.

[9] αὐτά, ὧν γευσάμενα τὰ ζῷα ἀποθνήσκει MSS; *the last five words would be appropriate if inserted after* ἔθος ἔχουσιν, *H* (1858).

[10] καὶ σῦκα μήν.

[11] ⟨οἱ⟩ *add. H.*

[12] ἐστιν, ἀναιρεῖται δὲ τὸν τρόπον τοῦτον διὰ τὸ πλῆθος.

αἰγῶν παρασχίσαντες τὴν πλευρὰν ἐς ὅσον καθεῖναι τὴν χεῖρα, ἐμπάττουσι [1] τῶν αὐτῶν ἀλεύρων,[2] προκεῖσθαι κακὸν ναὶ μὰ Δία δέλεαρ τοῖς προειρημένοις. ὅταν οὖν ἢ λέων ἢ πάρδαλις ἢ λύκος ἢ ἄλλο τι τοιοῦτον ἐντύχῃ καὶ γεύσηται, τέθνηκε παραχρῆμα. καὶ πᾶσα μὲν οὖν ἡ Ἀρμενία θηρίων ἀγρίων τροφός τε ἅμα καὶ μήτηρ ἐστίν, ἡ δὲ πεδιὰς ἔτι καὶ μᾶλλον ἡ πρὸς τῷ ποταμῷ.

32. Ἐν τῇ Κασπίᾳ γῇ λίμνην ἀκούω μεγίστην εἶναι, καὶ ἰχθῦς ἐν αὐτῇ γίνεσθαι μεγάλους, καὶ ὀξύρυγχοι καλοῦνται.[3] οὐκοῦν οἱ Κάσπιοι θηρῶσιν αὐτούς, καὶ διαπάσαντες ἁλσὶ καὶ ταρίχους ἐργασάμενοί τε καὶ ἀποφήναντες αὔους, ἐπισάξαντες καμήλοις κομίζουσιν ἐς Ἐκβάτανα. καὶ ποιοῦσιν ἄλειφα ἐκ τῶνδε τῶν ἰχθύων ἀφελόντες τὴν πιμελήν,[4] τῷ δὲ ἰχθυΐνῳ ἐλαίῳ χρίονται λιπαρῷ σφόδρα καὶ οὐ δυσώδει, τὰ δὲ ἔντερα ἐξέλκουσιν αὐτῶν καὶ ἕψουσι, καὶ ἐξ αὐτῶν ποιοῦσι κόλλαν καὶ μάλα γε ἐν χρείᾳ γίνεσθαι δυναμένην· συνέχει γὰρ πάντα ἐγκρατῶς, καὶ προσέχεται οἷς ἂν προσπλακῇ, καὶ ἰδεῖν ἐστι λαμπροτάτη. οὕτω δὲ συνέχει πᾶν ὅ τι ἂν συνδήσῃ τε καὶ συνάψῃ, ὡς καὶ δέκα ἡμερῶν αὐτὴν βρεχομένην μήτε λύεσθαι μήτε μὴν ἀφίστασθαι. ἀλλὰ καὶ τοὺς τὸν ἐλέφαντα χειρουργοῦντας [5] χρῆσθαί τε αὐτῇ καὶ τὰ ἔργα ἐκπονεῖν κάλλιστα.

[1] ἐμπλάττουσι. [2] κρέων.

[3] καλοῦνται κατὰ τὸ σχῆμα τοῦ προσώπου δηλονότι καὶ προϊέναι ἐς μῆκος καὶ ὀκτὼ πηχῶν.

[4] πιμελὴν καὶ τοῦ μὲν ταρίχου πιπράσκουσιν MSS, τοὺς . . . ταρίχους *Oud.*

[5] *Ges* : χειροῦντας.

slit in the side of a tame sheep or goat deep enough to admit a hand, and sprinkle in some of that self-same meal, and deadly indeed is the bait which is set before the above-mentioned animals. And so whenever a lion or a leopard or a wolf or other savage beast comes across the body and tastes it, it dies immediately. The whole country of Armenia is in fact the nurse and mother of wild animals, especially the plainlands bordering the river.[a]

The 'Oxyrhynchus' fish

32. I have heard that in the land of the Caspii there is a lake [b] of very wide extent, and that in it there occur large fishes which are called *Oxyrhynchi*.[c] Now the Caspii hunt them and after salting, pickling, and drying them, pack them on to camels and transport them to Ecbatana. And after removing the fat they make meal from these fish; with the oil, which is extremely rich and free from any evil smell, they anoint themselves; but the inwards they extract and boil, and therefrom they make a glue [d] which can be of great service, for it holds all objects together firmly, and sticks to whatever it has been attached to, and is very clear. And it holds all objects which it binds and unites, so tight that even if soaked in water for as much as ten days it will not dissolve or come away. Moreover workers in ivory use it and produce most beautiful pieces.

[a] The river Cyrus flows through the whole length of the Armenian plain.
[b] The Caspian Sea.
[c] 'Evidently a Sturgeon,' Thompson, *Gk. fishes*. This is not identical with the Nile fish of 10. 46.
[d] Isinglass.

33. Λέγει τις λόγος ἐν Κασπίοις ὄρνεον γίνεσθαι τὸ μὲν μέγεθος κατὰ τοὺς ἀλεκτρυόνας τοὺς μεγίστους, ποικίλον γε μὴν [1] καὶ πολυχροίᾳ διηνθισμένον. καὶ πέτεταί [2] γε ὕπτια [3] ὡς ἀκούω ὑποτεῖναν τῷ τραχήλῳ τὰ σκέλη καὶ οἷον ἀνέχον αὐτοῖς αὑτόν. κλαγγὴν δὲ προΐεσθαι σκυλακίου. ποιεῖσθαι δὲ τὴν πτῆσιν οὐκ ἐν ἀέρι βαθεῖ [4] ἀλλὰ περὶ τὴν γῆν, ἐλαφρίζειν ἐς ὕψος ἑαυτὸ [5] μὴ δυνάμενον.

Κάσπιος δὲ ἄρα καὶ οὗτος ὄρνις ἢ Ἰνδὸς μᾶλλον (λέγεται γὰρ καὶ ἐκείνῃ τὸ γένος οἱ καὶ ταύτῃ), καὶ εἴη τὸ μέγεθος κατὰ χῆνα ἄν. καὶ ἔχει κεφαλὴν πλατεῖαν μὲν λεπτὴν δέ, καὶ τὰ σκέλη οἱ μακρά. καὶ κεκραμένη χρόα οἱ καὶ μικτή· τὸ μὲν γὰρ νῶτον αὐτῷ πορφυροῖς ἠγλάϊσται, τὰ δὲ ὑπὸ τὴν γαστέρα [6] κόκκῳ γνησιωτάτῳ καὶ καλλίστῳ προσείκασται, κεφαλὴ δὲ καὶ δέρη λευκὰ ἄμφω. φθέγγεται δὲ κατὰ τὴν αἶγα.

34. Αἶγες δὲ Κάσπιαι γίνονται λευκαὶ ἰσχυρῶς, κεράτων [7] δὲ ἄγονοι, ⟨καὶ⟩ [8] μικραὶ τὸ μέγεθος καὶ σιμαί. [9] κάμηλοι δ᾽ ἀριθμοῦ [10] πλείους, αἳ μέγισται κατὰ τοὺς ἵππους τοὺς μεγίστους, εὔτριχες ἄγαν. ἁπαλαὶ γάρ εἰσι σφόδρα αἱ τούτων τρίχες, ὡς καὶ τοῖς Μιλησίοις ἐρίοις ἀντικρίνεσθαι τὴν μαλακότητα. οὐκοῦν ἐκ τούτων οἱ ἱερεῖς ἐσθῆτας [11] ἀμφιέννυνται καὶ οἱ τῶν Κασπίων πλουσιώτατοί τε καὶ δυνατώτατοι.

[1] γε μὴν τοῖς πτεροῖς.
[2] πέταται.
[3] ὕπτιον.
[4] βαθεῖαν.
[5] ἑαυτόν.
[6] τὸ δὲ ὑπὸ τῇ γαστρί.
[7] καὶ κεράτων.

A Caspian bird

33. There is a story that among the Caspii there occurs a bird as large as the largest cockerels, of variegated hue, and gay with many colours. And it flies, so I hear, upside down with its legs extended upwards beneath its neck, seeming to sustain itself by these means; and it utters a note like that of a puppy; and it flies not high up in the sky but along the ground, being unable to soar.

An Indian bird

The following bird also is a Caspian, or rather an Indian, bird, for its generic type is spoken of both in the latter and in the former connection, and it may be the size of a goose. It has a broad but shallow head and long legs; its colour is variegated, for its back is beautified with purple markings while its belly beneath is the colour of the purest and most splendid scarlet, and its head and throat are both white. It makes a sound like a goat.[a]

The Goats and Camels of the Caspii

34. The Goats of the Caspii are a pure white but grow no horns; they are small and snub-nosed. Their Camels are past numbering, and the largest are the size of the largest horses and have beautiful hair. For their hair is so fine that it can compare with Milesian wool for softness. Accordingly their priests and the wealthiest and most powerful of the Caspii clothe themselves in garments made from Camels' hair.

[a] These two birds have not been identified; they may even be legendary.

---

[8] ⟨καί⟩ *add. H.*
[9] *Ges* : οἶμαι.
[10] *Jac* : ἀριθμοῦνται.
[11] ἐσθῆτα.

35. Ἐν λόγοις Κρητικοῖς Ἀντήνωρ λέγει τῇ τῶν καλουμένων Ῥαυκίων [1] πόλει ἔκ τινος δαιμονίου προσβολῆς ἐπιφοιτῆσαι μελιττῶν σμῆνος, αἵπερ οὖν ᾄδονται [2] χαλκοειδεῖς, ἐγχριμπτούσας [3] δὲ ἄρα αὐτοῖς τὰ κέντρα εἶτα μέντοι πικρότατα λυπεῖν. ὧνπερ οὖν ἐκείνους τὴν προσβολὴν οὐ φέροντας ἀναστῆναι τῆς πατρίδος καὶ μέντοι καὶ ἐς χῶρον ἐλθεῖν ἄλλον, καὶ οἰκίσαι φιλίᾳ τῆς μητρίδος, ἵνα Κρητικῶς εἴπω, Ῥαῦκον,[4] εἰ [5] καὶ τοῦ χωρίου ὁ δαίμων ἤλαυνεν αὐτούς, ἀλλὰ γοῦν τελέως [6] ἀποσπασθῆναι τοῦ ὀνόματος οὐχ ὑπομείναντες. λέγει δὲ ὁ Ἀντήνωρ καὶ ἔτι κατὰ τὴν Ἴδην τὴν Κρῆσσαν ἐκείνου τοῦ γένους τῶν μελιττῶν εἶναι ἰνδάλματα, οὐ πολλὰ μέν, εἶναι δ' οὖν, καὶ πικρὰ [7] ἐντυχεῖν, ὡς ἐκεῖναι ἦσαν.

36. Καμήλου κρέας ἥδεται λέων ἐσθίων. καὶ τὸ μαρτύριον, Ἡρόδοτος λέγει ταῖς Ξέρξου καμήλοις ταῖς τὸν σῖτον φερούσαις ἐπιθέσθαι λέοντας. τὰ δὲ ἄλλα οὐκ ἐσίνοντο, οὐχ ὑποζύγιον, οὐκ ἄνθρωπον, ἢ δ' ὅς. ὀλίγα δὲ Ἡρόδοτος ᾔδει ἐξετάζων τροφὴν [8] λεόντων Θρᾳκίων· ἴσασι δὲ καὶ Ἄραβες ταῦτα, καὶ ὅσοι λεόντων καὶ καμήλων μητέρα τε ἅμα καὶ τροφὸν γῆν ἔχουσιν. οὐκ ἂν γοῦν θαυμάσαιμι εἰ φύσει τινὶ ἀπορρήτῳ λέων ἥδεται καμήλου κρέας καὶ μὴ θεασάμενος φαγεῖν, εἴ ποτε

[1] *Holstein*: Δραυκίων, Ῥακίων.
[2] καλοῦνται.
[3] ἐγχριπτούσας.
[4] *Ges*: Ῥᾶκον.
[5] Ῥ. ἐν αὐτῇ τῇ Κρήτῃ, εἰ.
[6] τελείως.
[7] πικράς.

35. Antenor in his *History of Crete* says that by way of an attack ordained of heaven a swarm of Bees, celebrated as copper-coloured, invaded the city of the people known as Rhaucii [a] and planting their stings in them, inflicted the most grievous pain. So as the people were unable to endure the Bees' attack they quitted their country and went to some other spot where through affection for their 'mother-city,' to use the Cretan idiom, they founded a second Rhaucus, since, even though the god drove them from their home, they could not endure to part utterly with the name. And Antenor states that there are still vestiges of this species of Bee on Mount Ida in Crete; they are not numerous, but they do still exist and are painful to encounter as the former were.

The Rhaucii expelled by Bees

36. The Lion delights to eat the flesh of Camels. Herodotus bears witness to this when he says [7. 125] that Lions fell upon the Camels of Xerxes which were carrying his provisions. But they did no damage to any other living beings, neither beast of burden nor man, so he says. But in his examination of the food of Thracian Lions Herodotus shows little knowledge. The Arabians however, and all whose country is at once the mother and the nurse of Lions, know these things. At any rate I should not be surprised if it were by some mysterious instinct that the Lion, in spite of having never seen one before, delights to eat the flesh of a Camel, if he chances to come across one. For a natural

Lion and Camels

[a] Of the two cities called 'Rhaucus' in Crete one may have lain between Cnossus and Gortyna, while the later foundation was on the eastern slopes of mt Ida.

[8] τροφὴν τήνδε καὶ τήνδε κατὰ τὴν ἡδονήν.

ἐντύχοι·[1] ἡ γὰρ φυσικὴ ἐπιθυμία καὶ τοὺς οὐκ ἰδόντας[2] ἐς τὴν τῆς τροφῆς ἐπιθυμίαν ἀναφλέγει.

37. Ἀμῶντες[3] ἄνθρωποι, τὸν ἀριθμὸν ἑκκαίδεκα,[4] τοῦ ἡλίου καταφλέγοντος δίψει[5] πιεζόμενοι ἕνα ἑαυτῶν ἀπέστειλαν ἐκ πηγῆς γειτνιώσης κομίσαι ὕδωρ. οὐκοῦν ὁ ἀπιὼν τὸ μὲν δρέπανον τὸ ἀμητικὸν διὰ χειρὸς εἶχε, τὸ δὲ ἀρυστικὸν ἀγγεῖον κατὰ τοῦ ὤμου ἔφερεν. ἐλθὼν δὲ καταλαμβάνει ἀετὸν ὑπό τινος ὄφεως ἐγκρατῶς τε καὶ εὐλαβῶς περιπλακέντα.[6] ἔτυχε δὲ ἄρα καταπτὰς μὲν ἐπ' αὐτὸν ὁ ἀετός, οὐ μὴν[7] τῆς ἐπιβουλῆς ἐγκρατὴς ἐγένετο, οὐδὲ (τοῦτο δὴ τὸ Ὁμηρικὸν) τοῖς ἑαυτοῦ τέκνοις τὴν δαῖτα ἐκόμισεν, ἀλλὰ τοῖς ἐκείνου[8] περιπεσὼν ἕρμασιν ἔμελλεν οὐ μὰ Δί' ἀπολεῖν ἀλλ' ἀπολεῖσθαι. εἰδὼς οὖν ὁ γεωργὸς[9] τὸν μὲν εἶναι Διὸς ἄγγελον καὶ ὑπηρέτην, εἰδώς γε μὴν κακὸν θηρίον τὸν ὄφιν, τῷ δρεπάνῳ τῷ προειρημένῳ διακόπτει τὸν θῆρα, καὶ μέντοι καὶ τῶν ἀφύκτων ἐκείνων εἱργμῶν τε καὶ δεσμῶν τὸν ἀετὸν ἀπολύει. ὁδοῦ μέντοι πάρεργον τῷ ἀνδρὶ ταῦτα καὶ δὴ διεπέπρακτο, ἀρυσάμενος δὲ τὸ ὕδωρ ἧκε, καὶ πρὸς τὸν οἶνον κεράσας ὤρεξε πᾶσιν, οἱ δὲ ἄρα ἔπιον[10] καὶ ἀμυστὶ καὶ πολλὰς ἐπὶ τῷ ἀρίστῳ. ἔμελλε δὲ καὶ αὐτὸς ἐπ' ἐκείνοις πίεσθαι· ἔτυχε γάρ πως[11] ὑπηρέτης κατ' ἐκεῖνο τοῦ καιροῦ ἀλλ' οὐ συμπότης ὤν. ἐπεὶ δὲ τοῖς χείλεσι τὴν κύλικα προσῆγεν, ὁ σωθεὶς ἀετὸς

[1] *Jac*: ἐντύχῃ MSS, *followed by* καὶ πρῶτον *del. H.*
[2] *Ges*: εἰδότας.
[3] *Reiske*: ἀλοῶντες.
[4] τὸν ἀρ. ἑκ. *in* MSS *after* πιεζόμενοι.
[5] δίψῃ.
[6] περιπλακέντα ἀποπνιγόμενον ἤδη.

appetite kindles the desire for a specific food even in those who have never seen it before.

An Eagle's gratitude

37. Some men, sixteen in all, reaping beneath a blazing sun and oppressed with thirst, despatched one of their number to fetch water from a spring near by. So the man went off with his reaping sickle in his hand and the pail for drawing water over his shoulder. On arrival he found an Eagle wrapped in the powerful grip of a snake. The Eagle happened to have swooped upon it but failed to achieve its design and could not, as in Homer [*Il.* 12. 219], carry their food to its young ones. Instead of that it fell into the serpent's coils and so far from killing was likely to be killed. So the husbandman knowing that the Eagle was the messenger and minister of Zeus and knowing too that the snake was an evil brute, cut the beast in two with the aforesaid sickle and released the Eagle from that inescapable grip that bound it. And yet all this was performed as a secondary purpose of the man's journey, and after drawing the water he returned, mixed it with the wine, and dispensed it to the company, whereupon they drained their cups at a single draught many times over at their luncheon. The man himself was intending to drink after the others, for he happened at that time to be rather their servant than their fellow at table. But when he raised the cup to his lips, the Eagle which

---

[7] *οὐ μὴν κρείττων γενόμενος οὐδέ.*

[8] *ταῖς ἐκείνου σπείραις.*

[9] *γεωργὸς ἢ ἀκούων.*

[10] *ἐξέπιον.*

[11] *πως after ἐκεῖνο in* MSS.

ζωάγρια ἐκτίνων οἱ καὶ κατὰ τύχην ἀγαθὴν ἐκείνου ἔτι διατρίβων περὶ τὸν χῶρον ἐμπίπτει τῇ κύλικι, καὶ ἐκταράττει αὐτήν, καὶ ἐκχεῖ τὸ ποτόν. ὁ δὲ ἠγανάκτησεν (καὶ γὰρ ἔτυχε διψῶν) καὶ λέγει 'εἶτα μέντοι σὺ ἐκεῖνος ὤν' (καὶ γὰρ τὸν ὄρνιν ἐγνώρισε) 'τοιαύτας ἀποδίδως τοῖς σωτῆρσι τὰς χάριτας; ἀλλὰ πῶς ἔτι ταῦτα καλά; πῶς δ' ἂν καὶ ἄλλος σπουδὴν καταθέσθαι θελήσειεν [1] ἔς τινα αἰδοῖ Διὸς χαρίτων ἐφόρου τε καὶ ἐπόπτου;' καὶ τῷ μὲν ταῦτα εἴρητο, καὶ ἐφρύγετο· ὁρᾷ δὲ ἐπιστραφεὶς τοὺς πιόντας ἀσπαίροντάς τε καὶ ἀποθνήσκοντας. ἦν δὲ ἄρα ὡς συμβαλεῖν ἐμημεκὼς ἐς τὴν πηγὴν ὁ ὄφις καὶ κεράσας αὐτὴν τῷ ἰῷ. ὁ μὲν οὖν ἀετὸς τῷ σώσαντι ἰσότιμον τῆς [2] σωτηρίας ἀπέδωκε τὸν μισθόν. λέγει δὲ Κράτης ὁ Περγαμηνὸς ὑπὲρ τούτων καὶ τὸν Στησίχορον ᾄδειν ἔν τινι ποιήματι οὐκ ἐκφοιτήσαντί [3] που ἐς πολλούς, σεμνόν τε καὶ ἀρχαῖον ὥς γε κρίνειν ἐμὲ τὸν μάρτυρα ἐσάγων.

38. Ἐν θαλάττῃ τῇ Κασπίᾳ [4] εἰσὶ νῆσοί φασι, καὶ γίνονται ἐν αὐταῖς ὄρνιθες διάφοροι μὲν καὶ ἄλλοι, εἷς δὲ εἰληχὼς τοιαύτην ἰδιότητα. εἶναι μὲν γὰρ κατὰ τοὺς χῆνας τὸ μέγεθός φασιν αὐτόν,[5] πόδας δὲ ἔχειν [6] ἐμφερεῖς γεράνῳ. καὶ τὰ μὲν νῶτα κοκκοβαφῆ καὶ σφόδρα ἀκράτως, τὰ δὲ ὑπὸ τὴν γαστέρα πράσινα· τὴν δέρην δὲ λευκὸν εἶναι, καί τινας καὶ ῥανίδας οἱονεὶ διασπαρείσας κροκοειδεῖς ἔχειν. μῆκος δὲ εἰληχέναι οὐ μεῖον

[1] *Bernhardy*: θελήσει.
[2] ἰσότιμον τῆς] ἀμοιβὴν τῆς ἰσοτίμου.
[3] *Ges*: εἰσφοιτήσαντι.

he had rescued and which, fortunately for him, was still lingering about the spot, to reward him for saving its life swooped upon the cup, dashed it from his hand, and spilt the drink. The man was annoyed, for he was indeed thirsty, and exclaimed 'So it is you' (for he recognised the bird), 'yet this is how you thank those who saved your life! I ask you, is this fair? And how should a man hereafter want to do a good turn to another from respect for Zeus who marks and watches over kind actions?' Such were his words and he felt parched. But turning round he saw the men who had drunk gasping and at the point of death. It seems, at a guess, that the snake had vomited into the spring and mingled the water with its poison. And so the Eagle repaid its saviour by similarly saving his life.

Crates of Pergamum says that Stesichorus also sings of this in a poem which has not, I think, reached a wide public, and he has cited, in my opinion, a weighty witness from ancient times.

38. In the Caspian Sea, they say, there are islands in which there occur birds of different species, but one species has this peculiarity. It is said to be the size of a goose, though its legs resemble those of a crane. Its back is an intense scarlet, while its belly below is green. The neck is white and has saffron-coloured dots as it were sprinkled over it. It

A bird from the Caspian Sea

[4] *Reiske*: τῆς Κασπίας.

[5] αὐτὸν ἀλλὰ καὶ τὸν εὐγενῆ χῆνα καὶ τοὺς ἄλλους ἰδεῖν ὑπερέχει.

[6] ἔχει.

πήχεων δύο, κεφαλὴν δὲ ἄρα λεπτήν τε ἅμα καὶ μακράν, τὸ ῥάμφος μέλαν· φωνήν τε ἀφιέναι ἐμφερῆ τοῖς βατράχοις.

39. Ἐν τῇ Πρασιακῇ [1] χώρᾳ (Ἰνδῶν δὲ αὕτη ἐστί) Μεγασθένης φησὶ πιθήκους εἶναι τῶν μεγίστων κυνῶν οὐ μείους, ἔχειν δὲ οὐρὰς πήχεων πέντε· προσπεφυκέναι δὲ ἄρα αὐτοῖς καὶ προκόμια καὶ πώγωνας καθειμένους καὶ βαθεῖς· καὶ τὸ μὲν πρόσωπον πᾶν εἶναι λευκούς, τὸ σῶμα δὲ μέλανας ἰδεῖν, ἡμέρους δὲ καὶ φιλανθρωποτάτους, καὶ τὸ τοῖς ἀλλαχόθι πιθήκοις συμφυὲς οὐκ ἔχειν τὸ κακόηθες.

40. Ἐν Ἰνδοῖς ἐστι χώρα περὶ τὸν Ἀσταβόραν [2] ποταμὸν ἐν τοῖς καλουμένοις Ῥιζοφάγοις. κατὰ τὴν τοῦ Σειρίου τοίνυν ἐπιτολὴν κωνώπων νέφη τινὰ ἐκπληκτικὰ καὶ οἷα [3] τὸν ἀέρα καταλαβεῖν ἐπιφανέντα εἶτα μέντοι ἐλύπησε πολλά. [4] κατὰ μέντοι τὴν λίμνην τὴν καλουμένην Ἀορατίαν [5] (Ἰνδῶν δὲ ἄρα καὶ αὕτη· πλησίον ⟨δέ⟩ [6] ἐστι τοῦ προειρημένου ποταμοῦ) τοῦτο [7] μὲν τὸ θηρίον τὸν κώνωπα ἐπιπολάζειν· ἔρημον δὲ καὶ εἶναι τὸν χῶρον καὶ καλεῖσθαι. τὴν δὲ αἰτίαν ἐκείνην Ἰνδοί φασιν οἱ κύκλῳ περιοικοῦντες, τὸν χῶρον τὸν προειρημένον οὐκ ἄνωθεν οὐδὲ ἐξ ἀρχῆς ἄγονον ἀνθρώπων γενέσθαι, σκορπίους δὲ ἐπιπολάσαι πλῆθος ἄμαχον, καὶ φαλαγγίων τινὰ

[1] *Schn*: Πραξιακῇ.
[2] *Gron*: Ἀσταβάραν, Ἐστα- *etc.* MSS, Ἀσταβόρραν *H.*
[3] *Jac*: οἷά τινα.
[4] τινα πολλά.
[5] Ἀορρατίαν L.
[6] ⟨δέ⟩ *add. H.*
[7] καὶ τοῦτο.

measures not less than two cubits; its head is narrow and long, its beak black, and its cry is like a frog's.[a]

Monkeys of Prasiaea

39. Megasthenes says that in the country of the Prasii (this is a part of India) there are Monkeys as large as the largest hounds, and that they have tails five cubits long. They have also forelocks and thick, pendent beards. Their face is completely white, whereas their body is black, and they are tame and very fond of human beings, and they have not the naturally mischievous temperament of Monkeys elsewhere.[b]

Population expelled by Mosquitoes, Scorpions, and Spiders

40. In India there is a region that lies about the river Astaboras[c] in the country of the *Rhizophagi* (root-eaters), as they are called. About the time of the rising of the Dog-star Mosquitoes, which appear in terrifying clouds such as to fill the sky, work widespread damage. It is about the lake called Aoratia[d] (this too is in India, not far from the aforesaid river) that these insects, the Mosquitoes, abound, and the district not only is but is called a desert. And the Indians who live round about give the following reason for it: the aforesaid district was not formerly or originally barren of human beings, but scorpions overran the country in numbers that defied resistance, and in addition there came a

[a] This 'reads like an imaginative account of the Flamingo' (Thompson, *Gk. birds*, p. 131).

[b] This is perhaps the *Presbytis johni* Fisch., Gossen § 239.

[c] The Astaboras (mod. Atbara) rises about Lat. 12, in Abyssinia, and flows N to join the Nile. Ael. appears to regard India as embracing NE Africa.

[d] Perhaps Lake Tana, not far from the sources of the river Atbara.

ἐπιφοιτῆσαι φοράν, φαλαγγίων δὲ ἃ καλοῦσι τετράγναθα. τεκεῖν δὲ ἄρα τὰ κακὰ ταῦτά φασιν ἀέρων[1] πονηρίαν. καὶ τέως[2] μὲν ἐγκαρτερεῖν τοὺς ἐκεῖθι τλημόνως τοῦ κακοῦ τὴν προσβολὴν καὶ φιλοπόνως ὑπομείναντας· ἐπεὶ δὲ ἦν παντελῶς ἄμαχον, καὶ διεφθείροντο ἡλικία πᾶσα, εἶτα μέντοι τελευτῶντες ὑπ' ἀπορίας τοῦ ἀμύνασθαι τὴν καταβολὴν τῆς ἐπιδημίας[3] τῆς προειρημένης ἐξέλιπον τὴν χώραν, καὶ ἐρήμην εἴασαν τὴν φίλην καὶ πρότερον ἀρίστην πατρίδα·[4] οὐχ ἁμαρτήσομαι δὲ ἴσως οὐδὲ μητρίδα εἰπὼν τὴν αὐτήν.

41. Μυῶν ἀρουραίων ἐπιφοίτησις καὶ στόλος οὐ μὰ τοὺς θεοὺς χρηστὸς τῶν ἐν Ἰταλίᾳ τινὰς ἐξήλασαν τῆς πατρῴας γῆς, καὶ φυγάδας ἀπέφηναν[5] δίκην αὐχμῶν ἢ κρυμῶν ἤ τινος ἀκαιρίας ὡρῶν ἑτέρας τὰ μὲν λήια κείροντες,[6] διακόπτοντες δὲ τὰς ῥίζας. τῇ Μηδικῇ δὲ ἐπιφοιτήσαντες στρουθῶν[7] φορά, ἐξήλασαν καὶ ἐκεῖνοι τοὺς κατοικοῦντας, διαφθείροντες τὰ σπέρματα καὶ ἀφανίζοντες αὐτά. βάτραχοι δὲ ἡμιτελεῖς πεσόντες ἐξ ἀέρος πολλοὶ Αὐταριάτας[8] μετῴκισαν[9] ἐς χῶρον ἕτερον. καὶ γένος μέντοι Λιβυστινόν, οὗ καὶ ἀνωτέρω μνήμην ἐποιησάμην, ἐπιφοιτησάντων αὐτοῖς λεόντων, εἶτα αὐτοὺς ἀναστῆναι τῆς πατρῴας γῆς ἐξενίκησαν.

[1] *Reiske*: ὄμβρων.
[2] *Jac*: πως.
[3] *Jac*: ἐπιμελείας.
[4] *Gow*: τὴν φίλην πρότερον καὶ πατρίδα ἀρίστην *corrupt H*.

crop of certain spiders which they call 'four-jawed.' Now they say that these plagues tainted the air. For a time the inhabitants courageously held out against the invading plague and stood their ground energetically, but when resistance became utterly impossible and all their men-folk were destroyed, then at length, being at their wits' end how to defend themselves against the attack of the aforesaid visitants, they abandoned the country, and left their cherished and once most kindly fatherland a desert. Perhaps I shall not be wrong if I say that it was not even their 'motherland.'[a]

41. The incursion of an army of Fieldmice, far from beneficial, I can assure you, drove certain people in Italy from their native country, and made them exiles, as a drought or frost or some other unseasonable event might have done, by shearing away the ears of corn and cutting through the roots. And a horde of Sparrows invaded Media and drove out the inhabitants by ruining and destroying the seeds. And half-formed Frogs fell in quantities from the sky causing the Autariatae[b] to emigrate to some other place. Further, a tribe in Libya, whom I have mentioned earlier on,[c] were compelled by an invasion of Lions to quit their native country.

A plague of Fieldmice

of Sparrows

of Frogs

of Lions

[a] Cp. Plato, *Rep.* 575 D.
[b] A tribe in Mysia.
[c] Ch. 27.

[5] ἀπέφηναν λυμαινόμενοι καὶ λήια καὶ φυτά.
[6] λήια κείροντες] διακείροντες.
[7] *Jac* : τύθων.
[8] *Schn* : Αὐτωριάτας.
[9] *Cas* : Ἰνδῶν μετῴκισαν.

42. Ἐν τῇ Βαβυλωνίᾳ γῇ γίνονται μύρμηκες, καὶ ἔχουσι τὸ παιδοποιὸν σῶμα ἐς τοὐπίσω μετεστραμμένον, ἀντίως τοῖς ἄλλοις καὶ ἔμπαλιν.

43. Πάρδαλις Καρικὴ καὶ Λυκιακὴ οὐκ ἔστι μὲν θυμική, οὐδὲ οἵα σφόδρα ἁλτικὴ εἶναι, τὸ σῶμα δὲ μακρά· τιτρωσκομένη δὲ καὶ δόρασι καὶ αἰχμαῖς ἀντίτυπός ἐστι, καὶ οὐ ῥᾳδίως τῷ σιδήρῳ εἴκει, τοῦτο δὴ τὸ Ὁμηρικὸν δρῶσα

ἦ ῥά τε καὶ περὶ δουρὶ πεπαρμένη οὐκ ἀπολήγει.

44. Ῥινοκέρωτος δὲ εἶδος γράφειν τρισέωλόν ἐστιν· ἴσασι γὰρ καὶ Ἑλλήνων πολλοὶ καὶ Ῥωμαίων τεθεαμένοι·[1] τὰ δὲ ἴδια αὐτοῦ ⟨τὰ⟩[2] κατὰ τὸν βίον εἰπεῖν οὐ χεῖρόν ἐστιν. ἐπ' ἄκρας τῆς ῥινὸς τὸ κέρας φέρει, ἔνθεν τοι ⟨καὶ⟩[3] κέκληται· καὶ ἔστι μὲν ὀξύτατον ἐπ' ἄκρου, σιδήρῳ δὲ τὸ καρτερὸν αὐτοῦ προσείκασται. ταῖς γε μὴν πέτραις[4] αὐτὸ παρατρίβων εἶτα ἐπιθήσει ἐλέφαντι ὁμόσε ἰών, τὰ δὲ ἄλλα οὐκ ὢν ἀξιόμαχος, διά τε τὸ ἐκείνου ὕψος καὶ τὴν ῥώμην τὴν τοῦ θηρὸς τὴν τοσαύτην. ὕπεισιν οὖν αὐτοῦ τὰ σκέλη, καὶ τὴν νηδὺν ὑποτέμνει τε καὶ ὑποσχίζει τῷ κέρατι· ὁ δὲ οὐ μετὰ μακρὸν[5] ἐκρυέντος οἱ τοῦ αἵματος κατολισθάνει. μάχη δὲ ῥινοκέρωτος πρὸς ἐλέφαντα ὑπὲρ τῆς νομῆς ἐστι, καὶ πολλοῖς γ' ἐλέφασιν[6] ἐντυχεῖν ἐστι τεθνεῶσι τὸν τρόπον τοῦτον. ἐὰν δὲ μὴ φθάσῃ ὁ ῥινόκερως δράσας

[1] οἱ τεθεαμένοι.
[2] ⟨τά⟩ add. H.
[3] ⟨καί⟩ add. H.
[4] ταῖς πέτραις γε μήν.
[5] Ges: μικρόν.
[6] γέ φασιν.

42. In Babylonia there occur Ants[a] with the generative part of their body turned in a backward direction, contrary to its position in Ants elsewhere.

Ants of Babylonia

43. The Leopard of Caria and Lycia is not fierce-tempered, nor of a kind that can leap high, though its body is long. But when wounded with pikes and spears it offers resistance and does not readily yield to the steel, behaving as Homer describes [*Il.* 21. 577]:

The Leopard of Caria

'Yet though pierced with a spear she does not cease.'[b]

44. A description of the shape and appearance of the Rhinoceros would be stale three times over, for there are many Greeks and Romans who know it from having seen it. But there is no harm in describing the characteristics of its way of life. It has a horn at the end of its nose, hence its name. The tip of the horn is exceedingly sharp and its strength has been compared to iron. Moreover it whets it on rocks and will then attack an Elephant in close combat, although in other respects it is no match for it because of the Elephant's height and immense strength. And so the Rhinoceros gets under its legs and gashes and rips up its belly from below with its horn, and in a short space the Elephant collapses from loss of blood. Rhinoceros and Elephant fight for possession of a feeding ground, and one may come across many an Elephant that has met its death in the above manner. If however the Rhinoceros is

The Rhinoceros

fight with Elephant

[a] These are fabulous.
[b] Add 'from her courage,' ἀλκῆς in l. 578.

τοῦτο, ἀλλὰ ὑποτρέχων πως [ὑποπεσόντος] [1] πιεσθῇ, περιβαλλόμενος [2] τὴν προβοσκίδα κατέχει καὶ πρὸς ἑαυτὸν ἕλκει, ἐμπίπτων δὲ τοῖς κέρασι κατακόπτει ὡς πελέκεσιν. εἰ γὰρ καὶ φορίνην ὁ ῥινόκερως ἔχει στερεὰν καὶ δυσδιακόντιστον, ἀλλ' ἡ βία τοῦ ἐμπίπτοντος μάλα καρτερά.

45. Ἀγριώτατον δὲ ἄρα ἦσαν τῶν ζῴων οἱ τῶν Αἰθιόπων ταῦροι οἱ [3] καλούμενοι σαρκοφάγοι. καί εἰσι μὲν τὸ μέγεθος τῶν παρὰ τοῖς Ἕλλησι διπλασίους, ὤκιστοι δὲ τὸ τάχος. εἰσὶ ⟨δὲ⟩ [4] πυρρότριχες, γλαυκοὶ τοὺς ὀφθαλμούς, καὶ ὑπὲρ τοὺς λέοντας οὗτοι. τὰ κέρατα δὲ τὸν μὲν ἄλλον χρόνον κινοῦσιν ὡς καὶ τὰ ὦτα, ἐν δὲ ταῖς μάχαις ἐγείρουσιν [5] αὐτὰ καὶ ἀναστήσαντες ἰσχυρῶς, [6] εἶτα οὕτω μάχονται· τὰ δὲ οὐ κλίνεται [7] ὑπὸ τοῦ θυμοῦ ἀνεστῶτα, φύσει ναὶ μὰ Δία θαυμαστῇ. ἄτρωτοι δέ εἰσι καὶ λόγχαις καὶ βέλει παντί· ὁ γάρ τοι σίδηρος [8] οὐκ εἰσδύεται· φρίξας γὰρ ὁ ταῦρος ἐκβάλλει αὐτὸν μάτην προσπεσόντα. ἐπιτίθεται δὲ καὶ ἵππων ἀγέλαις [9] καὶ θηρίων ἄλλων. οἱ τοίνυν νομεῖς ἐπαρκεῖν ταῖς ἑαυτῶν ἀγέλαις βουλόμενοι τάφρους [10] κρυπτὰς ἐργάζονται βαθείας, καὶ ταύταις αὐτοὺς ἐλλοχῶσιν· οἱ δὲ ὅταν ἐμπέσωσιν, ὑπὸ τοῦ θυμοῦ ἀποπνίγονται. κέκριται δὲ παρὰ τοῖς Τρωγλοδύταις τοῦτο τὸ ζῷον δικαίως ἄριστον· ἔχει μὲν γὰρ λέοντος τὴν ἀλκήν, τὴν δὲ ὠκύτητα ἵππου, ῥώμην δὲ ταύρου, σιδήρου δὲ κρεῖττόν ἐστι.

[1] [ὑποπεσόντος] *del. H*, ὑπ' ἐμπεσόντος *Schn.*
[2] περιβαλλόμενος ⟨ὁ ἐλέφας⟩ *add. Ges.*
[3] καί. [4] ⟨δέ⟩ *add. H.*

not quick enough to do as described but is crushed as it runs underneath, the Elephant slings its trunk round it, holds it fast, drags it towards itself, falls upon it, and with its tusks hacks it to pieces as with axes. For even though the Rhinoceros has a hide so strong that no arrow can pierce it, yet the might of its assailant is extremely powerful.

The flesh-eating Bull of Ethiopia

45. It seems that those Ethiopian Bulls which they call 'flesh-eaters' are the most savage of animals. They are twice the size of Bulls in Greece, and their speed is very great. Their hair is red, their eyes blue-grey, more so than the eyes of lions. In normal times they move their horns as they do their ears, but when fighting they raise them, making them stand strongly up, and so do battle; and once raised in passion owing to some truly wonderful natural cause their horns do not go aslant. No spear, no arrow can wound them: iron, you see, does not penetrate their hide, for the Bull raises its bristles and throws off the weapons showered upon it in vain. And it attacks herds of horses and also wild animals. Accordingly herdsmen who wish to protect their flocks dig deep concealed ditches and by these means ambush the Bulls. And when they fall into these ditches they are choked with rage. Among the Troglodytes this is judged to be the king of beasts, and rightly so, for it possesses the courage of a lion, the speed of a horse, the strength of a bull, and is stronger than iron.

---

[5] *Wesseling*: σπείρουσιν.
[6] αὐτοὺς ἰσχυρῶς.
[7] κλίνονται.
[8] σίδηρος ⟨αὐτοὺς⟩ οὐκ? *H.*
[9] ἀγέλαις καὶ ποίμναις.
[10] τάφρους αὐταῖς.

46. Λέγει Μνασέας ἐν τῇ Εὐρώπῃ Ἡρακλέους [1] ἱερὸν εἶναι καὶ τῆς τούτου γαμετῆς, ἣν ᾄδουσιν οἱ ποιηταὶ τῆς Ἥρας θυγατέρα. οὐκοῦν ἐν τῷ τοῦ νεὼ περιβόλῳ τιθασοὺς ὄρνιθας τρέφεσθαι πολλούς φησι, καὶ τοῦτο δέ, εἶναι ἀλεκτρυόνας τε καὶ ἀλεκτορίδας τούσδε τοὺς ὄρνεις.[2] νέμονται δὲ καὶ συναγελάζονταί σφισι κατὰ γένος, καὶ δημοσίας ἔχουσι τροφάς, καὶ τῶν θεῶν ἀναθήματά εἰσι τῶν προειρημένων. αἱ μὲν οὖν ἀλεκτορίδες ἐν τῷ τῆς Ἥβης [3] νέμονται νεῷ, οἱ δὲ ἐν Ἡρακλέους οἱ τῶνδε γαμέται. ὀχετὸς δὲ ἄρα ἀενάου [4] τε καὶ καθαροῦ ὕδατος διαρρεῖ μέσος. θῆλυς μὲν οὖν οὐδὲ εἷς ἐς Ἡρακλέους πάρεισιν· οἱ δὲ ἄρρενες, ὅταν ᾖ καιρὸς ἐπιθόρνυσθαι, ὑπερπέτονται τὸν ὀχετόν, εἶτα ὁμιλήσαντες ταῖς θηλείαις ἐπανίασιν ἐς τὰ σφέτερα αὖθις παρὰ [5] τὸν θεὸν ᾧ λατρεύουσι, καθηράμενοι τῷ διείργοντι τὰ γένη τῶν ὀρνίθων ὕδατι. τίκτεται οὖν, οἷα εἰκός, πρῶτον μὲν [6] ἐκ τῆς ὁμιλίας ᾠά· εἶτα ὅταν αὐτὰ θάλψωσι καὶ ἐκλέψωσι τοὺς νεοττοὺς αἱ μητέρες, τοὺς υἱεῖς οἱ ἄρρενες παρ᾽ ἑαυτοὺς ἄγουσι καὶ ἐκτρέφουσιν. αἱ δὲ ὄρνεις,[7] ἐκείναις [8] ἔργον ἐστὶ τρέφειν τὰς θυγατέρας.

[1] Εὐρώπῃ Διὸς Ἡ.
[2] ὄρνις.
[3] *Ges*: Ἥρας.
[4] ἀεννάου.
[5] *Abresch*: περί.
[6] πρῶτα.
[7] ὄρνις.
[8] κἀκείναις.

Cockerels and Hens in the temples of Heracles and Hebe

46. Mnaseas in his work *On Europe* says that there is a temple to Heracles and to his spouse whom poets celebrate as the daughter of Hera. Now they say that in the precincts of these temples a large number of tame birds are kept, adding that these birds are cockerels and hens. They feed and consort together according to their sex, are fed at the public expense, and are consecrated to the aforesaid gods. The hens feed in the temple of Hebe while their mates feed in the temple of Heracles. And a never-failing channel of clear water flows between them. Now on the one hand not a single hen ever appears in the temple of Heracles. On the other hand at the season of mating the cockerels fly across the channel and after consorting with the hens return again to their own quarters at the side of the god whom they serve, cleansed by the water that separates the sexes. And so to begin with, as a natural result of this union eggs are laid; later on when the hens have warmed them and hatched the chicks, the cockerels carry off the male birds and rear them, while the hens make it their business to rear their daughters.

EPILOGUE

ΕΠΙΛΟΓΟΣ

Ὅσα μὲν οὖν σπουδή τε ἐμὴ καὶ φροντὶς καὶ πόνος καὶ ἐς τὸ πλέον μαθεῖν καὶ ἐν τοῖσδε ἡ γνώμη προχωροῦσα ἀνίχνευσέ τε καὶ ἀνεῦρε, δοκίμων τε ἀνδρῶν καὶ φιλοσόφων ἀγώνισμα θεμένων τὴν ἐπ' αὐτοῖς ἐμπειρίαν, καὶ δὴ λέλεκταί μοι, ὡς οἷόν τε ἦν εἰπεῖν, μὴ παραλείποντι ἅπερ ἔγνων μηδὲ βλακεύοντι, ὡς ἀλόγου τε καὶ ἀφώνου ἀγέλης ὑπεριδόντι καὶ ἀτιμάσαντι, ἀλλὰ κἀνταῦθα ἔρως με σοφίας ὁ σύνοικός τε καὶ ὁ συμφυὴς ἐξέκαυσεν. οὐκ ἀγνοῶ δὲ ὅτι ἄρα [1] καὶ τῶν ἐς χρήματα ὁρώντων ὀξὺ καὶ τεθηγμένων ἐς τιμάς τε καὶ δυνάμεις τινὲς καὶ πᾶν τὸ φιλόδοξον δι' αἰτίας ἕξουσιν, εἰ τὴν ἐμαυτοῦ σχολὴν κατεθέμην ἐς [2] ταῦτα, ἐξὸν καὶ ὠφρυῶσθαι καὶ ἐν ταῖς αὐλαῖς ἐξετάζεσθαι καὶ ἐπὶ μέγα προήκειν πλούτου. ἐγὼ δὲ ὑπέρ τε ἀλωπέκων καὶ σαυρῶν καὶ κανθάρων καὶ ὄφεων καὶ λεόντων καὶ τί δρᾷ πάρδαλις καὶ ὅπως πελαργὸς φιλόστοργον καὶ ὅτι ἀηδὼν εὔστομον καὶ πῶς φιλόσοφον [3] ἐλέφας καὶ εἴδη ἰχθύων καὶ γεράνων ἀποδημίας καὶ δρακόντων φύσεις καὶ τὰ λοιπὰ ὅσα ἥδε ἡ συγγραφὴ πεπονημένως ἔχει καὶ φυλάττει, περιέρχομαι· ἀλλὰ οὔ μοι φίλον

[1] ἄρα ὅτι. [2] καὶ εἰς.
[3] θυμόσοφον *Ges.*

EPILOGUE

All that my own application, reflection, and labour to augment my knowledge, all that the advance of understanding in these studies (as eminent scholars vied with each other in acquainting themselves with these matters) have traced out and discovered—all this I have now set down to the best of my ability. I have not through idleness omitted anything that I have learnt, as though animals, void of reason and of speech, were beneath my notice and to be despised, but here as elsewhere I have been fired by that love of knowledge which in me is inherent and innate. I am well aware that among those who keep a sharp look-out for money, or who are keen in the pursuit of honours and influence and all that brings reputation, there are some who will blame me for devoting my leisure to these studies, when I might have given myself airs and appeared in palaces and attained to considerable wealth. I however occupy myself with foxes and lizards and beetles and snakes and lions, with the habits of the leopard, the affectionate nature of the stork, the melodiousness of the nightingale, the sagacity of the elephant, and the shapes of fishes and the migrations of cranes and the various species of serpents, and so on—everything which in this account of mine has been carefully got together and observed. But it is no pleasure to me to be numbered among your rich men and to be compared with them. But if I exert myself and desire some-

σὺν [1] τοῖσδε τοῖς πλουσίοις ἀρίθμεῖσθαι καὶ πρὸς ἐκείνους ἐξετάζεσθαι, εἰ δὲ ὧν καὶ ποιηταὶ σοφοὶ καὶ ἄνδρες φύσεως ἀπόρρητα ἰδεῖν τε ἅμα καὶ κατασκέψασθαι δεινοὶ καὶ συγγραφεῖς τῆς [2] πείρας ἐς τὸ μήκιστον προελθόντες ἑαυτοὺς ἠξίωσαν, τούτων τοι καὶ ἐμαυτὸν ἁμωσγέπως ἕνα πειρῶμαι ἀριθμεῖν καὶ ἐθέλω, δῆλον ὡς ἀμείνων ἐμαυτῷ σύμβουλός εἰμι τῆς ἐξ ἐκείνων κρίσεως. βουλοίμην γὰρ ἂν μάθημα ἓν γοῦν πεπαιδευμένον περιγενέσθαι μοι ἢ τὰ ᾀδόμενα τῶν πάνυ πλουσίων χρήματά τε ἅμα καὶ κτήματα. καὶ ὑπὲρ μὲν τούτων ἱκανὰ νῦν. οἶδα δὲ ὅτι καὶ ἐκεῖνα οὐκ ἐπαινέσονταί τινες, εἰ μὴ καθ' ἕκαστον τῶν ζῴων ἀπέκρινα μου [3] τὸν λόγον, μηδὲ ἰδίᾳ τὰ ἑκάστου εἶπον ἀθρόα, ἀνέμιξα δὲ καὶ τὰ ποικίλα ποικίλως, καὶ ὑπὲρ πολλῶν διεξῆλθον, καὶ πῇ μὲν ἀπέλιπον τὸν περὶ τῶνδε λόγον τῶν ζῴων, πῇ δὲ ὑπέστρεψα ὑπὲρ τῆς αὐτῶν φύσεως ἕτερα εἴρων. ἐγὼ δὲ πρῶτον μὲν τὸ ἐμὸν ἴδιον οὔκ εἰμι τῆς ἄλλου κρίσεώς τε καὶ βουλήσεως δοῦλος, οὐδέ φημι δεῖν ἕπεσθαι ἑτέρῳ, ὅποι μ' ἂν ἀπάγῃ· δεύτερον δὲ τῷ ποικίλῳ τῆς ἀναγνώσεως τὸ ἐφολκὸν θηρῶν καὶ τὴν ἐκ τῶν ὁμοίων βδελυγμίαν ἀποδιδράσκων, οἱονεὶ λειμῶνά τινα ἢ στέφανον ὡραῖον ἐκ τῆς πολυχροίας, ὡς ἀνθεσφόρων τῶν ζῴων τῶν πολλῶν, ᾠήθην δεῖν τήνδε ὑφᾶναί τε καὶ διαπλέξαι τὴν συγγραφήν. εἰ δὲ τοῖς θηρατικοῖς καὶ ἓν ζῷον εὑρεῖν δοκεῖ πως εὐερμία, ἀλλὰ τό γε τῶν τοσούτων οὐ τὰ ἴχνη, οὐδὲ τὰ μέλη συλλαβεῖν ἐγώ φημι γενναῖον, ⟨ἀλλ'⟩ [4] ὁπόσα ἡ φύσις ἔδωκέ τε αὐτοῖς καὶ ὅσων ἠξίωσεν ἀνιχνεῦσαι. τί πρὸς ταῦτα

[1] ἐν ? *H.* [2] *Schn* : ἐκ. [3] μοι. [4] ⟨ἀλλ'⟩ *add. Ges.*

how to count myself one of that company to which learned poets, and men clever at detecting and probing the secrets of nature, and writers who have attained the greatest experience, claim to belong, it is obvious that my own counsel is better than the judgment of those men. For I would rather attain to expert knowledge in at least one branch than to the belauded riches and possessions of your wealthiest men. So enough of this for the present.

I am aware too that some will express disapproval because I have not in my discourse kept each creature separate by itself, and have not said in its own place all that is to be said about each, but have mixed the various kinds like a varied pattern in the course of describing a great number, at one point dropping the narrative about such-and-such animals, at another going back and stringing together other facts about their nature. Now in the first place, speaking for myself, I am no slave to another's judgment and will: I maintain that it is not my duty to follow another's lead wherever it may take me. And in the second place, since I was aiming to attract through the variety of my reading matter, and since I flee from the tedium arising from monotony, I felt that I ought to weave the tissue of this narrative of mine so as to resemble a meadow or a chaplet beautiful with its many colours, the many creatures, as it were, contributing their flowers. And although hunters regard the finding of even one animal as a piece of luck, I maintain that there is nothing splendid in finding the tracks or capturing the bodies of such a multitude of animals, whereas to track down the faculties which nature has seen fit to bestow upon them—that is splendid.

Κέφαλοί τε καὶ Ἱππόλυτοι καὶ εἴ τις ἐν ὄρεσιν ἀγρίοις θηρία μετελθεῖν δεινὸς ἕτερος ἢ αὖ πάλιν τῶν ἐν ὑδροθηρίαις δεινῶν [1] Μητρόδωρος ὁ Βυζάντιος ἢ Λεωνίδης ὁ τούτου παῖς ἢ Δημόστρατος ἢ ἄλλοι τινὲς θηραταὶ ἰχθύων οἱ δεινότατοι, πολλοὶ ναὶ μὰ Δία; καὶ γραφικοὶ δὲ ἄνδρες, μέγα αὐτοὺς φρονεῖν ἀνέπειθεν ἢ ἵππος γραφεὶς κάλλιστα, ὡς Ἀγλαοφῶντα, ἢ νεβρός, ὡς Ἀπελλῆν, ἢ [2] πλασθὲν βοΐδιον, ὡς Μύρωνα, ἢ ἄλλο τι. εἰ δὲ εἰς τὰ τῶν τοσούτων ἐκδεικνύει καὶ ὑπ' αὐγὰς ἄγει καὶ ἤθη καὶ πλάσεις καὶ σοφίαν καὶ ἀγχίνοιαν καὶ δικαιοσύνην καὶ σωφροσύνην καὶ ἀνδρείαν καὶ στοργὴν καὶ εὐσέβειαν θηράσας, πῶς οὐκ ἤδη καὶ θαυμάσαι ἄξιος; ἥκων δὲ ἐνταυθοῖ τοῦ λόγου καὶ πάνυ ἄχθομαι, εἰ ζῴων μὲν εὐσέβειαν ἀλόγων ᾄδομεν, ἀνθρώπων δὲ ἀσέβειαν [3] ἐλέγχομεν. καὶ τοῦτο μὲν οὐκ ἐνταῦθα ἀποδείξομεν, ἐκεῖνο δὲ προσέτι εἰπεῖν δικαιότατον, οὗπερ οὖν καὶ ἐναρχόμενος τῶνδε τῶν λόγων μνήμην ἐποιησάμην, εἰ ταῦτα εἶπον, ὅσα πάντες, ἢ οἵ γε πλεῖστοι, οὔπω δίκαιον αἰτιᾶσθαι· ζῷα γὰρ αὐτὸς ἄλλα πλάσαι οὐκ ἠδυνάμην, ὅτι δὲ ἔγνων πολλὰ ἐπεδειξάμην. ἤδη μέντοι καὶ εἶπόν τινα, ὧν οὐκ ἄλλος εἶπε διά γε

[1] ὑδροθηρίᾳ οἶδεν (*or* ᾔδει ὡς ἢ) *most* MSS, ἐνυδροθηριῶν M.
[2] ἢ τό.
[3] *Ges*: εὐσέβειαν.

[a] Cephalus and Hippolytus are examples drawn from mythology; C. with his dog Laelaps, which no quarry could escape, joined in the pursuit of the Teumessian Vixen, which none could catch. Dog and Vixen were changed into stone by Zeus.—Hippolytus, son of Theseus and Hippolyte, and a votary of the virgin Artemis, spent his days hunting; see Euripides' *Hippolytus*.

EPILOGUE

What have they to say to this, your Cephaluses and Hippolytuses,[a] and all the others so skilful in the chase upon the wild mountains, or again, among those who were skilled in fishing, Metrodorus of Byzantium, or his son Leonidas, or Demostratus, or any others who were past masters at the catching of fish? And there were many such, god knows! Painters too: the picture of a horse consummately drawn fills them with pride, as it did Aglaophon;[b] or the picture of a fawn, as it did Apelles; or his statue of a calf, as it did Myron;[c] or take any other work of art. But when one man displays and brings forth to the light of day his researches into the habits, the forms, the sagacity, the shrewdness, the justice, the temperance, the bravery, the affection, the filial piety of such a great number of animals, he cannot fail to claim immediate respect. Having reached this point in my discourse I am distressed that while praising the filial piety of unreasoning animals, I have to accuse men of the reverse. I shall not here enlarge on this subject, but this much I have every right to add—indeed I mentioned this point at the beginning of this treatise: it is not fair to censure me for repeating what all, or at any rate most, writers have said already. After all I could not create other animals, though I have given evidence that I have known a great many. Yet I have in fact mentioned certain characteristics

[b] Aglaophon, of Thasos, painter, early in 5th cent. B.C.; father of Polygnotus and Aristophon; was the first to depict *Nike* as winged.

[c] Myron, famous sculptor, of the first half of the 5th cent. B.C.; worked chiefly in bronze. His *Discobolus* and *Athena and Marsyas* survive in copies.

τῆς πείρας τῆσδε αὐτὸς ἐλθών· φίλη δὲ ἡ ἀλήθειά μοι τῇ τε ἄλλῃ καὶ ἐνταῦθα οὐχ ἥκιστα. ὅπως δὲ αὐτὰ εἶπον καὶ σὺν ὅσῳ πόνῳ, τό τε εὐγενὲς τῆς λέξεως ὁποῖον καὶ τῆς συνθήκης, τῶν τε ὀνομάτων καὶ τῶν ῥημάτων τὸ κάλλος, ὁπόσοις ἂν μὴ χρήσωμαι πονηροῖς κριταῖς, ἐκεῖνοι εἴσονται.

which no other writer who has attempted the work on my scale has mentioned. But I prize truth in all spheres, most of all in this, and critics who handle me without malice will realise the quality of my work, the labour it cost, the dignity of its style and composition, and the propriety of the words and phrases employed.

INDEXES:

I. GREEK

References to the passages in which a Greek word occurs are given under the English equivalent in INDEX II, *English.*

INDEX: GREEK

INDEX: GREEK

θύννος tunny
θύον citrus
θώς jackal

ἶβις ibis
ἱερὰ νόσος epilepsy
ἱέραξ falcon, hawk
— θαλάττιος flying-fish
— πελάγιος sea-hawk
ἴκτερος jaundice
ἰκτῖνος kite
ἰξός bird-lime
ἰουλίς rainbow-wrasse
ἰπνός lantern
ἱππόκαμπος sea-horse
ἱππομανές hippomanes
ἵππος horse; ἵ. θήλεια mare
— ποτάμιος hippopotamus
ἵππουρος hippurus (fly)
ἶρις iris
ἰσχάς fig, dried
ἰσχίου πόνος sciatica
ἰτέα willow-tree
ἴυγξ wryneck
ἰχθῦς fish
ἰχνεύμων ichneumon
ἰχώρ serum
ἴωψ minnow

καλαμίνθη νοτερά water-mint
καλαμοδύτης reed-warbler
κάλαμος (i) reed (ii) cane (iii) fishing-rod
καλλίωνυμος star-gazer (fish)
κάμηλος camel
κάμπη caterpillar
κανθαρίς blister-beetle
κάνθαρος (i) beetle (ii) scarab
— θαλάττιος black sea-bream
κάπρος caprus (fish)
κάραβος crayfish
καρίς prawn
καρκινάς hermit-crab
καρκίνος crab
καρτάζωνος cartazonus (= rhinoceros)
καρτόν leek, chopped
κάρυον nut
καρχαρόδοντα, τὰ saw-toothed animals
καστορίδες, αἱ sea-calves
κάστωρ beaver
κατρεύς manâl pheasant
κατώβλεπον gnu
καύσων dipsas
κεγχρηΐς kestrel
κεγχρίς ortolan
κέγχρος millet
κέδρος cedar
κεκρύφαλος reticulum
κεμάς pricket
κεντρίνης (i) dipsas (ii) spiny dog-fish
κεντρίς dipsas
κέρας horn
κεράστης cerastes
κερκίων mynah
κερκόρωνος mynah
κερχνῄς kestrel
κεστρεύς mullet, grey
κέφαλος mullet, grey
κήλας adjutant (bird)
κημός, muzzle, horse's
κῆπος kepos (monkey)
κηρύλος ceryl
κῆρυξ trumpet-shell, whelk
κῆτος sea-monster, cetacean
κηφήν drone
κίγκλος wagtail
κιθαρῳδός harper (fish)
κιννάβαρι vermilion
κιννάμωμον cinnamon
κίρκη circe (bird)
κίρκος falcon
κίττα jay
κιττός ivy
κίχλη (i) thrush (ii) wrasse
κλαδαρόρυγχος clapperbill
κλύσμα clyster
κνίδη nettle
κόγχη mussel, shellfish
κόκκυξ (i) cuckoo (ii) piper (fish)
κολίας Spanish mackerel
κόλλα glue
κολοιός (i) jackdaw (ii) little cormorant
κόνικλος rabbit
κόνυζα fleabane
κορακῖνος crow-fish
κόραξ raven
κορίαννον coriander
κοροκόττας corocottas
κορυδαλλός crested lark
κόρυδος lark
κορώνη (i) crow (ii) κ. ἐναλία shear-water, little Manx
κόσκινον sieve
κόσσυφος (i) blackbird (ii) κ. θαλάττιος wrasse
κότινος olive, wild
κοττάνη, *see* 12. 43*n*.
κοχλίας (i) snail (ii) κ. θαλάττιος sea-snail

INDEX: GREEK

INDEX: GREEK

II. ENGLISH

INDEX: ENGLISH

INDEX: ENGLISH

INDEX: ENGLISH

Blindness, asp causes **3**. 33; lizard cured of **5**. 47
Blister-beetle (κανθαρίς) **9**. 39
Blood-letter (αἱμόρρους), snake, effects of bite **15**. 13; *also* **15**. 18
Blow-hole (αὐλός), of dolphin and whale **2**. 52; (φυσητήρ) of porpoise **5**. 4
Blue-grey fish (γλαῦκος), paternal instincts **1**. 16
Blue Tit (αἴγιθος), and ass **5**. 48
Boar, Wild (ὗς ἄγριος), *see* Pig, Wild
Boasting, Greek characteristic **5**. 49
Boccalis (βώκκαλις), bird **13**. 25
Bocchoris, king of Egypt, and Mneuis **11**. 11; *also* **12**. 3
Boeotia, no moles in **17**. 10; partridges in **3**. 35
Bones, as fuel **12**. 34; of lion **4**. 34
Boreas, sons of **11**. 1; *also* **5**. 45
Bosphorus, Thracian, crabs in **7**. 24; pearl-oysters in **15**. 8
Box-tree (πύξος), honey from **5**. 42
Boy, loved by asp **4**. 54; — dolphin **6**. 15; **8**. 11; — goose **5**. 29; — -horse **6**. 44; — jackdaw **1**. 6; — snake **6**. 63
Brahmins (Βραχμᾶνες), as historians **16**. 20; and hoopoe **16**. 5; and parrots **13**. 18
Bream, *see* Black Sea-bream
Brenthus (βρένθος), bird, and sea-mew **5**. 48
Britain, pearl-oysters from **15**. 8
Britannicus, son of emperor Claudius, poisoned **5**. 29
Bronze (χαλκός) **16**. 6
Bubastus, fishes at **12**. 29
Bucephala, in India **16**. 3
Bucephalus, horse of Alexander the Great **6**. 44
Buck-thorn (ῥάμνος), charm against sorcery **1**. 35
Budini, their sheep **3**. 32; **16**. 33
Bull (ταῦρος), angry **4**. 48; blood of, medicinal **11**. 35; as body-guard **7**. 46; of Chaonia **12**. 11; flesh-eating bulls of Ethiopia **17**. 45; and golden eagle **2**. 39; horns **2**. 20; hunted **17**. 26; and lion **5**. 48; lungs as bait **14**. 25; performing **7**. 4; and raven **2**. 51; **5**. 48; self-training **6**. 1; wild **15**. 15; and wolf **5**. 19. *See also* Apis, Cattle, Mneuis, Onuphis
Bumble-bee [?] (ἀνθηδών) **15**. 1
Buprestis (βούπρηστις), kills cows **6**. 35
Burial customs, of ants **6**. 43
Busiris, people of **10**. 28
Bustard (ὠτίς), and dogs **5**. 24; and fox **6**. 24; and horses **2**. 28
Butter (βούτυρον) **9**. 54; **13**. 7
Buzzard (τριόρχης) **12**. 4
Byzantium, dolphins at **8**. 3; whelks at **7**. 32

Cabbage (κράμβη) **9**. 39; kind of sea-weed **13**. 3
Cabbage-caterpillar (κραμβίς) **9**. 39
Cabiri, gods of Samothrace **15**. 23
Caecinus, river **5**. 9*n*.
Caeneus, changes his sex **1**. 25
Calf (μόσχος), flesh of, for fishes **12**. 1; a freak **11**. 40; sacrificed to Dionysus **12**. 34; and wolves **8**. 14
Calingae, Indian people **16**. 18
Callias, Athenian **3**. 42
Callimachus, Athenian Polemarch **7**. 38
Calypso **15**. 28
Cambyses, king of Persia, outrages in Egypt **10**. 28
Camel (κάμηλος), anatomy of **10**. 3; of the Caspii **17**. 34; castrated **4**. 55; drinks muddy water **17**. 7; and horses **3**. 7; **11**. 36; and incest **3**. 47; lions eat **17**. 36; longevity of **4**. 55; mating of **6**. 60; races **12**. 34; *also* **5**. 50(i)
Campylinus, river **3**. 4
Cane (κάλαμος), wine from **13**. 8
Cannibalism, *see* Fish, Hippopotamus, Pig, Wolf
Canobus, helmsman of Menelaus **15**. 13
Cappadocia, bees in **5**. 42
Caprus (κάπρος), fish **10**. 11
Car, son of Zeus and Creta **12**. 30
Caria, fishing in **13**. 2; leopards in **17**. 43; mercenaries from **12**. 30
Cariscus, river **16**. 33
Carmania, dogs of **3**. 2
Carmel, mt **5**. 56
Carp (κυπρῖνος) **14**. 23, 26
Cartazonus (καρτάζωνος), 'Indian unicorn' (rhinoceros) **16**. 20
Caspian Sea, birds on islands in **17**. 38; sturgeon in **17**. 32
Caspii, foxes among the **17**. 17; their goats and camels 34; horses and cattle 17; plagued by mice 17; and

INDEX: ENGLISH

INDEX: ENGLISH

INDEX: ENGLISH

Toes, animals with **11**. 37
Torpedo (*νάρκη*), fish, flesh of, as depilatory **13**. 27; numbing action **1**. 36; **9**. 14; and silphium **5**. 37; *also* **11**. 37; **14**. 3
Tortoise (*χελώνη χερσαία*), causes Aeschylus's death **7**. 16; in Egypt and Arabia **17**. 3; in India **16**. 14; male and female **15**. 19; and partridge **4**. 5; its remedy against poison **3**. 5; **6**. 12; shell of **14**. 17; and viper **6**. 12; *also* **11**. 37
— -stone (*χελωνία*) **4**. 28
Touchstone (*βάσανος*), for gold **3**. 13
Trachis **9**. 27
Transmigration of soul **12**. 7
Trapezus, in Pontus, honey in **5**. 42
Tree, evergreen in India **13**. 18; grafting of **9**. 37; killed by sting-ray **2**. 36; **8**. 26; moon-fish and **15**. 4
Tree-medick (*κύτισος*) **6**. 42; **16**. 32
Triplets, Indian sheep and goats bear **4**. 32; lions bear **4**. 34
Triton, at Tanagra **13**. 21; *also* **14**. 28
Troad, mouse worshipped in **12**. 5
Trochus (*τροχός*), fish **13**. 20
Troglodytes (*Τρωγλοδύται*), of Arabia **6**. 10(i); eat snakes **9**. 44; tortoises and vipers in their country **17**. 3; *also* **17**. 45
Troy, country round **10**. 37; tomb of Memnon near **5**. 1; Trojan War **1**. 1; — cause of **11**. 27
Trumpet (*σάλπιγξ*), compared to braying of ass **10**. 28; *also* **12**. 21
— -shell (*κῆρυξ*), in Indian Ocean **16**. 12; (*στρόμβος*) **11**. 37; **15**. 8
Tunny (*θύννος*), blood of, as depilatory **13**. 27; t.-fishing in Euxine **15**. 5, 6; habits **15**. 3; how caught **13**. 16; one-eyed **9**. 42; strength **13**. 17; and whale **16**. 18. *See also* Pelamyd
—, Great (*ὄρκυνος*), *see* Great Tunny
Turbot (*ῥόμβος*) **14**. 3
Turtle (*χελώνη θαλαττία*), and eggs **5**. 52; eyes of, as jewels **4**. 28; in Indian Ocean **16**. 17; *see also* **9**. 41 noto *d*; (*χ. ποταμία*) in Ganges **12** 41; of India **16**. 14
Turtle-dove (*τρυγών*), continence **10**. 33; loquacity of, proverbial **12**. 10; and pigeon, pyrallis, and greenfinch **5**. 48; and pomegranate **6**. 46; and pyrallis **4**. 5; and raven and falcon **6**. 45; and sorcerers **1**. 35; white, sacred **10**. 33; *also* **13**. 25
Twilight (*λυκόφως*) **10**. 26
Twins, she-ass never bears **10**. 28
Typhlops (*τυφλώψ* ' blind-eyes '), lizard **8**. 13
Typho, and ass **10**. 28; as crocodile **10**. 21
Tyrants, expelled **5**. 10; five notorious **5**. 15
Tyro **11**. 18
Tyrrhenian islands **13**. 17

Udad (*αἴξ ἄγριος Λιβυκός*) **14**. 16
Ulcers, from scorpion's dung **8**. 13
Unicorn (*μονόκερως ἵππος, ὄνος*), horn of, protects from poison **3**. 41; *also* **13**. 25; **15**. 15. *See also* Cartazonus
Urine (*οὖρον*), of lynx **4**. 17; none in birds of prey **4**. 20; retention of, cured **11**. 18; snake-bite stops flow **17**. 4; spider's bite stops flow **6**. 26

Vaccaei, Spanish tribe, funeral customs **10**. 22
Veneti, and jackdaws **17**. 16
Vermilion (*κιννάβαρι*) **4**. 46(i)
Vervain (*ἀριστερεών*), charm against sorcery **1**. 35
Vibo, Gulf of, tunny in **15**. 3
Vicetia **14**. 8
Vine (*ἄμπελος*), first cultivated by Icarius **7**. 28; gum of, and crane **6**. 46; a kind of sea-weed **13**. 3; sacred to Heracles **6**. 40
Vinegar (*ὄξος*) **14**. 21
Viper (*ἔχις, ἔχιδνα*), bite of **1**. 54; **10**. 9; in Egypt and Arabia **17**. 3; and human spittle **2**. 24; male and female **1**. 24; mates with moray **1**. 50; **9**. 66; none in Clarus **10**. 49; poison of, and wasps **5**. 16; and tortoise **6**. 12; and young **1**. 24; **15**. 16
Viviparous animals **2**. 52
Vulpanser, *see* Egyptian Goose
Vulture (*γύψ*), and falcon **2**. 42; feathers of, and snakes **1**. 45; feeds on corpses **2**. 46; **10**. 22; and fowls **5**. 50(i); impregnated by wind **2**. 46; killed by perfumes **3**. 7; **4**. 18; no male birds **2**. 46; and pomegranate **6**. 46; and Romulus **10**. 22; sacred to Hera *ib.*; *also* **12**. 4. *See also* Aegypius

INDEX: ENGLISH

III. CLASSIFIED CATALOGUE OF FAUNA, FLORA, ETC.

1. MAMMALS

Acomys *Mus cahirinus*
Ampelus (leopard), perh. *Felis serval*
Antelope *Bubalis mauretanica*
Ass *Equus asinus*
Aurochs *Bos bonasus*
Baboon *Cynocephalus babuin*
Bat *Vespertilio serotinus*
Bear *Ursus arctos*
Beaver *Castor fiber*
Blind-rat *Spalax typhlus*
Boar *Sus scrofa*
Bull *Bos taurus*
Camel *Camelus bactrianus*
Cartazonus *Rhinoceros indicus*
Cat, domestic *Felis domestica*; wild *F. catus*
Chimpanzee *Troglodytes niger*
Corocottas *Hyaena crocuta*
Cow *Bos femina, Vacca*
Deer *Cervus elaphus*
Dog *Canis familiaris*
Dolphin *Delphinus delphis*
Dugong *Halicore dugong*
Elephant *Elephas africanus* and *E. indicus*
Elk, *see* Tarandus
Fawn, *see* Deer
Field-mouse, gen. *Mus silvaticus*
Fox *Canis vulpes*
Gazelle *Antilope dorcas*
Gibbon *Hylobates hulok*
Gnu *Catoblepas gnu*
Goat *Capra hircus*
Gorilla *Troglodytes gorilla*
Hare *Lepus timidus*
Hedgehog *Erinaceus europaeus*
Hippopotamus *H. amphibius*
Horse *Equus caballus*
Hunuman *Semnopithecus entellus*
Hyena *Hyaena striata*
Ibex *Ovis lervia*
Ichneumon *Herpestes ichneumon*
Jackal *Canis aureus*
Jerboa *Dipus aegypticus*
Kepos (monkey) *Cercopithecus pyrrhonotus*
Killer Whale *Orca gladiator*
Leopard *Felis pardus*
Lion *Felis leo*
Lynx *Felis lynx*
Mandrill *Cynocephalus maimon*
Mantichore, fabulous
Marmot *Arctomys bobac*
Marten *Mustela martes*
Mole *Spalax typhlus*
Monkey, *see* Baboon, Chimpanzee, Gibbon, Kepos, Mandrill, Sphinx
Mouse *Mus musculus*
Mule *Mulus*
Onocentaura, *see* Chimpanzee
Otter *Lutra vulgaris*
Pangolin *Maris longicauda*
Panther *Felis pardus panthera*
Pig, gen. *Sus*
Porcupine *Hystrix cristata*
Pricket, *see* Deer
Rabbit *Lepus cuniculus*
Rat *Epimys norwegicus?*
Reindeer *Rangifer tarandus*
Rhinoceros *Rhinoceros indicus*
Roe-deer *Cervus capreolus*
Satyr (monkey), *see* Gibbon
Sea-calf = ? Walrus, *Odobaenas rosmarus*
Seal *Phoca vitulina*
Sheep *Ovis aries*
Shrew-mouse *Sorex araneus*
Sphinx (ape) *Cercopithecus Diana*
Tarandus *Alces malchis?*
Tiger *Felis tigris*
Udad *Ovis lervia*
Unicorn, fabulous
Warthog *Phaeochoerus aethiopicus*
Whale *Balaena biscayensis*
Wolf *Canis lupus*
Yak *Poephagus grunniens*

2. Birds

Adjutant *Leptopilus argala*
Aegypius, perh. Lämmergeier, *q.v.*
Asterias (i) perh. Starling, *q.v.*; (ii) Golden Eagle, *q.v.*
Beccafico *Sylvia atricapilla*
Bee-eater *Merops apiaster*
Blackbird *Turdus merula*
Blue Tit *Parus cyanus*
Boccalis, unidentified
Brenthus, unidentified
Bustard *Otis tarda*
Buzzard *Buteo vulgaris*
Ceryl, unidentified
Chaffinch *Fringilla coelebs*
Chicken, Cock *Gallus gallinaceus*
Cinnamon bird, fabulous
Circe, unidentified
Clapperbill *Pluvianus aegyptius*
Corn-crake[?] *Rallus crex*
Crane *Grus cinereus*
Crested Lark *Alauda cristata*
Crow *Corvus corone*
Cuckoo *Cuculus canorus*
Dabchick *Podiceps ruficollis*
Dove (i) *Crocopus chlorogaster* **16.** 2; (ii) *Columba palumbus*
Dove-killer *Astur palumbarius*
Duck *Anas boschas*
Duck-killer, sp. *Aquila*
Eagle, sp. *Aquila*
Egyptian Goose *Chenalopex aegyptiacus*
Egyptian Plover *Pluvianus aegyptius*
Falcon, gen. *Falco*
Francolin *Tetras francolinus*
Goatsucker *Caprimulgus europaeus*
Golden Eagle *Aquila chrysaetus*
Golden Oriole *Oriolus galbula*
Goldfinch *Carduelis elegans*
Goose *Anser cinereus*
Greenfinch *Fringilla chloris*
Guinea-fowl *Numida meleagris*
Halcyon *Alcedo ispida*
Harpe, perh. Sea-hawk, *q.v.*
Hawk, gen. *Accipiter*
Heron *Ardea cinerea*
Heron, Buff-backed *Ardea bubulcus*
Hoopoe *Upupa epops*
Hornbill, sp. *Bucero*
Ibis White *Tantalus aethiopicus*; Black *Falcinellus igneus*
Jackdaw *Corvus monedula*
Jay *Garrulus glandarius*
Kestrel *Falco tinnunculus*
Kite *Milvus ictinus*
Lämmergeier *Gypaëtus barbatus*
Lark *Alauda arvensis*
Little Cormorant *Phalacrocorax pygmaeus*
Manâl Pheasant *Lophophorus impeyanus*
Marsh Tit *Parus palustris*
Merlin *Falco aesalon*
Mermnus, perh. *Buteo desertorum*
Mynah *Gracula religiosa*
Night-hawk *Strix uralensis?*
Nightingale *Daulias luscinia*
Ocypterus *Accipiter nisus*
Orion, fabulous
Orites, perh. *Falco sacer*
Ortolan *Emberiza hortulana*
Ostrich *Struthio camelus*
Owl *Athene noctua*
Owl, Little Horned *Strix scops*
Pappus, unidentified
Parrot *Palaeornis cyanocephalus*
Partridge *Perdix graeca* (or *saxatilis*)
Partridge-catcher, perh. *Astur brevipes*
Peacock *Pavo cristatus*
Pelican *Pelicanus crispus*
Pheasant *Phasianus colchicus*
Pigeon *Columba palumbus*
Purple Coot *Porphyrio veterum*
Pyrallis, unidentified
Quail *Coturnix vulgaris*
Raven *Corvus corax*
Reedwarbler [?] *Acrocephalus arundinaceus*
Ring-dove *Columba palumbus*
Robin *Erithacus rubecula*
Rock-dove *Columba livia*
Roller *Coracias garrulus*
Rose-coloured Pastor *Pastor roseus*
Ruff *Machetes pugnax*
Salpinx, unidentified
Sand-partridge *Ammoperdix Bonhami*
Sea-eagle *Pandion haliaëtus*
Sea-hawk *Megalestris catarractes*
Sea-mew *Larus canus*
Seagull, gen. *Larus*
Shearwater *Puffinus kuhli*
—, Little Manx *P. yelkuan*
Siren (ii) *Serinus hortulanus*
Siskin *Fringilla spinus*
Skua, Great, *see* Sea-hawk
Sparrow *Passer domesticus*

INDEX: FAUNA, FLORA, ETC.

Spindalus, unidentified
Starling *Sturnus vulgaris*
Stone-curlew *Charadrius oedicnemus*
Stork *Ciconia alba*
Swallow *Hirundo rustica*
Swan *Cygnus olor*
Syrian Nuthatch *Sitta syriaca*
Thrush *Turdus musicus*
Titmouse *Parus major*
Turtle-dove *Turtur communis*
Vulture *Gyps fulvus*
Wagtail, sp. *Motacilla*
Wide-wing, sp. *Circus*
Woodpecker *Picus martius*
Wryneck *Yunx torquilla*

3. Reptiles

Acontias *Zamenis gemonensis*
Amphisbaena *Typhlops vermicularis?*
Asp *Naia haie*
Basilisk, fabulous
Blood-letter *Vipera latastei?*
Cerastes *Cerastes cornutus*
Chameleon *Chamaeleo vulgaris*
Chelydrus *Tropidonotus tessellatus*
Cobra, *see* Asp
Crocodile *Crocodilus vulgaris*; Gangetic *Gavialis gangeticus*; Indian *C. palustris*
—, Land- *Psammosaurus griseus*
Dipsas *Vipera prester*
Gecko *Platydactylus mauretanicus*
Lizard (i) *Lacerta viridis*; (ii) gen. *Varanus* **16**. 41
Melanurus, *see* Dipsas
Pareas *Coluber longissimus*, or *Aesculapii*
Prester, *see* Dipsas
Purple Snake *Dryophis intestinalis*
Python *Python molurus*, or *P. cebae?*
Salamander *Salamandra maculosa*
Sepedon, unidentified
Sêps *Vipera macrops*
Snake, generic term
Thermuthis, *see* Asp
Tortoise *Testudo graeca*
Turtle (i) *Thalassochelys caretta*; (ii) perh. *Trionyx gangeticus*. *See also* **16**. 14*n*.
Typhlops *Pseudopus pallasi*
Viper *Vipera aspis*
Water-snake, *see* Chelydrus

4. Amphibia

Frog *Rana agilis*; *R. graeca*, **3**. 37
Toad *Bombinator pachypus*

5. Fishes

Adonis, unidentified
Anchovy *Engraulis encrasicholus*
Anthias, unidentified
Aulopias *Thynnus alalonga?*
Basse *Lupus labrax*
Black Sea-bream *Cantharus lineatus*
Blue-grey, unidentified
Caprus, unidentified
Carp *Cyprinus carpio*
Cat-fish *Parasilurus Aristotelis*
Charax, unidentified
Chromis, perh. *Umbrina cirrhosa*
Conger-eel *Conger vulgaris*
Crane-fish, perh. *Regalecus Banksi*
Crow-fish (i) *Chromis castanea*; (ii) unidentified, **14**. 23, 26
Dog-fish *Mustelus laevis*
Eel *Anguilla vulgaris*
Etna-fish, unidentified
Fishing-frog *Lophius piscatorius*
Flounder *Pleuronectes flesus*
Flying-fish *Exocoetus volitans*
Flying Gurnard *Dactylopterus volitans*
Fox-shark *Alopecias vulpes*
Garfish *Belone acus*
Gilthead *Chrysophrys aurata*
Globe-fish *Diodon hystrix*
Gnawer, perh. *Alopecias vulpes*
Goby, sp. *Gobius*
Grayling *Thymallus vulgaris*
Great Sea-perch *Polyprion cernium*
Great Tunny *Thynnus thynnus*
Gurnard, sp. *Trigla*
Hake *Gadus merluccius*
Hammer-headed Shark *Zygaena malleus*
Harper, sp. *Chaetodon*
Hepatus, unidentified
Horned Ray *Cepaloptera giorna*
Horse-mackerel *Caranx trachurus*
Hyena-fish, unidentified
John Dory *Zeus faber*
Leopard-fish, unidentified
Mackerel *Scomber scomber*
Maeotes, unidentified
Maigre *Sciaena aquila*
Maltha, unidentified

6. Insects

7. Arachnida

INDEX: FAUNA, FLORA, ETC.

INDEX: FAUNA, FLORA, ETC.

15. Metals and Minerals

IV. AUTHORS CITED

INDEX: AUTHORS CITED

Aristoxenus, Greek of Asia Minor, doctor and medical writer, fl. about the beginning of the Christian era **8.** 7

Aristoxenus, of Tarentum, 4th cent. B.C., son of Spintharus (*q.v.*) and pupil of Aristotle, wrote on musical theory **2.** 11

Artemon, date and identity uncertain **12.** 38

Autocrates, of Athens, 5th/4th cent. B.C., wrote tragedies and comedies **12.** 9

Bacchylides, of Ceos, 5th cent. B.C., nephew of Simonides (*q.v.*), wrote epinician odes, dithyrambs, hymns, paeans, etc. **6.** 1

Callias, of Syracuse, 4th/3rd cent. B.C., his history of Agathocles, Tyrant of S. (316–289), was regarded as too favourable **16.** 28

Callimachus, of Cyrene, *c.* 305–*c.* 240 B.C., employed in the library of Alexandria, wrote hymns and other poems in a great variety of metres, also prose works on birds, rivers, etc. **6.** 58; **9.** 27; **15.** 28

Callisthenes, of Olynthus, nephew of Aristotle and historiographer to Alexander the Great on his expedition **16.** 30

Charmis, of Massilia, not certainly identified with a famous doctor of the time of Nero, mid-1st cent. A.D. **5.** 38

Chios, historians of **16.** 39

Cleanthes, Stoic philosopher, 331–232 B.C., succeeded his master Zeno as head of the Stoic school at Athens **6.** 50

Clearchus, of Soli, 3rd cent. B.C., wrote on philosophy, natural history, painting, and biographical works **12.** 34

Cleitarchus, of Alexandria, 3rd cent. B.C., wrote an untrustworthy account of Alexander the Great **17.** 2, 22–3, 25

Crates, 'of Pergamum', b. at Mallus in Cilicia Pedias, 2nd cent. B.C., Stoic philosopher and head of the library at Pergamum, wrote upon Homer and other Greek poets, and on the Attic dialect **17.** 9, 37

Cratinus, 5th cent. B.C., the older contemporary and rival in Comedy of Aristophanes and Eupolis **12.** 10

Cretan histories **5.** 2

Ctesias, of Cnidus, late 5th cent. B.C., wrote a history of Persia, where he spent some years as doctor to Artaxerxes, and a work on India **3.** 3; **4.** 21, 26–7, 46(i), 52; **5.** 3; **7.** 1; **16.** 31, 42; **17.** 29; p. xvi

Damon, of Athens, 5th cent. B.C., taught Socrates, wrote on music, cited with approval by Plato **2.** 11

Demetrius, end of 5th cent. B.C., writer of Old Comedy **12.** 10

Democritus, of Abdera, *c.* 460–361 B.C., philosopher and a man of immense learning in the physical and other sciences **5.** 39; **6.** 60; **9.** 64; **12.** 16–20; p. xvi

Demostratus, 2nd cent. A.D. (?), Roman Senator, wrote on fishes and divination **13.** 21; **15.** 4, 9, 19; *epil.*; p. xx f., xxiii f.

Dinolochus, perh. 5th cent. B.C., Sicilian writer of comedies **6.** 51

Dinon, of Colophon, 4th cent. B.C., wrote a history of Persia **17.** 10

Diocles, of Carystus, 4th cent. B.C., contemporary of Aristotle, wrote on anatomy, physiology, and allied subjects **17.** 15

Egyptian histories **16.** 39

Empedocles, of Acragas in Sicily, 5th cent. B.C., philosopher, statesman, and poet **9.** 64; **12.** 7; **16.** 29

Epicharmus, fl. 5th cent. B.C., of Sicilian origin, writer of Comedy **13.** 4 (ter); *also* **6.** 51

Epicrates, of Ambracia, 4th cent. B.C., one of the earliest writers of Middle Comedy **12.** 10

Epimenides, of Crete, perh. 6th cent. B.C., author of religious and mystical works **12.** 7

Eratosthenes, of Cyrene, 3rd cent. B.C., head of the Alexandrian library, wrote on ancient Comedy, astronomy, mythology, chronology, geography, and philosophy **7.** 48

Ethiopian histories **2.** 21; **5.** 49

Eudemus, date and place of origin unknown, seemingly a writer on

INDEX: AUTHORS CITED

Theophrastus, of Eresus in Lesbos, *c.* 370–*c.* 285 B.C., pupil of Aristotle whom he succeeded as head of the Lyceum at Athens, wrote on philosophy, botany, and other sciences **3**. 17, 32, 35, 37, 38; **5**. 27, 29; **7**. 7; **9**. 15, 27, 37, 64; **10**. 35; **11**. 40; **12**. 36; **15**. 16, 26

Theopompus, of Chios, 4th cent. B.C., pupil of Isocrates, wrote a continuation of Thucydides and a history of Philip of Macedon **5**. 27; **11**. 40; **17**. 16

Timaeus, date and identity uncertain, perhaps the authority on mineral drugs cited by Celsus 5. 22. 7 and Plin. *HN* 1. 34 **17**. 15

Tyrtaeus, of Sparta, 7th cent. B.C., elegiac poet, wrote war-songs and political verse **6**. 1

Xenophon, of Athens, *c.* 430–*c.* 354 B.C., disciple of Socrates, served as cavalry officer under Cyrus II against Artaxerxes, wrote historical works (*Anabasis*, *Hellenica*), memoirs of Socrates, and on horsemanship and hunting **2**. 11; **6**. 25, 43; **8**. 3; **13**. 24 (bis)

Zenothemis, date uncertain, wrote a *Περίπλους* in verse, containing 'tales of wonder' **17**. 30

PRINTED IN GREAT BRITAIN BY
RICHARD CLAY AND COMPANY, LTD.,
BUNGAY, SUFFOLK.

THE LOEB CLASSICAL LIBRARY

VOLUMES ALREADY PUBLISHED

Latin Authors

AMMIANUS MARCELLINUS. Translated by J. C. Rolfe. 3 Vols. (*3rd Imp., revised.*)
APULEIUS: THE GOLDEN ASS (METAMORPHOSES). W. Adlington (1566). Revised by S. Gaselee. (*8th Imp.*)
S. AUGUSTINE: CITY OF GOD. 7 Vols. Vol. I. G. E. McCracken.
ST. AUGUSTINE, CONFESSIONS OF. W. Watts (1631). 2 Vols. (Vol. I. *7th Imp.*, Vol. II. *6th Imp.*)
ST. AUGUSTINE, SELECT LETTERS. J. H. Baxter. (*2nd Imp.*)
AUSONIUS. H. G. Evelyn White. 2 Vols. (*2nd Imp.*)
BEDE. J. E. King. 2 Vols. (*2nd Imp.*)
BOETHIUS: TRACTS and DE CONSOLATIONE PHILOSOPHIAE. Rev. H. F. Stewart and E. K. Rand. (*6th Imp.*)
CAESAR: ALEXANDRIAN, AFRICAN and SPANISH WARS. A. G. Way.
CAESAR: CIVIL WARS. A. G. Peskett. (*6th Imp*)
CAESAR: GALLIC WAR. H. J. Edwards. (*11th Imp.*)
CATO: DE RE RUSTICA; VARRO: DE RE RUSTICA. H. B. Ash and W. D. Hooper. (*3rd Imp.*)
CATULLUS. F. W. Cornish; TIBULLUS. J. B. Postgate; PERVIGILIUM VENERIS. J. W. Mackail. (*13th Imp.*)
CELSUS: DE MEDICINA. W. G. Spencer. 3 Vols. (Vol. I. *3rd Imp. revised*, Vols. II. and III. *2nd Imp.*)
CICERO: BRUTUS, and ORATOR. G. L. Hendrickson and H. M. Hubbell. (*3rd Imp.*)
[CICERO]: AD HERENNIUM. H. Caplan.
CICERO: DE FATO; PARADOXA STOICORUM; DE PARTITIONE ORATORIA. H. Rackham (With De Oratore. Vol. II.) (*2nd Imp.*)
CICERO: DE FINIBUS. H. Rackham. (*4th Imp. revised.*)
CICERO: DE INVENTIONE, etc. H. M. Hubbell.
CICERO: DE NATURA DEORUM and ACADEMICA. H. Rackham. (*3rd Imp.*)
CICERO: DE OFFICIIS. Walter Miller. (*7th Imp.*)
CICERO: DE ORATORE. 2 Vols. E. W. Sutton and H. Rackham. (*2nd Imp.*)
CICERO: DE REPUBLICA and DE LEGIBUS; SOMNIUM SCIPIONIS. Clinton W. Keyes. (*4th Imp.*)
CICERO: DE SENECTUTE, DE AMICITIA, DE DIVINATIONE. W. A. Falconer. (*6th Imp.*)
CICERO: IN CATILINAM, PRO FLACCO, PRO MURENA, PRO SULLA. Louis E. Lord. (*3rd Imp. revised.*)

CICERO: LETTERS TO ATTICUS. E. O. Winstedt. 3 Vols. (Vol. I. *7th Imp.*, Vols. II. and III. *4th Imp.*)
CICERO: LETTERS TO HIS FRIENDS. W. Glynn Williams. 3 Vols. (Vols. I. and II. *4th Imp.*, Vol. III. *2nd Imp. revised.*)
CICERO: PHILIPPICS. W. C. A. Ker. (*4th Imp. revised.*)
CICERO: PRO ARCHIA, POST REDITUM, DE DOMO, DE HARUSPICUM RESPONSIS, PRO PLANCIO. N. H. Watts. (*3rd Imp.*)
CICERO: PRO CAECINA, PRO LEGE MANILIA, PRO CLUENTIO, PRO RABIRIO. H. Grose Hodge. (*3rd Imp.*)
CICERO: PRO CAELIO, DE PROVINCIIS CONSULARIBUS. PRO BALBO. R. Gardner.
CICERO: PRO MILONE, IN PISONEM, PRO SCAURO, PRO FONTEIO. PRO RABIRIO POSTUMO, PRO MARCELLO, PRO LIGARIO, PRO REGE DEIOTARO. N. H. Watts. (*3rd Imp.*)
CICERO: PRO QUINCTIO, PRO ROSCIO AMERINO, PRO ROSCIO COMOEDO, CONTRA RULLUM. J. H. Freese. (*3rd Imp.*)
CICERO: PRO SESTIO, IN VATINIUM. R. Gardner.
CICERO: TUSCULAN DISPUTATIONS. J. E. King. (*4th Imp.*)
CICERO: VERRINE ORATIONS. L. H. G. Greenwood. 2 Vols. (Vol. I. *3rd Imp.*, Vol. II. *2nd Imp.*)
CLAUDIAN. M. Platnauer. 2 Vols. (*2nd Imp.*)
COLUMELLA: DE RE RUSTICA. DE ARBORIBUS. H. B. Ash, E. S. Forster and E. Heffner. 3 Vols. (Vol. I. *2nd Imp.*)
CURTIUS, Q.: HISTORY OF ALEXANDER. J. C. Rolfe. 2 Vols. (*2nd Imp.*)
FLORUS. E. S. Forster and CORNELIUS NEPOS. J. C. Rolfe. (*2nd Imp.*)
FRONTINUS: STRATAGEMS and AQUEDUCTS. C. E. Bennett and M. B. McElwain. (*2nd Imp.*)
FRONTO: CORRESPONDENCE. C. R. Haines. 2 Vols. (*3rd Imp.*)
GELLIUS, J. C. Rolfe. 3 Vols. (Vol. I. *3rd Imp.*, Vols. II. and III. *2nd Imp.*)
HORACE: ODES and EPODES. C. E. Bennett. (*14th Imp. revised.*)
HORACE: SATIRES, EPISTLES, ARS POETICA. H. R. Fairclough. (*9th Imp. revised.*)
JEROME: SELECTED LETTERS. F. A. Wright. (*2nd Imp.*)
JUVENAL and PERSIUS. G. G. Ramsay. (*8th Imp.*)
LIVY. B. O. Foster, F. G. Moore, Evan T. Sage, and A. C. Schlesinger and R. M. Geer (General Index). 14 Vols. (Vol. I. *5th Imp.*, Vol. V. *4th Imp.*, Vols. II.–IV., VI. and VII., IX.–XII. *3rd Imp.*, Vol. VIII., *2nd Imp. revised.*)
LUCAN. J. D. Duff. (*4th Imp.*)
LUCRETIUS. W. H. D. Rouse. (*7th Imp. revised.*)
MARTIAL. W. C. A. Ker. 2 Vols. (Vol. I. *5th Imp.*, Vol. II. *4th Imp. revised.*)
MINOR LATIN POETS: from PUBLILIUS SYRUS TO RUTILIUS NAMATIANUS, including GRATTIUS, CALPURNIUS SICULUS, NEMESIANUS, AVIANUS, and others with "Aetna" and the "Phoenix." J. Wight Duff and Arnold M. Duff. (*3rd Imp.*)

Ovid: The Art of Love and Other Poems. J. H. Mozley. (*4th Imp.*)
Ovid: Fasti. Sir James G. Frazer. (*2nd Imp.*)
Ovid: Heroides and Amores. Grant Showerman. (*7th Imp.*)
Ovid: Metamorphoses. F. J. Miller. 2 Vols. (Vol. I. *11th Imp.*, Vol. II. *10th Imp.*)
Ovid: Tristia and Ex Ponto. A. L. Wheeler. (*4th Imp.*)
Persius. Cf. Juvenal.
Petronius. M. Heseltine, Seneca Apocolocyntosis. W. H. D. Rouse. (*9th Imp. revised.*)
Plautus. Paul Nixon. 5 Vols. (Vol. I. *6th Imp.*, II. *5th Imp.*, III. *4th Imp.*, IV. and V. *2nd Imp.*)
Pliny: Letters. Melmoth's Translation revised by W. M. L. Hutchinson. 2 Vols. (*7th Imp.*)
Pliny: Natural History. H. Rackham and W. H. S. Jones. 10 Vols. Vols. I.–V. and IX. H. Rackham. Vols. VI. and VII. W. H. S. Jones. (Vol. I. *4th Imp.*, Vols. II. and III. *3rd Imp.*, Vol. IV. *2nd Imp.*)
Propertius. H. E. Butler. (*7th Imp.*)
Prudentius. H. J. Thomson. 2 Vols.
Quintilian. H. E. Butler. 4 Vols. (Vols. I. and IV. *4th Imp.*, Vols. II. and III. *3rd Imp.*)
Remains of Old Latin. E. H. Warmington. 4 vols. Vol. I. (Ennius and Caecilius.) Vol. II. (Livius, Naevius, Pacuvius, Accius.) Vol. III. (Lucilius and Laws of XII Tables.) (*2nd Imp.*) (Archaic Inscriptions.)
Sallust. J. C. Rolfe. (*4th Imp. revised.*)
Scriptores Historiae Augustae. D. Magie. 3 Vols. (Vol. I. *3rd Imp. revised*, Vols. II. and III. *2nd Imp.*)
Seneca: Apocolocyntosis. Cf. Petronius.
Seneca: Epistulae Morales. R. M. Gummere. 3 Vols. (Vol. I. *4th Imp.*, Vols. II. and III. *3rd Imp.*)
Seneca: Moral Essays. J. W. Basore. 3 Vols. (Vol. II. *4th Imp.*, Vols. I. and III. *2nd Imp. revised.*)
Seneca: Tragedies. F. J. Miller. 2 Vols. (Vol. I. *4th Imp.* Vol. II. *3rd Imp. revised.*)
Sidonius: Poems and Letters. W. B. Anderson. 2 Vols. (Vol. I. *2nd Imp.*)
Silius Italicus. J. D. Duff. 2 Vols. (Vol. I. *2nd Imp.* Vol. II. *3rd Imp.*)
Statius. J. H. Mozley. 2 Vols. (*2nd Imp.*)
Suetonius. J. C. Rolfe. 2 Vols. (Vol. I. *7th Imp.*, Vol. II. *6th Imp. revised.*)
Tacitus: Dialogues. Sir Wm. Peterson. Agricola and Germania. Maurice Hutton. (*7th Imp.*)
Tacitus: Histories and Annals. C. H. Moore and J. Jackson. 4 Vols.. (Vols. I. and II. *4th Imp.* Vols. III. and IV. *3rd Imp.*)
Terence. John Sargeaunt. 2 Vols. (Vol. I. *8th Imp.*, Vol. II. *7th Imp.*)
Tertullian: Apologia and De Spectaculis. T. R. Glover. Minucius Felix. G. H. Rendall. (*2nd Imp.*)
Valerius Flaccus. J. H. Mozley. (*3rd Imp. revised.*)

VARRO: DE LINGUA LATINA. R. G. Kent. 2 Vols. (*3rd Imp. revised.*)
VELLEIUS PATERCULUS and RES GESTAE DIVI AUGUSTI. F. W. Shipley. (*2nd Imp.*)
VIRGIL. H. R. Fairclough. 2 Vols. (Vol. I. *19th Imp.*, Vol. II. *14th Imp. revised.*)
VITRUVIUS: DE ARCHITECTURA. F. Granger. 2 Vols. (Vol. I. *3rd Imp.*, Vol. II. *2nd Imp.*)

Greek Authors

ACHILLES TATIUS. S. Gaselee. (*2nd Imp.*)
AELIAN: ON THE NATURE OF ANIMALS. 3 Vols. Vols. I. and II. A. F. Scholfield.
AENEAS TACTICUS, ASCLEPIODOTUS and ONASANDER. The Illinois Greek Club. (*2nd Imp.*)
AESCHINES. C. D. Adams. (*3rd Imp.*)
AESCHYLUS. H. Weir Smyth. 2 Vols. (Vol. I. *7th Imp.*, Vol. II. *6th Imp. revised.*)
ALCIPHRON, AELIAN, PHILOSTRATUS LETTERS. A. R. Benner and F. H. Fobes.
ANDOCIDES, ANTIPHON, Cf. MINOR ATTIC ORATORS.
APOLLODORUS. Sir James G. Frazer. 2 Vols. (*3rd Imp.*)
APOLLONIUS RHODIUS. R. C. Seaton. (*5th Imp.*)
THE APOSTOLIC FATHERS. Kirsopp Lake. 2 Vols. (Vol. I. *8th Imp.*, Vol. II. *6th Imp.*)
APPIAN: ROMAN HISTORY. Horace White. 4 Vols. (Vol. I. *4th Imp.*, Vols. II.–IV. *3rd Imp.*)
ARATUS. Cf. CALLIMACHUS.
ARISTOPHANES. Benjamin Bickley Rogers. 3 Vols. Verse trans. (*5th Imp.*)
ARISTOTLE: ART OF RHETORIC. J. H. Freese. (*3rd Imp.*)
ARISTOTLE: ATHENIAN CONSTITUTION, EUDEMIAN ETHICS, VICES AND VIRTUES. H. Rackham. (*3rd Imp.*)
ARISTOTLE: GENERATION OF ANIMALS. A. L. Peck. (*2nd Imp.*)
ARISTOTLE: METAPHYSICS. H. Tredennick. 2 Vols. (*4th Imp.*)
ARISTOTLE: METEOROLOGICA. H. D. P. Lee.
ARISTOTLE: MINOR WORKS. W. S. Hett. On Colours, On Things Heard, On Physiognomies, On Plants, On Marvellous Things Heard, Mechanical Problems, On Indivisible Lines, On Situations and Names of Winds, On Melissus, Xenophanes, and Gorgias. (*2nd Imp.*)
ARISTOTLE: NICOMACHEAN ETHICS. H. Rackham. (*6th Imp. revised.*)
ARISTOTLE: OECONOMICA and MAGNA MORALIA. G. C. Armstrong; (with Metaphysics, Vol. II.). (*4th Imp.*)
ARISTOTLE: ON THE HEAVENS. W. K. C. Guthrie. (*3rd Imp. revised.*)
ARISTOTLE: ON THE SOUL. PARVA NATURALIA, ON BREATH. W. S. Hett. (*2nd Imp. revised.*)

ARISTOTLE: ORGANON—Categories, On Interpretation, Prior Analytics. H. P. Cooke and H. Tredennick. (*3rd Imp.*)
ARISTOTLE: ORGANON—Posterior Analytics, Topics. H. Tredennick and E. S. Forster.
ARISTOTLE: ORGANON—On Sophistical Refutations.
On Coming to be and Passing Away, On the Cosmos. E. S. Forster and D. J. Furley.
ARISTOTLE: PARTS OF ANIMALS. A. L. Peck; MOTION AND PROGRESSION OF ANIMALS. E. S. Forster. (*4th Imp. revised.*)
ARISTOTLE: PHYSICS. Rev. P. Wicksteed and F. M. Cornford. 2 Vols. (Vol. I. *2nd Imp.*, Vol. II. *3rd Imp.*)
ARISTOTLE: POETICS and LONGINUS. W. Hamilton Fyfe; DEMETRIUS ON STYLE. W. Rhys Roberts. (*5th Imp. revised.*)
ARISTOTLE: POLITICS. H. Rackham. (*4th Imp. revised.*)
ARISTOTLE: PROBLEMS. W. S. Hett. 2 Vols. (*2nd Imp. revised.*)
ARISTOTLE: RHETORICA AD ALEXANDRUM (with PROBLEMS. Vol. II.). H. Rackham.
ARRIAN: HISTORY OF ALEXANDER and INDICA. Rev. E. Iliffe Robson. 2 Vols. (*3rd Imp.*)
ATHENAEUS: DEIPNOSOPHISTAE. C. B. Gulick. 7 Vols. (Vols. I.–IV., VI. and VII. *2nd Imp.*, Vol. V. *3rd Imp.*)
ST. BASIL: LETTERS. R. J. Deferrari. 4 Vols. (*2nd Imp.*)
CALLIMACHUS: FRAGMENTS. C. A. Trypanis.
CALLIMACHUS, Hymns and Epigrams, and LYCOPHRON. A. W. Mair; ARATUS. G. R. Mair. (*2nd. Imp.*)
CLEMENT of ALEXANDRIA. Rev. G. W. Butterworth. (*3rd Imp.*)
COLLUTHUS. Cf. OPPIAN.
DAPHNIS AND CHLOE. Thornley's Translation revised by J. M. Edmonds; and PARTHENIUS. S. Gaselee. (*4th Imp.*)
DEMOSTHENES I.: OLYNTHIACS, PHILIPPICS and MINOR ORATIONS. I.–XVII. AND XX. J. H. Vince. (*2nd Imp.*)
DEMOSTHENES II.: DE CORONA and DE FALSA LEGATIONE. C. A. Vince and J. H. Vince. (*3rd Imp. revised.*)
DEMOSTHENES III.: MEIDIAS, ANDROTION, ARISTOCRATES, TIMOCRATES and ARISTOGEITON, I. AND II. J. H. Vince (*2nd Imp.*)
DEMOSTHENES IV.–VI.: PRIVATE ORATIONS and IN NEAERAM. A. T. Murray. (Vol. IV. *3rd Imp.*, Vols. V. and VI. *2nd Imp.*)
DEMOSTHENES VII.: FUNERAL SPEECH, EROTIC ESSAY, EXORDIA and LETTERS. N. W. and N. J. DeWitt.
DIO CASSIUS: ROMAN HISTORY. E. Cary. 9 Vols. (Vols. I. and II. *3rd Imp.*, Vols. III.–IX. *2nd Imp.*)
DIO CHRYSOSTOM. J. W. Cohoon and H. Lamar Crosby. 5 Vols. (Vols. I.–IV. *2nd Imp.*)
DIODORUS SICULUS. 12 Vols. Vols. I.–VI. C. H. Oldfather. Vol. VII. C. L. Sherman. Vols. IX. and X. R. M. Geer. Vol. XI. F. Walton. (Vol. I. *3rd Imp.*, Vols. II.–IV. *2nd Imp.*)
DIOGENES LAERTIUS. R. D. Hicks. 2 Vols. (*5th Imp.*).
DIONYSIUS OF HALICARNASSUS: ROMAN ANTIQUITIES. Spelman's translation revised by E. Cary. 7 Vols. (Vols. I.–V. *2nd Imp.*)

Epictetus. W. A. Oldfather. 2 Vols. (*3rd Imp.*)
Euripides. A. S. Way. 4 Vols. (Vols. I. and IV. *7th Imp.*, Vol. II. *8th Imp.*, Vol. III. *6th Imp.*) Verse trans.
Eusebius: Ecclesiastical History. Kirsopp Lake and J. E. L. Oulton. 2 Vols. (Vol. I. *3rd Imp.*, Vol. II. *5th Imp.*)
Galen: On the Natural Faculties. A. J. Brock. (*4th Imp.*)
The Greek Anthology. W. R. Paton. 5 Vols. (Vols. I.–IV. *5th Imp.*, Vol. V. *3rd Imp.*)
Greek Elegy and Iambus with the Anacreontea. J. M. Edmonds. 2 Vols. (Vol. I. *3rd Imp.*, Vol. II. *2nd Imp.*)
The Greek Bucolic Poets (Theocritus, Bion, Moschus). J. M. Edmonds. (*7th Imp. revised.*)
Greek Mathematical Works. Ivor Thomas. 2 Vols. (*3rd Imp.*)
Herodes. Cf. Theophrastus: Characters.
Herodotus. A. D. Godley. 4 Vols. (Vol. I. *4th Imp.*, Vols. II. and III. *5th Imp.*, Vol. IV. *3rd Imp.*)
Hesiod and The Homeric Hymns. H. G. Evelyn White. (*7th Imp. revised and enlarged.*)
Hippocrates and the Fragments of Heracleitus. W. H. S. Jones and E. T. Withington. 4 Vols. (Vol. I. *4th Imp.*, Vols. II.–IV. *3rd Imp.*)
Homer: Iliad. A. T. Murray. 2 Vols. (*7th Imp.*)
Homer: Odyssey. A. T. Murray. 2 Vols. (*8th Imp.*)
Isaeus. E. W. Forster. (*3rd Imp.*)
Isocrates. George Norlin and LaRue Van Hook. 3 Vols. (*2nd Imp.*)
St. John Damascene: Barlaam and Ioasaph. Rev. G. R. Woodward and Harold Mattingly. (*3rd Imp. revised.*)
Josephus. H. St. J. Thackeray and Ralph Marcus. 9 Vols. Vols. I.–VII. (Vol. V. *4th Imp.*, Vol. VI. *3rd Imp.*, Vols. I.–IV. and VII. *2nd Imp.*)
Julian Wilmer Cave Wright. 3 Vols. (Vols. I. and II. *3rd Imp.*, Vol. III. *2nd Imp.*)
Lucian. A. M. Harmon. 8 Vols. Vols. I.–V. (Vols. I. and II. *4th Imp.*, Vol. III. *3rd Imp.*, Vols. IV. and V. *2nd Imp.*)
Lycophron. Cf. Callimachus.
Lyra Graeca. J. M. Edmonds. 3 Vols. (Vol. I. *5th Imp.* Vol. II *revised and enlarged*, and III. *4th Imp.*)
Lysias. W. R. M. Lamb. (*3rd Imp.*)
Manetho. W. G. Waddell: Ptolemy: Tetrabiblos. F. E. Robbins. (*3rd Imp.*)
Marcus Aurelius. C. R. Haines. (*4th Imp. revised.*)
Menander. F. G. Allinson. (*3rd Imp. revised.*)
Minor Attic Orators (Antiphon, Andocides, Lycurgus, Demades, Dinarchus, Hypereides). K. J. Maidment and J. O. Burtt. 2 Vols. (Vol. I. *2nd Imp.*)
Nonnos: Dionysiaca. W. H. D. Rouse. 3 Vols. (*2nd Imp.*)
Oppian, Colluthus, Tryphiodorus. A. W. Mair. (*2nd Imp.*)
Papyri. Non-Literary Selections. A. S. Hunt and C. C. Edgar. 2 Vols. (*2nd Imp.*) Literary Selections. (Poetry). D. L. Page. (*3rd Imp.*)

Parthenius. Cf. Daphnis and Chloe.
Pausanias: Description of Greece. W. H. S. Jones. 5 Vols. and Companion Vol. arranged by R. E. Wycherley. (Vols. I. and III. *3rd Imp.*, Vols. II., IV. and V. *2nd Imp.*)
Philo. 10 Vols. Vols. I.–V.; F. H. Colson and Rev. G. H. Whitaker Vols. VI.–IX.; F. H. Colson. (Vols. I–II., V.–VII., *3rd Imp.*, Vol. IV. *4th Imp.*, Vols. III., VIII., and IX. *2nd Imp.*)
Philo: two supplementary Vols. (*Translation only.*) Ralph Marcus.
Philostratus: The Life of Appollonius of Tyana. F. C. Conybeare. 2 Vols. (Vol. I. *4th Imp.*, Vol. II. *3rd Imp.*)
Philostratus: Imagines; Callistratus: Descriptions. A. Fairbanks. (*2nd Imp.*)
Philostratus and Eunapius: Lives of the Sophists. Wilmer Cave Wright. (*2nd Imp.*)
Pindar. Sir J. E. Sandys. (*8th Imp. revised.*)
Plato: Charmides, Alcibiades, Hipparchus, The Lovers, Theages, Minos and Epinomis. W. R. M. Lamb. (*2nd Imp.*)
Plato: Cratylus, Parmenides, Greater Hippias, Lesser Hippias. H. N. Fowler. (*4th Imp.*)
Plato: Euthyphro, Apology, Crito, Phaedo, Phaedrus. H. N. Fowler. (*11th Imp.*)
Plato: Laches, Protagoras, Meno, Euthydemus. W. R. M. Lamb. (*3rd Imp. revised.*)
Plato: Laws. Rev. R. G. Bury. 2 Vols. (*3rd Imp.*)
Plato: Lysis, Symposium Gorgias. W. R. M. Lamb. (*5th Imp. revised.*)
Plato: Republic. Paul Shorey. 2 Vols. (Vol. I. *5th Imp.*, Vol. II. *4th Imp.*)
Plato: Statesman, Philebus. H. N. Fowler; Ion. W. R. M. Lamb. (*4th Imp.*)
Plato: Theaetetus and Sophist. H. N. Fowler. (*4th Imp.*)
Plato: Timaeus, Critias, Clitopho, Menexenus, Epistulae. Rev. R. G. Bury. (*3rd Imp.*)
Plutarch: Moralia. 14 Vols. Vols. I.–V. F. C. Babbitt. Vol. VI. W. C. Helmbold. Vol. VII. P. H. De Lacy and B. Einarson. Vol. X. H. N. Fowler. Vol. XII. H. Cherniss and W. C Helmbold. (Vols. I.–VI. and X. *2nd Imp.*)
Plutarch: The Parallel Lives. B. Perrin. 11 Vols. (Vols. I., II., VI., VII., and XI. *3rd Imp.*, Vols. III.–V. and VIII.–X. *2nd Imp.*)
Polybius. W. R. Paton. 6 Vols. (*2nd Imp.*)
Procopius: History of the Wars. H. B. Dewing. 7 Vols. (Vol. I. *3rd Imp.*, Vols. II.–VII. *2nd Imp.*)
Ptolemy: Tetrabiblos. Cf. Manetho.
Quintus Smyrnaeus. A. S. Way. Verse trans. (*3rd Imp.*)
Sextus Empiricus. Rev. R. G. Bury. 4 Vols. (Vol. I. *4th Imp.*, Vols. II. and III. *2nd Imp.*)
Sophocles. F. Storr. 2 Vols. (Vol. I. *10th Imp.* Vol. II. *6th Imp.*) Verse trans.

STRABO: GEOGRAPHY. Horace L. Jones. 8 Vols. (Vols. I., V., and VIII. *3rd Imp.*, Vols. II., III., IV., VI., and VII. *2nd Imp.*)
THEOPHRASTUS: CHARACTERS. J. M. Edmonds. HERODES, etc. A. D. Knox. (*3rd Imp.*)
THEOPHRASTUS: ENQUIRY INTO PLANTS. Sir Arthur Hort, Bart. 2 Vols. (*2nd Imp.*)
THUCYDIDES. C. F. Smith. 4 Vols. (Vol. I. *5th Imp.*, Vols. II. and IV. *4th Imp.*, Vol. III., *3rd Imp. revised.*)
TRYPHIODORUS. Cf. OPPIAN.
XENOPHON: CYROPAEDIA. Walter Miller. 2 Vols. (Vol. I. *4th Imp.*, Vol. II. *3rd Imp.*)
XENOPHON: HELLENICA, ANABASIS, APOLOGY, and SYMPOSIUM. C. L. Brownson and O. J. Todd. 3 Vols. (Vols. I. and III *3rd Imp.*, Vol. II. *4th Imp.*)
XENOPHON: MEMORABILIA and OECONOMICUS. E. C. Marchant (*3rd Imp.*)
XENOPHON: SCRIPTA MINORA. E. C. Marchant. (*3rd Imp.*)

IN PREPARATION

Greek Authors

ARISTOTLE: HISTORY OF ANIMALS. A. L. Peck.
PLOTINUS: A. H. Armstrong.

Latin Authors

BABRIUS AND PHAEDRUS. Ben E. Perry.

DESCRIPTIVE PROSPECTUS ON APPLICATION

London — WILLIAM HEINEMANN LTD
Cambridge, Mass. — HARVARD UNIVERSITY PRESS